Topics in Safety, Risk, Reliability and Quality

Volume 32

More information about this series at http://www.springer.com/series/6653

Alec Groysman

Corrosion Problems and Solutions in Oil Refining and Petrochemical Industry

Springer

Alec Groysman
The Israeli Society of Chemical Engineers and Chemists
Association of Engineers and Architects in Israel
Tel Aviv
Israel

ISSN 1566-0443 ISSN 2215-0285 (electronic)
Topics in Safety, Risk, Reliability and Quality
ISBN 978-3-319-45254-8 ISBN 978-3-319-45256-2 (eBook)
DOI 10.1007/978-3-319-45256-2

Library of Congress Control Number: 2016948810

Printed on acid-free paper

This Springer imprint is published by Springer Nature
The registered company is Springer International Publishing AG
The registered company address is: Gewerbestrasse 11, 6330 Cham, Switzerland

One advantage of writing to yourself is that you know at least somebody's reading.

Paraphrase by Franklin P. Jones (1908–1980), an American journalist

To my great, wise, intelligent, bright
and smart wife Olga for regular support,
endurance, understanding and assistance
in creating and writing this book,
and also to my lovely children Sasha,
Anat, Tal, and beautiful grandchildren
Yonatan and Ido who sometimes take
my first two books and look at wonderful
corrosion pictures inside

Preface

I invite you to start our journey into the amazing world of crude oil, fuels, and corrosion problems and solutions at oil refineries and petrochemical plants. Look around. Vehicles, computers, modern sources of energy, materials, such as medicines, different goods from polymers, cosmetics, to name a few. The source of all these materials and energy is crude oil and products of its processing. Crude oil was formed in the depths of the Earth during millions of years. Since ancient times, people have used bitumen and other compounds accompanying crude oil or producing from it for waterproof, lubricating axles, and medical treatment.

The modern world depends severely on fuels which are obtained from crude oil in refineries.

The first oil refinery was built at Ploieşti, Romania, in 1856–1857. About 60 refineries were built in 1860s in the USA. Then in the beginning of the twentieth century, refineries were erected like 'mushrooms after the rain.' Nowadays, more than 700 oil refineries function all over the world and use about 150 different types of crude oil. Most technological processes of elaboration of crude oil were created in the twentieth century.

The word 'petroleum' means 'rock oil' from the Latin 'petra' (rock or stone) and 'oleum' (oil). Therefore, 'crude oil,' or simply 'crude,' is synonym to 'petroleum'.

We should also differentiate between oil refining and petrochemical industries. Oil refining industry produces the following products from crude oil: liquefied petroleum gas (LPG), naphtha, gasoline, kerosene (jet fuel), gas oil (diesel fuel), fuel oil, lubricating oils, paraffin wax, asphalt (bitumen), coke, and sulfur. Petrochemical industry produces olefins and aromatics. Then, these chemicals are used for manufacturing solvents, polymers, paints, medicines, fertilizers, etc. Our comfort life, health, and we can safely say 'lifes pan' are linked to them. Oil refineries and petrochemical plants are firmly connected because the former produce raw materials for the latter. Both are considered as typical chemical plants.

Development of numerous vehicles on the land, sea, and air and demands of chemical, pharmaceutical, and other industries resulted in intensive development of oil refining and petrochemical industry in the twentieth–twenty-first centuries. Metallic equipment and constructions contact crude oils, petroleum products and fuels, solvents, water, atmosphere, and soil. All processes with participation of aggressive substances occur in metallic equipment at temperatures from −196 °C to +1400 °C and pressures from vacuum to 1000 bar.

Oil refineries and petrochemical plants represent also a high hazard industry with media which are flammable, explosive, toxic to human health, or harmful to the environment. The combination of numerous factors makes refinery equipment very vulnerable to a variety of corrosion phenomena that can lead to serious accidents.

On the one hand, oil refining and petrochemical industry has accumulated large experience. On the other hand, the introduction of new technologies, materials, and strict requirements to the quality of fuels and to the reduction of environmental pollution state new problems to safe functioning of equipment and constructions.

In order to understand and to solve corrosion problems in refinery and petrochemical units, corrosion and materials specialist should learn diverse physicochemical processes which are the basis of production of fuels and other chemicals. During my long carrier in oil refining and petrochemical plants, above 3000 corrosion events were analyzed and the reasons were defined. It was established that people are responsible in 65–85 % of corrosion cases. Using proper corrosion management, it is possible to diminish them.

In spite of many conferences, publications, researches, reports, and achievements in refining and petrochemical corrosion control and monitoring, a number of corrosion problems is increasing in the last 20 years because of four factors: the first—the introduction of new processes; the second—some universities and colleges removed corrosion courses they had before in the engineering curricula; the third—corrosion engineers in most of oil refineries and petrochemical plants were replaced with consultants; the fourth—corrosion specialists retire and are not replaced.

There are many ways to avoid or control corrosion hazard: selection of corrosion resistant or suitable materials, correct design, use of anti-corrosive chemicals, control of technological parameters, use of coatings, cathodic protection, and, what is very important, inspecting and controlling at all stages of application of these actions.

Interesting event happened to corrosion scientist C. Edeleanu who suggested anodic protection in 1954. He moved into industry from academy and did not take part in corrosion conferences during 15 years. Attending corrosion conference and listening to all presentations after such a long period he exclaimed: "Nothing changed." New generations of engineers come and face the problems which were solved and even documented. There are good books, but most new information is dispersed in the literature or is present in the heads of specialists.

In this book, considering corrosion cases at different units, I tried to unify and allocate them according to appropriate systems and phenomena. You will find description of processing conditions, materials of constructions, history and service period, visual examination and findings, characterization of failure phenomenon, causes of failure and its explanation, solutions, and practical recommendations. My experience is given, and the last literature data as much as possible is included. I hope that reading this book will enrich your knowledge and help in your understanding, experience, and job.

Shekhaniya, Israel Alec Groysman

Acknowledgements

I am extremely grateful to my friend and colleague Dr. Boris Feldman (IKA Laboratories, Israel) who carried out metallurgical analysis of all the samples during our long mutual work and elucidated many metallurgical problems. I am sincerely thankful to Roman Kotlyar for his kind proofreading.

Contents

About the Author

Dr. Alec Groysman graduated in 1973 from the Chemico-Technological University named after Mendeleev in Moscow. He received his Ph.D. in physical chemistry and corrosion in 1983 in Moscow. He has experience in corrosion and protection from corrosion from 1976 in the oil refining and petrochemical industry.

He deals with kinetics and thermodynamics of corrosion processes, online corrosion monitoring, choice and use of corrosion inhibitors, coating systems, selection of appropriate alloys for corrosive conditions, and failure analysis.

He has special interests in corrosion education in oil and petrochemical industry and in the searching of relationships between corrosion, art, history, and philosophy.

His first book 'Corrosion for Everybody' published by Springer in 2010 received the innovation award winner of Materials Performance Readers' choice in 2012 year in the USA.

His second book 'Corrosion in Systems for Transportation and Storage of Petroleum Products and Biofuels' was published by Springer in 2014.

He is a lecturer of the courses 'Materials and Standards in Oil and Gas Engineering' and 'Corrosion and Corrosion Control' in the Technion (Haifa) in Israel.

Abbreviations

0.1 M solution	0.1 Molar (volume concentration) solution. 0.1 mole of substance in one liter of solution
AE	Acoustic emission
AFNOR	Association Française de Normalisation (France)
ANSI	American National Standards Institute
API	American Petroleum Institute
API RP	American Petroleum Institute Recommended Practice
ASI	Advanced Study Institute
ASME	The American Society of Mechanical Engineers
ASTM	American Society for Testing and Materials (ASTM International)
BS	British Standard
BS&W	Basic sediment and water (the quantity of sediments and water in crude oil)
BTEX	Benzene, Toluene, Ethyl benzene, Xylene
BTX	Benzene, Toluene, Xylene
CCR	Continuous catalytic reforming
CDA	Copper Development Association
CFU/ml or CFU/g	Colony-forming units per milliliter of liquid or gram of deposit; an estimate of viable bacterial or fungal numbers
Cl SCC	Chloride stress corrosion cracking
CLO	Clarified oil (bottom of the main column at FCCU)
COC	Cycles of concentration represents the accumulation of dissolved salts in the recirculating cooling water
cP	Centipoise, the unit of dynamic viscosity. $1\ cp = 10^{-2}\ P = 10^{-3}\ Pa{\cdot}s = 1\ mPa{\cdot}s$
DAF	Dissolved air flotation
DCU	Delayed coking unit
DEA	Diethanolamine

DIN	Deutsches Institut Fur Normung E.V. (German National Standard)
EC	Eddy current
ECTFE	Halar ECTFE (ethylenechlorotrifluoroethylene)
EDS	Energy-dispersive spectroscopy
ED-XRF	Energy-dispersive X-ray fluorescence
EEMUA	The Engineering Equipment and Materials Users' Association
EIP	Emulsion inversion point
EN	European Norm, European Standard
EPA	US Environmental Protection Agency
ER	Electrical resistance
ETFE	Tefzel ETFE (ethylene tetrafluoroethylene)
FBE	Fusion bonded epoxy
FCCU	Fluid catalytic cracking unit
FEP	Teflon FEP (fluorinated ethylene propylene)
FHWA	Federal Highway Administration
FRP (GFRP, GRP)	Fiberglass-reinforced plastic
FTIR	Fourier transform infrared spectroscopy
GFRP (see GRP, FRP)	Glass-fiber-reinforced plastic
GOST	Gosstandard (Russia)
Gr.	Grade
GRP (see GFRP, FRP)	Glass-reinforced plastic
HAB	Heterotrophic aerobic bacteria
HAGO	Heavy atmospheric gas oil
HAZ	Heat-affected zone
HB	Hardness Brinell
HC	Hydrocarbons
HDPE	High-density polyethylene
HDS	Hydrodesulphurizer
HE	Hydrogen embrittlement
HIC	Hydrogen-induced cracking
HRC	Hardness Rockwell
HSS	Heat stable salts
HV	Hardness Vickers
HVGO	Heavy vacuum gas oil
IFI	Industrial Fasteners Institute
IOB	Iron-oxidizing bacteria
IPA	Isopropanol
IPSC	European Commission in Institute for the Protection and Security of the Citizen in Italy
ISO	International Organization for Standardization
IUPAC	International Union of Pure and Applied Chemistry
JRC	Joint Research Center
kPa	Kilo Pascal

KWA	Ken Wilcox Associates, Inc
LAGO	Light atmospheric gas oil
LC50	Lethal concentration 50
LCO	Light cycle oil
LPG	Liquefied petroleum gas
m/sec	Meter per second
MDEA	Methyldiethanolamine
MDPE	Medium-density polyethylene (density 926–940 g/dm^3)
MEA	Monoethanolamine
MFL	Magnetic flux leakage
mm L^{-1}	Millimole per liter (volume concentration)
MPa	Megapascal (a unit of pressure); 1 MPa = 1,000,000 Pa. Standard atmospheric pressure is 101,325 Pa
mpy	Mils per year (corrosion rate unit). 1 mpy = 0.025 mm/y (millimeter per year)
NA	Naphthenic acids
NAC	Naphthenic acid corrosion
NACE	National Association of Corrosion Engineers International
NAN	Naphthenic acid number
NDT	Non-destructive technique
NFPA	National Fire Protection Association
NLPA	National Leak Prevention Association
O/W	Emulsion in which water is the continuous phase
OSHA	US Occupational Safety and Health Administration
P	Pressure
Pa	Pascal (a unit of pressure)
PE	Polyethylene
PEC	Pulsed eddy current
PEI	Petroleum Equipment Institute
PFA	Teflon PFA (perfluoroalkoxy)
PI	Plant Information
PP	Polypropylene
Ppb	Parts per billion; weight concentration; 1 mg of substance (solute) in 1,000,000,000 mg (1000 kg) of solution
ppm	Parts per million; weight concentration; 1 mg of substance (solute) in 1,000,000 mg (1 kg) of solution
Ppmv	Parts per million by volume; 1 mg of substance (solute) in 1,000,000 ml of solution
PTB	Pounds of salt per thousand barrels of crude oil (1 pound = 0.453592 kg; 1 barrel = 159 liters of crude oil)
PVC	Polyvinyl chloride
PWHT	Post-weld heat treatment

RPM	Revolutions per minute (a measure of the frequency of rotation)
S&W	Sediment and water (the quantity of sediments and water in crude oil)
SCC	Stress corrosion cracking
SCE	Saturated calomel electrode
Sch	Schedule
sec	Second
SEM	Scanning electron microscope
SHE	Standard hydrogen electrode
SOHIC	Stress-oriented hydrogen-induced cracking
SP	Standard practice
SRB	Sulfate-reducing bacteria
SRU	Sulfur recovery unit
SS	Stainless steel
SSC	Sulfide stress cracking
SSPC	Steel Structures Painting Council (The Society for Protective Coatings)
T	Temperature
TAN	Total acid number
TBC	Total bacteria count
TEMA	Tubular Exchanger Manufacturers Association
TFE	Teflon TFE (tetrafluoroethylene)
TGT	Tail gas treating
TM	Test methods
TPC	Total plate count
TSP	Trisodium phosphate
UL	Underwriters Laboratory Inc.
UNESCO	United Nations Educational, Scientific and Cultural Organization
UNS	Unified Numbering System (for Metals and Alloys)
US	Ultrasound
UV	Ultraviolet
V	Volt, unit for electric potential
VOC	Volatile organic compounds
Vol%	Volume percent
W/O	Emulsion in which oil is the continuous phase
Wt%	Weight percent
XRF	X-ray fluorescence
μs/cm	Micro-siemens per centimeter (electrical conductivity)

Chapter 1
Process Units in Oil Refineries and Petrochemical Plants

Before we proceed, we need to understand.
Kapitsa Sergey Petrovich (1928–2012), a Russian physicist.

Abstract Main process units and their functions in oil refineries and petrochemical plants are described. Among them are atmospheric and vacuum crude distillation, fluid catalytic cracking and hydrocracking, hydroprocessing (including hydrodesulfurization), and catalytic reforming (continuous catalytic regeneration and isomerization) units. Supporting units include amine treating, sulfur recovery, tail gas treatment, sour water stripper, and wastewater facilities. Three units at petrochemical plant (benzene-toluene-xylene; toluene and para-xylene; and phthalic anhydride) are described too.

An oil refinery includes many unit operations and unit processes. A unit operation is a basic step in a process. Like evolution of our planet, life and technology, oil refining industry has been developing with increasing complexity since its foundation in 1859. Unit operations involve physicochemical processes. Physical transformations, such as desalting, distillation, filtration, and evaporation, relate to separation. Chemical transformations, such as isomerization, hydrogenation, oxidation, and polymerization, relate to chemical reactions. These unit operations are connected into process. The main goal of refinery units is to extract useful substances from crude oil. Oil refinery and petrochemical plant are two 'living organisms' which are tightly connected. The dream of the creator of the Periodic Table of elements, Dmitri Mendeleev is, that any chemical plant should be wasteless and is realized in the refinery: a product obtained in one unit is the raw material for the other unit. Nearly all wastes are utilized. Excluding is some gases (mostly CO_2) and water vapor emitted into the atmosphere. Even nowadays CO_2 utilization can be applied in oil refining and petrochemical industries. There are associated facilities, such as cooling water system, power station (with water treatment and steam providing), and units related to the protection of the environment and people (the utilization of hydrocarbon wastes, purification of wastewater and emitted gases, and deodorization). To sum up, any oil refinery is a very complicated alive "organism" (see Appendix A). Each oil refinery has its own

A. Groysman, *Corrosion Problems and Solutions in Oil Refining and Petrochemical Industry*, Topics in Safety, Risk, Reliability and Quality 32,
DOI 10.1007/978-3-319-45256-2_1

unique processing scheme which is determined by the process equipment available, crude oil characteristics, operating costs, and product demand [1]. There are no refineries absolutely identical in their operations but most corrosion problems and solutions may be similar. We will describe shortly main units and processes where corrosion problems occur more often.

1.1 Crude Distillation Unit

Crude distillation unit is a "heart" of any oil refinery and consists of a preheat train, a desalter, a preflash drum, a furnace, an atmospheric and vacuum distillation columns. Not at once crude oil coming from a storage tank or a transportation pipeline is distilled. Crude oil is prepared by means of settling, then is treated in desalters to remove dissolved salts. Then crude oil is heated in furnaces, and the resultant liquid-vapor mixture flows via a transfer line to the flash zone of the preflash drum, then passes through the furnace, and then liquid crude enters into atmospheric distillation column. This column is the "main organ" of any crude distillation unit where crude is distilled and various petroleum products are obtained. These products are sent to other units for further treatment to obtain useful fuels or other substances as raw materials for petrochemical, chemical and pharmaceutical plants. The heaviest product, residual bottom, flows to the vacuum distillation column where is distilled under vacuum to form also valuable petroleum products.

Desalter. Crude oil is an emulsion which contains various small amounts of water and salts. These salts in the presence of water cause corrosion, fouling and poison the catalysts in processing units downstream of the crude distillation unit. Desalting is the process of diluting the salt amount with fresh water and applying electric fields and special chemicals (demulsifiers and surfactants). Thus, the function of the desalter is to extract the water soluble salts (more than 90 wt%), suspended solids (corrosion products, soil, silt, and sand) from the crude oil into the aqueous phase (lower layer in the desalter). The heavy aqueous phase containing dissolved salts, H_2S and suspended solids is sent to a sour water stripper. The light crude oil phase (containing small amounts of H_2S and traces of water and salts) is withdrawn to the intermediate section of the preheat train.

Preheat train is made up of many heat exchangers with the crude usually processed on the tube-side. Then crude enters the furnace. Following the changes in crude temperature, the preheat train is usually divided into cool section (upstream of the desalter), intermediate (between the desalter and the preflash), and hot (downstream of the preflash).

Furnace is a fired heater where the crude oil is brought at the desired inlet temperature of the preflash drum or the distillation column. Usually crude units contain furnaces before the preflash, the atmospheric and the vacuum columns. Fuel oil, gas oil or natural gas are used as fuels in furnaces for the heating crude oil.

Preflash drum (column). The preflash separates lighter hydrocarbon components of the crude oil before it enters the furnace. This process was suggested by Brugma A. J. in 1941 [2]. The use of preflash reduces the risk of a two-phase flow in the hot section of the train. While the gas phase bypasses the furnaces and enters the distillation column directly, the liquid phase is further preheated in the hot end of the preheat train before entering the furnace. If the crude oil feed to an atmospheric column is heavy and has nearly no light hydrocarbons, it is possible to avoid using the preflash.

Atmospheric distillation column. Crude is fractionated at atmospheric pressure in this column into petroleum products (named also primary products, distillates, fractions, or cuts). Fractional continuous distillation (named also rectification) is a physicochemical process in which numerous hydrocarbons are separated according to their boiling temperatures, and new different chemical mixtures are obtained. These combinations of chemical compounds (petroleum products) are withdrawn from the column according to their boiling temperatures (see Appendices A, B). Gases and naphtha are withdrawn from the distillation column head (overhead). Heavier fractions gasoline, kerosene, light and heavy gas oil, and fuel oil are withdrawn from different sections of the column. These petroleum products from gases to viscous liquids are not finished fuels and have a different fate. Some of them are used for inner needs at refinery. Others are sent for further treatment to obtain finished products. The third group is the raw material for petrochemical industry.

If we ask a layman how many petroleum products he knows, he can say gasoline, kerosene, fuel oil. Really, over 2000 products have different specifications. Only about 40 types of gasoline are produced by refineries [1]. The residual bottom after atmospheric distillation is sent to the next unit.

Vacuum distillation unit. Atmospheric residue is distilled in vacuum column to obtain useful petroleum products. By lowering the pressure the boiling point of hydrocarbons is decreased and their destruction is minimized. Thus, unconditioned atmospheric bottoms is transformed into light hydrocarbons, light vacuum gas oil and heavy vacuum gas oil. The lightest fraction together with steam are taken off the top of the distillation column. The vacuum residue is taken off the bottom of the column and sent to a visbreaker, cocker or deasphalting unit for further processing.

1.2 Cracking

The developing of the automobile industry at the beginning of the twentieth century encouraged erection of oil refineries. The first automobile engines used ethanol as fuel. Use of 'straight-run' gasoline for automobile engines was insufficient. At the first refineries, only kerosene fraction was used for lightning and heating. It was necessary to solve what to do with atmospheric residue. The solution was found by the creation of thermal and later catalytic cracking of heavy fractions with the aim to obtain high quality gasoline. The cracking is the process in which long-chain hydrocarbons are broken into simpler molecules of light hydrocarbons. This process is realized in thermal and catalytic cracking. The latter includes fluid catalytic cracking (FCC) and hydrocracking.

Thermal cracking. Two American chemists Burton W. M. and Humphreys R. E. developed in 1911–1913 the process of destructive distillation of crude oil heated in a still under pressure [3]. More fairly, the first thermal cracking method was invented by the Russian engineer-polymath, scientist and architect Shukhov V. G. and chemist Gavrilov S. in 1891. The use of these stills allowed to double the production of gasoline from crude oils.

The most prevailing thermal (noncatalytic) processes are visbreaking (mild) and coking (severe) thermal cracking. A visbreaker is intended for reducing the residue produced in the distillation unit and increasing the yield of more valuable diesel fuel and heating oil. A visbreaker thermally cracks large hydrocarbon molecules in the residue to diminish its viscosity (therefore, it was named '*vis*breaker') and to produce small amounts of LPG and gasoline.

Catalytic cracking. It is the process of converting high-molecular weight hydrocarbon fractions of crude oils in the presence of catalyst to more valuable gasoline, olefinic gases, and other products. A catalyst allows a lower reaction temperatures to be used. In 1937 the Burton thermal cracking process was superseded by catalytic cracking, but it is still in use today to produce diesel fuel. The cracking process produces coke which remains on catalyst particles and lowers its activity. To maintain the catalyst activity, it is necessary to regenerate the catalyst by burning off this coke with air. As a result, the catalyst is continuously moved from reactor to regenerator and back to reactor.

Fluid catalytic cracking (FCC) unit is intended for producing high octane gasoline. Unlike atmospheric and vacuum distillations, which are physicochemical separation processes, FCC is a chemical process that uses a catalyst to create new, smaller molecules from larger hydrocarbon molecules (gas oil, heavy fuel oil and residues). Major FCC products are gasoline, diesel oil, heating oil, cycle oil, and olefinic hydrocarbons. The FCC process employs a catalyst in the form of fine particles which behave as a fluid when aerated with a vapor. The fluidized catalyst is circulated continuously between the reaction zone and the regeneration zone. The FCC process vaporizes and breaks the long-chain molecules of the high-boiling hydrocarbon liquids into much shorter molecules by contacting the feedstock, at high temperature and moderate pressure, with a fluidized powdered catalyst. Light hydrocarbons contain also remains of water vapor, hydrogen sulfide, ammonia, and hydrogen cyanide. FCC units were implemented into the practice in 1942.

Hydrocracking is a type of catalytic cracking in the presence of hydrogen at elevated temperature and pressure. Hydrocracking was invented in Germany in 1915 and commercially used in 1927 for producing gasoline from lignite. The main products in hydrocracking are gasoline, kerosene, and diesel fuel. Light hydrocarbons can contain also H_2S, NH_3, and water vapor.

1.3 Hydroprocessing

Hydroprocessing is a process to catalytically stabilize petroleum products and/or remove objectionable elements from feedstocks by reacting them with hydrogen. Stabilization usually involves converting unsaturated hydrocarbons (olefins and unstable diolefins) to paraffins. Higher-value products are obtained: gasoline, kerosene, and diesel fuel. Objectionable elements removed by hydroprocessing include sulfur, nitrogen, oxygen, halides, and trace metals. When the process is employed specifically for sulfur removal it is usually called hydrodesulfurization (HDS). Hydroprocessing is also named hydrotreating, hydroconversion, hydrorefining, hydrocracking, desulfurization, or hydrodesulfurization. Feedstocks can vary from light naphtha and gas oil to heavy vacuum residue. Common to all hydroprocessing units is the formation of H_2S and NH_3 resulting from the reaction of hydrogen with S- and N-containing organic compounds. The removal of sulfur is necessary for either processing in downstream units where the sulfur can contaminate the catalyst or for improving fuel quality. The "heavier" hydrocarbon feedstocks require higher temperature and pressure and contain higher concentrations of sulfur and nitrogen that produce the highest quantities of H_2S and NH_3. The conversion of any chlorides in the feed to HCl also takes place. The reactor effluent is a mixture of hydrocarbons, H_2, H_2S, NH_3, and possibly HCl and H_2O. The source and upstream processing of hydrogen can have a significant impact on corrosion and fouling in hydroprocessing units since chlorides can contaminate these hydrogen streams. If the hydrogen is not scrubbed of chlorides prior to injecting into the hydroprocessing unit feed stream, it may contain HCl. Additional sources of chloride may be organochlorine substances, organic and inorganic chlorides in the hydrocarbon feed stream (see Sect. 3.1.1.1).

Hydrodesulfurization (HDS) is a catalytic chemical process using for removing sulfur from petroleum products, such as naphtha, gasoline, kerosene, diesel fuel, and fuel oil. The purpose of removing sulfur is to reduce the sulfur dioxide emissions that result from burning fuels. Another important reason for removing sulfur is that it, even in extremely low concentrations, poisons the noble metal catalysts. S-organic compounds react with hydrogen gas to form hydrocarbons and H_2S. Then H_2S is converted into elemental sulfur by the Claus process at sulfur recovery unit.

1.4 Catalytic Reforming

Catalytic Reforming is a catalytic process in which high octane gasoline (rich in aromatics) is produced from naphtha. This process was developed by the American chemist Haensel V. in the 1940s using a catalyst containing platinum and therefore received the name *Plat*forming process. Most catalysts contain platinum or rhenium on a silica or silica-alumina support base. Fresh catalyst is chlorinated before use and is the source of corrosion problems by chlorides appearing in reformate. The dehydrogenation also produces significant amounts of byproduct hydrogen gas, which is fed into other refinery processes, i.e. hydrocracking. In addition to a gasoline blending stock, reformate is the main source of aromatic compounds (BTXE) and is sent to petrochemical plant for their extraction.

Continuous Catalytic Regeneration (CCR) Reforming. Platforming units were non-regenerative or semi-regenerative, that is periodically shutdown to regenerate the catalyst. This regeneration includes burning off catalyst coke and reconditioning the catalyst's active metals. CCR was developed in 1971. The process employs continuous catalyst regeneration in which catalyst continuously removed from the last reactor, regenerated in a controlled environment, and then transferred back to the first reactor.

Isomerization unit is intended for transformation of linear molecules of hydrocarbons to branched molecules possessing by higher octane number for blending into gasoline or feed to alkylation units.

1.5 Supporting Processes

Amine Treating Unit. Acid gases mostly containing H_2S are absorbed by aqueous amine solution. The amine is regenerated by removing H_2S, which is fed to a sulfur plant (named also sulfur recovery unit).

Sulfur Recovery Unit (SRU). Hydrogen sulfide is converted to elemental sulfur by partial oxidation with air at this unit.

The process was invented by the chemist Carl Friedrich Claus in England in 1883 and bears his name. Sulfur is very important product in the arsenal of manufactured goods of oil refineries because it is a raw material for chemical and pharmaceutical industry. Sulfur is used alone as fertilizer and pesticide but mostly for manufacturing sulfuric acid, rubber, fertilizers, medicines, and cosmetics.

The oxidation of hydrogen sulfide is not complete, sulfur vapor and SO_2 gas remain in the Claus process. CS_2 and COS can be also present in remaining gases. The emissions of sulfur compounds to the atmosphere is strongly restricted by environmental regulations. In order to minimize the releasing of noxious sulfur mixture, it is utilized at the tail gas treating unit.

Tail Gas Treatment (TGT) reduces sulfur vapor, SO_2 and other sulfur contaminants contained in the tail gas from the Claus process to H_2S, absorbs it in the absorbing aqueous amine solution and returns it to the Claus process, thereby achieving a high efficiency of sulfur recovery.

Sour Water Stripper (SWS). The main purpose of sour water stripper is to remove H_2S and NH_3 from the process water so that the treated water can be disposed off in an environmentally acceptable condition. The H_2S and NH_3 are sent to the SRU. The FCC sour water is different from the sour water from the other units because it can contain phenols and cyanides.

Wastewater facility is intended for collection and treating of wastes forming at the units of oil refineries and petrochemical plants. These wastes contain mixture (emulsion) of crude oil, petroleum products, fuels, solvents and other organic substances, aqueous solutions, and suspended solids. The wastewater facility consists of API separators, dissolved gas flotation (DGF), and biotreatment units. At last, separated organic phase is sent to refinery units for retreatment and producing of useful products. Separated water is reused or disposed.

Wastewater facilities as a rule serve separately for treatment of wastes forming at oil refinery units and petroleum plant.

1.6 Petrochemical Plant

Benzene-Toluene-Xylene (BTX) unit separates aromatics from hydrocarbon mixtures by sulfolane extraction. Sulfolane is considered as stable substance and can be reused many times. However, when using long time it eventually degrades into acidic products. The following methods were developed to regenerate spent sulfolane: vacuum and steam distillation, back extraction, adsorption, and ion exchange.

TPX unit separates toluene and para-xylene from a mixture of xylene isomers.

Phthalic Anhydride unit produces solid phthalic anhydride by the catalytic oxidation of ortho-xylene by atmospheric oxygen at high temperature.

References

1. Gary JH, Handwerk GE (2001) Petroleum refining, 4th edn. Marcel Dekker, New York
2. Brugma AJ (1941) The Brugma process. Ref Nat Gasoline Manuf 20:86–88
3. Speight JG (2006) The chemistry and technology of petroleum, 4th edn. CRC Press, New York

Recommended Literature

4. Antos GJ, Aitani AM (eds) (2004) Catalytic naphtha reforming, 2nd edn. Marcel Dekker, New York
5. Jones DSJ, Pujado PP (eds) (2006) Handbook of petroleum processing. Springer, New York
6. Hudec P (2011) FCC catalyst—key element in refinery technology. In: Presented at the 45th international petroleum conference, June 13, 2011, Bratislava, Slovak Republic
7. Giddens PH (1955) Standard oil company (Indiana): oil pioneer of the middle west
8. Coletti F, Hewitt JF (eds) (2015) Crude oil fouling: deposit characterization, measurements, and modeling. Gulf Professional Publishing, Houston

Chapter 2
Physicochemical Properties of Crude Oils

The great aim of education is not pure knowledge but its application.
Herbert Spencer (1561–1626), an English philosopher.

Abstract Physicochemical properties and components of crude oils affecting their corrosiveness are described. Corrosiveness of crude oils is defined by water and salt content, total sulfur content, total acid number, microorganisms, and conditions (temperature, flow regime, etc.). Not all sulfur compounds and acids are corrosive to metals. Hydrogen sulfide is most corrosive among sulfur substances. The corrosiveness of crude oil containing water can be determined by a combination of three properties: the type of emulsion formed between oil and water, the wettability of the steel surface, and the corrosiveness of aqueous phase in the presence of oil. A case study with analysis and solution is given.

In order to understand corrosion problems and solutions in oil refining and petrochemical industry, we will describe physicochemical characteristics of crude oils and their corrosiveness. Other media, such as water (cooling water, boiler feed water, extinguishing water), steam, different gases and chemicals also can participate in corrosion of equipment at refining and petrochemical units. Crude oil types and their physicochemical properties are described in Appendix B. First, we will describe constituents of crude oils and components related to their corrosiveness.

2.1 Physicochemical Composition of Crude Oils

Crude oil is a mixture of numerous liquid hydrocarbons containing dissolved gases, water, and salts. Crude oils vary widely: from almost solid and heavy that sink in water up to light materials that float on water. Crude oils are, in fact, emulsions—drops of aqueous solution dispersed throughout the continuous hydrocarbon phase. Interstitial or connate water is always present in crudes. These water drops vary from almost fresh to saturated aqueous solutions of salts and are the main cause of

A. Groysman, *Corrosion Problems and Solutions in Oil Refining and Petrochemical Industry*, Topics in Safety, Risk, Reliability and Quality 32,
DOI 10.1007/978-3-319-45256-2_2

Table 2.1 Crude oil constituents [2, 3]

Constituent	Chemical type
Hydrocarbons:	
Paraffinic (Alkanes)	Straight chain; branched chain
Naphthenic	Alkyl cyclopentanes; alkyl cyclohexanes
Aromatic	Alkyl benzenes; aromatic naphthenic fluorenes; polynuclear aromatics
Dissolved gases	Nitrogen (N_2); carbon dioxide (CO_2)
Sulfur compounds	Elemental sulfur (S_8), hydrogen sulfide (H_2S)[a], mercaptans; organic sulfides, disulfides and polysulfides; thiophenes and benzothiophenes; sulfones
Organic nitrogen compounds	Pyridine, quinoline
Organic oxygen compounds	Carboxylic acids (including naphthenic acids)[b], alcohols, phenols[b], aldehydes, ketones, esters, ethers, oxyacids
Organic metallic compounds	Porphyrins
Colloidal particles	Asphaltenes; resins; paraffin waxes
Surfactants	Sulfonic acids, sulfonates, sodium napthenates
Metals	Vanadium, nickel[c], iron[c], aluminum, sodium, potassium, calcium, copper
Water (S&W[d] or BS&W[d])[e]	Fresh or saline
Solids	Sand, dirt, silt, soil dust, mud, corrosion products (metals' oxides, sulfides, salts)

[a]Hydrogen sulfide is present as dissolved gas
[b]They are surfactants
[c]They are present in porphyrins
[d]S&W—sediment and water; as previously called BS&W—bottoms sediment and water [2]
[e]Microorganisms can be present in crude oils

crude corrosiveness. Crude oils in addition to hydrocarbons can include also compounds containing sulfur (S), nitrogen (N), oxygen (O), and metals (Table 2.1). All these contaminants may be present in crude oils as dissolved gases, liquids and solids or distinctive phases. Microorganisms also can be present in crudes in active or dormant state. Every crude contains about the same kinds of compounds, but in different proportions. As a result, crude oil differ by their corrosiveness.

2.2 Corrosiveness of Crude Oils

Chemistry of crude oils is so complicated and there are so many factors influencing corrosiveness of crudes that it is difficult and even in most cases impossible to predict their corrosiveness according to physicochemical composition. Usually corrosiveness of crude oils is defined by total acid number (TAN), total sulfur content, water and salt content, and microorganisms. The combination of these parameters and components influences corrosion by different manner and differently at stages of preparation, transportation, storage, and processing crude oils at refineries. Both TAN value and total sulfur content do not always define corrosiveness of crudes. For instance, H_2S and alkanethiols are corrosive to carbon steel, while thiophenes are not corrosive. Therefore, it was suggested speciation of corrosive (H_2S and alkanethiols) and non-corrosive (thiophenes) organic sulfur compounds in crudes using physicochemical analytical methods with the aim to define potential corrosiveness of crudes [1]. The problem is complicated by the fact that H_2S and alkanethiols reacting with iron form iron sulfide scale which under particular conditions can be protective (see Fig. 3.1).

Sometimes crudes containing relatively high water content can inhibit corrosion and, vice versa, crudes may be corrosive at relatively low water concentrations. Only experiment can respond on the question about corrosiveness of crudes. Usually crude oils are not corrosive under pipeline transportation conditions. However, water can drop-out and accumulate at low velocities and under stagnant conditions. As a result, corrosive contaminants in crude are dissolved in separate liquid aqueous phase and cause corrosion of inner surface of pipelines. It is possible to use model for determination the velocities at which water could drop out of crude oil as a function of the crude oil density and viscosity [4]. Water is less likely to drop out at lower velocities when entrained in heavier crudes compared to light crudes. These velocities are well below normal operating velocities on most transmission pipelines. The deposition of sediments and water are minimized in highly turbulent flow.

The low corrosiveness of most crude oils (including oil sands) in transmission pipelines is due to the fact that corrosive (water, dissolved salts) and erosive (sand, mud) constituents are largely removed before transportation [5]. Quality specification for crude oil transporting in pipelines requires that basic sediment and water (BS&W) content should be <0.5 vol% because most corrosion problems are caused by dropping out the above mentioned corrosive contaminants of the crudes [6–10]. Corrosive sulfur compounds (excluding H_2S) and organic acids (including naphthenic acids) do not cause corrosion in pipelines at operating temperatures of transportation, usually at 5–50 °C. They can cause corrosion under refinery processing conditions at temperatures >190 °C (naphthenic acids) and >230 °C (sulfur compounds).

Water, salts, H_2S and microorganisms containing in crudes as contaminants can cause corrosion during storage in tanks or transportation in tankers. In the absence of water, the crude oil is noncorrosive in tanks and tankers, even some types of crude oils possess inhibitive properties. It is not easy to predict an extent of crude oil corrosiveness because of many complicated factors. The methods of determination of crude oil corrosiveness were developed and they will be considered in the next section.

2.3 Determination of Crude Oil Corrosiveness

The corrosiveness of crude oil containing water can be determined by a combination of three properties:

(a) The type of emulsion formed between oil and water (the emulation of the oil and water).
(b) The wettability of the steel surface.
(c) The corrosiveness of aqueous phase in the presence of oil.

These three properties are described below.

(a) The type of emulsion formed between oil (O) and water (W). Water and oil are immiscible but they can form emulsion under certain conditions. There are two kinds of emulsion: O/W and W/O. W/O emulsion (in which oil is the continuous phase) has low electric conductivity and is thus less corrosive; whereas O/W (in which water is the continuous phase) has high electric conductivity and, hence, can be more corrosive. The percentage of water at which W/O converts to O/W is known as the emulsion inversion point (EIP). EIP can be determined by measuring the electrical conductivity of the emulsion. At and above the EIP, a continuous (separate) phase of water or free water is present. Therefore, there is a potential for corrosion.
(b) The wettability of the steel surface. Whether water phase can cause corrosion in the presence of oil depends on whether the surface is oil-wet (hydrophobic) or water-wet (hydrophilic). Because of higher electrical resistance, an oil-wet metal surface is not susceptible to corrosion, but a water-wet surface is. It is important also the geometry type of metal surface on which water droplets are separated from crude oils: horizontal, sloping, or vertical. Horizontal metal surfaces, on which water droplets tend to be retained, are more prone to corrosion than sloping or vertical surfaces [11].
(c) Influence of impurities in oil phase on corrosiveness of aqueous phase. Dissolution of corrosive contaminants (H_2S, chloride salts, HCl—see Sect. 2.3.1, light organic acids) from crude oils may alter the corrosiveness of the aqueous phase. Crude oils can be classified as corrosive, neutral, or inhibiting crudes. This classification is based on the corrosiveness of the aqueous phase in the presence of crude oil [12–14].

Corrosion problems related to crude oils in addition to low velocity flow and stagnation conditions, can occur later in oil refineries if desalters do not function properly.

It is important to emphasize that crude oil supply patterns are often changing as production from traditional supply basins decline and new sources of supply come on-stream. Corrosiveness of crudes can also change respectively. Sometimes groundless allegations appear in mass media, for instance, about increased corrosiveness of new unconventional crudes. In this case, it is important to prove by means of experiment that these new crudes do not represent increased corrosion risk compared to conventional crudes. For instance, the bitumen produced in oil sands operations must be processed or mixed with lighter hydrocarbons for transport in pipelines. Some bitumen is upgraded to synthetic crude oil and called *synbit*. Bitumen mixed with light Canadian condensate is referred to as *dil*uted *bit*umen, or *dilbit* (see Appendix B). Mixtures of the two are sometimes referred to as *syndilbit* [15]. Corrosion rates of steel used for pipelines (API 5L X-65) in different types of crudes (dilbit, synbit and conventional crudes) changed in 0.001–0.03 mm/year (so low values!) both in aqueous and oil phases [15]. There was no correlation between physicochemical properties of crudes (TAN, sulfur and water content) and corrosiveness. Weight loss (WL) method of determination of corrosiveness of crudes has long exposure time and low reproducibility. Coupons (used in WL method) can sometimes gain weight due to adhesive product buildup. Radioactive tracer technology was used to measure high temperature crude corrosiveness with very high sensitivity during 1 h to avoid thermal degradation of organic acids [16]. The disadvantage of radioactive method is that it requires complicated equipment and skillful personnel.

Example of corrosion problems described above concerning crude oil at oil refinery is shown in the following case study.

2.4 Case Study

Conditions. Crude oil after desalter flowed in the shell at 140–190 °C and 12 bar.

Material of construction. Carbon steel A516 Gr.70. Wall thickness of the shell—12.5 mm.

Service period before the failure: 2 years. The heat exchanger was filled with crude oil and has not been worked during 8 months among 2 years of service.

Visual examination and findings. Deposits and pits of 6–7 mm in deepness were found on inner surface of the lower part of the shell (Fig. 2.1).

Corrosion pitting rate growth was 3–3.5 mm/year. Analysis of salt content in crude oil after the desalter showed that during 2 years these values changed in the range 2–6 ptb when allowable maximum concentration is 1 ptb.

(a) (b) (c) (d)

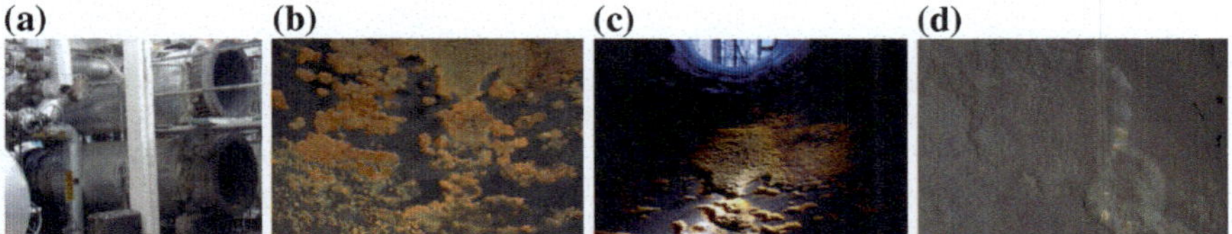

Fig. 2.1 **a** General view of failed two heat exchangers. **b–d** Inner surface of corroded heat exchanger shell

X-ray fluorescence (XRF) analysis. The deposits formed inside the shell of the heat exchanger contained Fe, S, Ca, Na, Si, and Cl.

Chemical analysis. Deposits (inorganic part) mostly consisted of FeS and $Fe_2(SO_4)_3$.

Failure phenomenon: *Under deposit acid corrosion* [17].

Cause of failure and its explanation. Crude oil was in stagnation and did not flow 8 months through the heat exchanger. Deposits from the crude precipitated on the lower part of the heat exchanger shell. Hydrogen sulfide contained in crude oil reacted with iron forming iron sulfide. Ingress of air (oxygen) into the shell caused oxidation of sulfides to sulfates. Presence of water in crude oil resulted in hydrolysis of ferric sulfate $Fe_2(SO_4)_3$ salts and as a result pH decreased to 1–2 (reactions 2.1 and 2.2):

$$Fe_2(SO_4)_{3(aq)} \xrightarrow{\text{Water}} 2Fe^{3+}_{(aq)} + 3SO^{2-}_{4(aq)} \tag{2.1}$$

$$Fe^{3+}_{(aq)} + H_2O_{(l)} \rightarrow H^+_{(aq)} + (FeOH)^{2+}_{(aq)} \tag{2.2}$$

Thus, corrosion occurred under acid deposits of ferric sulfate salts.

Solutions and recommendations.

(a) In no case not to leave crude oil without flow in heat exchanger shell. To fill shell with gas oil for the stand-by period.
(b) To repair pits by welding.

References

1. Ávila BMF et al (2014) Speciation of organic sulfur compounds using comprehensive two-dimensional gas chromatography coupled to time-of-flight mass spectrometry: a powerful tool for petroleum refining. Fuel 126:188–194
2. Manning FS, Thompson RE (1995) Oilfield processing of petroleum: vol two: crude oil. Penn Well Publishing Company, Tulsa
3. Groysman A (2014) Corrosion in systems for transportation and storage of petroleum products and biofuels. Springer, Dordrecht
4. NACE Standard SP0208 (2008) Internal corrosion direct assessment methodology for liquid petroleum pipelines. NACE International, Houston, TX, USA
5. Larsen KR (2013) New standard determines corrosivity of crude oils. Mater Perform 52(2):18
6. Been J et al (2011) Development of a test protocol for the evaluation of underdeposit corrosion inhibitors in large diameter crude oil pipelines. Paper 11263 presented at the NACE CORROSION conference, NACE International, Houston, TX, USA
7. Been J, Place TD, Holm M (2010) Evaluating corrosion and inhibition under sludge deposits in large diameter crude oil pipelines. Paper 10143 presented at the NACE CORROSION conference, NACE International, Houston, TX, USA
8. Place TD, Holm MR, Cathrea C, Ignacz T (2009) Understanding and Mitigating large Pipe Underdeposit Corrosion. Mater Perform 48(1):54–60

9. Salama MM (2000) Influence of sand production on design and operations of piping systems. Paper 00080 presented at the NACE CORROSION conference, NACE International, Houston, TX, USA
10. Larsen KR (2013) Managing corrosion of pipelines that transport crude oils. Mater Perform 52(5):28–35
11. ASTM D3603–14 (2014) Standard test method for rust-preventing characteristics of steam turbine oil in the presence of water (horizontal disk method). ASTM International, West Conshohocken, PA, USA
12. ASTM G205–10 (2012) Standard guide for determining corrosivity of crude oils. ASTM International, West Conshohocken, PA, USA
13. TM0172-2001 (2001) Determining corrosive properties of cargoes in petroleum product pipelines. NACE International, Houston, TX, USA
14. ASTM D665–14e1 (2014) Standard test method for rust-preventing characteristics of inhibited mineral oil in the presence of water. ASTM International, West Conshohocken, PA, USA
15. McIntyre DR et al (2014) Laboratory tests comparing the corrosivity of dilbit and synbit with conventional crudes under pipeline conditions. Paper 3824 presented at the NACE CORROSION conference, NACE International, Houston, TX, USA
16. Eberle DC, Wall CM, Treuhaft MB (2006) Crude oil corrosivity measurement using radioactive tracer technology. Presentation at the 2006 spring meeting & 2nd global congress on process safety, 9th topical conference on refinery processing, AIChE, USA
17. Groysman A (2010) Corrosion for everybody. Springer, Dordrecht

Recommended Literature

18. Hatch LF, Matar S (1977) From hydrocarbons to petrochemicals, part 2. Hydrocarb Process 56(6):189–194
19. Matar S, Hatch LF (1994) Chemistry of petrochemical processes. Gulf Publishing Company, Houston
20. Dowd WT, Kuuskraa VA (1988) A technical and economic assessment of domestic heavy oil. Final report. DOE report, DOE/BC/10840-1
21. Rossini F (1960) Hydrocarbons in petroleum. J Chem Educ 37(11):554–561
22. Nelson WL (1958) Petroleum refinery engineering, 4th edn. McGraw-Hill Book Co, New York

Chapter 3
Physicochemical Basics of Corrosion at Refineries' Units

Only fools and charlatans know everything and understand everything.
Anton Chekhov (1860–1904), a Russian writer.

Abstract Physicochemical mechanisms of metal corrosion at refineries' units are described. Low- ($T < 100$ °C) and high- ($T > 200$ °C) temperature corrosion problems are analyzed. Corrosion by hydrochloric acid and its prevention, electrochemical mechanism of corrosion in acid, neutral and alkali aqueous solutions of electrolytes are explained. Main factors affecting corrosion rate and its intensity are temperature, pressure, flow regime, and media. The following chemical compounds and mixtures are analyzed from corrosiveness: water, air, hydrogen sulfide, substances containing nitrogen, phenols, polythionic acids, organochlorine and organic chloride compounds, aluminum chloride, sulfuric acid, and alkaline solutions (sodium hydroxide, sodium carbonate, ammonium hydroxide, amines, sour water, and spent caustic).

Corrosion of metallic equipment and structures at refineries' units occur in different media and phases under very wide conditions. It is possible to divide corrosion at refineries' and petrochemical units into five groups:

(a) Low temperature ($T < 100$ °C) corrosion in the presence of electrolytes, usually water and aqueous solutions of electrolytes, such as dissolved corrosive gases (e.g., HCl and H_2S), or dissolved salts (e.g., NaCl and Na_2SO_4). We can add ammonia (NH_3) to this list but we should remember that ammonia is a 'double-faced' towards metals. For iron, ammonia is a corrosion inhibitor, whereas for copper, it is a very aggressive substance which can dissolve it and induce SCC. Really H_2S, NaCl and Na_2SO_4 are also 'Janus-faced' towards carbon steel corrosion. At some definite H_2S concentrations and conditions, it can form protective passive film on carbon steel surface (see Fig. 3.1). Salts NaCl and Na_2SO_4 at concentrations up to $\sim$3 wt% in water accelerate carbon steel corrosion, but while adding very large concentrations (>26 wt%) they cause salting-out of dissolved oxygen, and corrosion of carbon steel is nil [1].

A. Groysman, *Corrosion Problems and Solutions in Oil Refining and Petrochemical Industry*, Topics in Safety, Risk, Reliability and Quality 32,
DOI 10.1007/978-3-319-45256-2_3

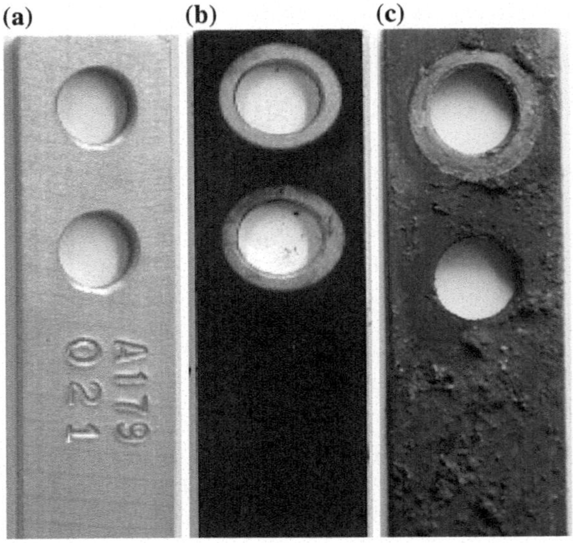

Fig. 3.1 Carbon steel coupon: **a** original, **b** with protective smooth thin tenacious iron sulfide film (5–20 μm in thickness), **c** with thick fragile iron sulfide non-protective layer (50–100 μm in thickness)

(b) High temperature (T > 200 °C) corrosion by non-electrolytes (usually gaseous H_2S and H_2; naphthenic acid corrosion; hot ash corrosion; oxidation by oxygen in furnaces).
(c) Intermediate temperature (100 °C < T < 200 °C) corrosion which can occur either in the presence of electrolytes or non-electrolytes depending on substances and conditions.
(d) Corrosion in natural environments: in the atmosphere, soil, and natural waters as equipment and constructions at oil refineries and petrochemical plants can contact them.
(e) Specific corrosion phenomena: acid dew point corrosion, corrosion under thermal insulation, boiler feedwater corrosion, and steam condensate corrosion.

The first group (T < 100 °C) occurs according to electrochemical mechanism: corrosion in acid, neutral and alkali aqueous solutions. Corrosion in aqueous amine solutions (in amine treating units) belongs to this group. The second group (T > 200 °C) falls into non-electrochemical mechanism, i.e. sulfidic corrosion, naphthenic acid corrosion, hydrogen damages, hot ash (or fuel ash), and oxidation (by oxygen) corrosion. We should mention that this is conditional division and is made only for convenience to describe by known and accepted corrosion mechanisms. Some corrosion phenomena, i.e. hydrogen damages, can occur both at low and high temperatures (see Sect. 4.3). In the third interjacent group which occurs between 100 and 200 °C, the presence of electrolytes in water and increase of pressure expands the existence of liquid aqueous phase above 100 °C and realization of electrochemical mechanism at higher temperatures. For instance, aqueous solutions of sulfuric acid (H_2SO_4) can exist in liquid state up to 160–180 °C depending on its concentration. The situation can be even more complicated when we consider corrosion in hot hydrocarbons (i.e. gasoline, kerosene or gas oil at ~100–290 °C). Rectification of crude oil in atmospheric distillation column proceeds in the presence of steam which is added to the bottom of the column to facilitate fractionation of hydrocarbons. Water (H_2O) can be in the form of either steam or liquid condensate in hydrocarbon organic phase at 100–290 °C. Because of the polarity of H_2O molecules, they are adsorbed on metal surface and it is very difficult to remove them from it. Rough non-uniform surface of metal helps water molecules retain under hydrocarbon phase on the metal surface, and corrosion can occur in a thin electrolytic water layer similar to atmospheric corrosion. If cations (i.e. Fe^{2+} and Fe^{3+}) are present in aqueous phase, hydrates ($FeCl_n \cdot mH_2O$) are firmly held on metal surface. The difference between corrosion in hydrocarbons (e.g., in petroleum products and fuels) and atmospheric corrosion is that the atmosphere is the reservoir of water and dissolved oxygen for occurring of electrochemical mechanism in thin layer of water (electrolyte) on metal surface. Hydrocarbon organic phase (mixed with steam) is the supplier of water, hydrogen sulfide, sometimes dissolved oxygen, for the occurring of electrochemical mechanism on metal surface. Really adsorbed water layer can remain on metal surface till 300 °C.

The fourth (corrosion in natural environments) and fifth (specific corrosion phenomena, such as acid dew point corrosion and corrosion under thermal insulation) groups are described in detail in [1, 2]. However, corrosion cases and their analysis will be given in this book too (see Chaps. 5, 6).

3.1 Low-Temperature Corrosion

Most low-temperature corrosion problems in oil refineries and petrochemical plants are not caused by hydrocarbons that are processed but by various inorganic compounds, such as H_2O, H_2S, HCl, NH_3, NaOH, H_2SO_4, Na_2CO_3, and dissolved oxygen. Organic acids that are contained or occasionally formed in hydrocarbons and amines using as neutralizers and corrosion inhibitors in the overhead of distillation systems may also promote low-temperature corrosion (see Sects. 8.2.1 and 8.2.2).

There are two sources of compounds causing corrosion: contaminants contained in crude oil and process chemicals. Prolonged use of aqueous solutions of amines for the absorption of acid gases H_2S and CO_2 at amine treating units can result in formation of corrosive heat stable amine salts (see Sect. 4.4). Use of sulfolane for a long time also can cause formation of destructive corrosive products (acids) at petrochemical plant (see Sects. 5.4.1.4 and 5.4.1.5).

Hydrochloric acid takes a special place in corrosion problems at refinery and petrochemical units. Hence, we will describe corrosion by hydrochloric acid and its mechanism, then corrosion by other compounds.

Corrosion by hydrochloric acid and its prevention. Hydrochloric acid appears in crude oil and petroleum products as a result of hydrolysis of calcium and magnesium chloride salts ($CaCl_2$, $MgCl_2$), organochlorine (RCl; R—radical, i.e. methyl CH_3– or ethyl C_2H_5–) compounds, and aluminum chloride ($AlCl_3$) catalyst (Appendix C). In addition, some chlorides can appear from chemicals used in crude oil production. Most corrosive contaminants are removed during preliminary treating of crude oils in the fields and then during settling in storage tanks. However, some contaminants (mostly chloride salts and additional water) can pick up in marine tankers and pipelines. Inorganic salts ($CaCl_2$, $MgCl_2$, and NaCl) remain in any case in small amounts in crude oil after desalting. The hydrolysis of the first two salts is greater at higher temperatures, especially at 120–150 °C. Sodium chloride is stable to 760 °C [3]. Hydrogen chloride can primarily evolve in crude preheat furnaces. Preferable and recommended salt content in crude oil after desalter is below 1 PTB (pound of salt per thousand barrels of crude), corresponding to ~3.5–4 ppm HCl. This acid remaining in crude oil after desalter enters distillation column and goes up with light hydrocarbons, water vapor and H_2S into upper part (overhead) of the column. Corrosion by hydrochloric acid is a major problem in the overhead of crude distillation columns. It can occur also in reforming and hydrotreating units. Organochlorine compounds (RCl) are used in reforming units to regenerate reformer catalyst. Hydrogen chloride is stripped off the catalyst if moisture is present in the reformer feed. Chlorides are formed as a result of hydrolysis of organochlorine catalyst (Appendix C). This situation causes corrosion in reforming units, and then in hydrotreating units that use excess hydrogen (containing traces of HCl) from the reformer. In addition, reformate (containing even several ppm chlorides) from the CCR unit as a feedstock to petrochemical plants can cause corrosion of different equipment, from piping and pumps to tanks at BTX and TPX units (see Sect. 5.4). Chloride scavengers must be installed at CCR unit for complete removing of chlorides from reformate. Hydrogen gas containing traces of HCl should be passed through a sodium carbonate solution scrubber for complete HCl removing. Malfunction (insufficient absorption) of a scrubber results in severe corrosion of carbon steel piping and valves (see Sect. 5.5.1.8).

How does hydrogen chloride cause corrosion of carbon steel equipment in the overhead of atmospheric distillation column? Dry HCl gas is not corrosive. In the overhead of distillation column at T > 100 °C (usually 120–130 °C), HCl gas and water vapor (H_2O) mixed with light hydrocarbons are not corrosive to carbon steel. However, HCl becomes corrosive at initial water dew point (T ≤ 100 °C) when contacts liquid water condensate. Usually this happens in air coolers but can take place before, even in upper part of the distillation column under faulty operation. Therefore, it is very important to maintain temperature in the overhead of distillation column 20–30 °C more than water dew point at atmospheric pressure. Depending on concentration of HCl, values of pH of water condensate can be in the range 2.5–4 (Table 3.1). During desalter fault and the presence of organochlorine compounds in crude oil pH values may be even less.

Intensive corrosion of carbon steel begins at pH < 4 [4]. This means that the acid water condensate should be neutralized (see Sect. 8.2.1.1). This is realized by injection of two chemicals, ammonia and amine, into the overhead vapor line of the

Table 3.1 Affect of HCl concentration in water condensate on pH

HCl, ppm	5	10	15	20	30	50	70	100
pH	3.86	3.56	3.39	3.26	3.08	2.86	2.72	2.56

crude oil distillation column to maintain the pH = 5.5–6.5 of the condensate. Why not neutral value pH = 7? Three reasons exist for this:

(a) At higher pH (≥7) deposits (fouling) of ammonium chloride (NH_4Cl) and amine chloride (RNH_3Cl) salts can be formed which represent not less danger for occurring acid localized under deposit corrosion. These vulnerable locations are trays in distillation column, header boxes of air coolers, heat exchanger shells and baffles of tube-sheets.
(b) Corrosion of carbon steel in sour water can increase at pH ≥ 7 (see Sect. 3.1.1 and Appendix D).
(c) In addition to neutralizers (ammonia and amine), corrosion inhibitors (amines and other organic compounds) are injected into the overhead vapor line. Corrosion inhibitors function well in pH = 5.5–6.5 region.

Usually ammonia is used as 25 wt% aqueous solution NH_4OH. This solution quickly neutralizes acid solution while amines neutralize relatively smoother. If overdosage of ammonia solution occurs, pH rapidly raises above 7 which can cause ammonium chloride deposits. In order to prevent such situation, in addition to ammonia solution, amine is injected too. Because of safety and environmental problems which ammonia application brings (see Appendix E), so called 'low salt' amines are injected alone, which do not cause formation of amine chloride salt deposits.

Electrochemical mechanism of corrosion in acid solutions. When carbon steel equipment in the overhead of distillation column contacts aqueous solution of acids HCl or H_2S, iron corrodes according to electrochemical mechanism which includes two reactions, anodic and cathodic, and they occur simultaneously [1]. Anodic dissolution of iron—exit of positive metallic ions (Fe^{2+}) into solution and releasing of electrons on the metal surface:

$$Fe_{(s)} \rightarrow Fe^{2+}_{(aq)} + 2e^- \quad (3.1)$$

These electrons move along the metal surface to adjacent area (with higher electric potential) where $H^+_{(aq)}$ cations (existing as a result of dissociation of acids) are collecting, and cathodic process occurs:

$$2H^+_{(aq)} + 2e^- \rightarrow H_{2(g)} \quad (3.2)$$

As $H^+_{(aq)}$ cations in water immediately combine with $H_2O_{(l)}$ molecules forming hydronium ions $H_3O^+_{(aq)}$, more correctly to write cathodic reaction (3.2) in the form:

$$2H_3O^+_{(aq)} + 2e^- \rightarrow H_{2(g)} + 2H_2O_{(l)} \quad (3.3)$$

Both (3.2) and (3.3) records of cathodic reaction are equivalent. Really iron in contact with any acid (i.e. H_2S, H_2SO_3, H_2SO_4, H_3PO_4, HNO_3, CH_3COOH) corrodes according to this electrochemical mechanism, and complete reaction may be written in ionic (3.4) or molecular (3.5) forms:

$$Fe_{(s)} + 2H^+_{(aq)} \rightarrow Fe^{2+}_{(aq)} + H_{2(g)} \quad (3.4)$$

$$Fe_{(s)} + 2HCl_{(aq)} \rightarrow FeCl_{2(aq)} + H_{2(g)} \quad (3.5)$$

The result is the appearance of dissolved iron salts in solution or undissolved salt layers (in the case of H_2S and H_3PO_4) on iron surface.

Electrochemical corrosion mechanism in neutral and alkali aqueous solutions of electrolytes. Water is a weak electrolyte. This means that there are no many H^+ ions which can take part in cathodic reaction when iron contacts water. Another component, dissolved oxygen in water, takes part in uptake of electrons on the metal surface:

$$O_{2(g)} + 2H_2O_{(l)} + 4e^- \rightarrow 4OH^-_{(aq)} \quad (3.6)$$

The cathodic reaction (3.6) prevails in neutral and alkali aqueous solutions of electrolytes than the reaction (3.3) because concentration of $O_{2(g)}$ is 1000 greater than that of $H_3O^+_{(aq)}$. Such electrochemical mechanism with cathodic reaction (3.6) occurs when metallic equipment contacts water, soil or atmosphere of relatively high humidity. Both anodic dissolution of metal and cathodic reaction (reduction of O_2 or H_3O^+) proceed simultaneously. Anodic reaction cannot proceed without cathodic reaction and vice versa. Avoiding cathodic reaction, we can stop also anodic dissolution of metal, and as a result corrosion. Removing dissolved oxygen from water or aqueous solution we eliminate cathodic reaction (3.6) and thus also stop anodic (corrosion!) reaction (3.1). This is the basics of deaeration of water when preparing for boilers (boiler feed water). In addition to HCl, numerous chemical substances influence low-temperature corrosion, and they will be analyzed in the next section.

3.1.1 Factors Influencing Low-Temperature Corrosion

Main factors affecting corrosion rate and its intensity are temperature, pressure, flow regime, and media. Changes in conditions and process parameters can also influence the effect of all these factors.

Temperature. Temperature has complicated effect on corrosion. Solubility of aggressive gases (O_2, H_2S, and CO_2) in water and in liquid hydrocarbons decreases with increase of temperature and at high temperatures concentration of these gases in liquids is very low. However, the rates of corrosion reactions increase with temperature growth. This situation leads to the position that corrosion rate as a function of temperature passes through maximum (at 60–80 °C) in open aqueous systems [1]. At T ~ 100 °C corrosion rate of carbon steel in aqueous solutions is close to zero in open systems because solubility of oxygen is nil, and there is no main 'actor' (dissolved oxygen O_2) in cathodic reaction, and as a result anodic reaction (corrosion) is also depressed. Increase of temperature also results in better dissociation of water molecules and decrease of pH and thus to acidification. For instance, for pure water, pH = 7.0 at 25 °C and pH = 4 at 90 °C.

Solubility of water in liquid hydrocarbons increases with increase of temperature. Decrease of temperature causes separation into two phases: aqueous and hydrocarbon.

Considering the conditions in the overhead of distillation column, where the temperature is 120–130 °C at 1 bar, there is only one vapor phase, containing light hydrocarbons, water vapor (H_2O), HCl, and H_2S gases. Piping from the upper part of distillation column to the air cooler has no insulation because light hydrocarbons flowing inside the pipe should be cooled. Injection of ammonia in the form of gas (NH_3) or liquid solution (NH_4OH) at T > 100 °C cannot result in condensation of water vapor. However, when T < 100 °C, liquid water containing two acids, HCl and H_2S, are formed.

Ammonia and neutralizing amine neutralize this acid condensate solution to pH = 5.5–6.5 (required range). Side effect of the injection of ammonium and amine compounds is that they react with chlorides, and solid ammonium chloride (NH_4Cl) and amine chloride $[RNH_3]^+Cl^-$ salts can be formed and deposit on metal surface. These salts are not corrosive under dry conditions. However, the environment is not dry and salts are hydrolyzed in the presence of water with formation of strong $HCl_{(aq)}$ acid (see Appendix C). Local appearance of HCl under salt deposits lead to pitting corrosion. Therefore, it is important to wash and remove these acid deposits (see Sect. 8.2.1.2).

Influence of pressure. Increase of pressure causes increasing of concentration of H_2S and CO_2 in aqueous and hydrocarbon phases. Thus, pH decreases and aqueous phase becomes more acidic and corrosive. Liquid hydrocarbon phase also becomes more corrosive because of greater concentration of acidic gases in it.

Flow regime. Flow regime of liquid and gaseous phases influence corrosion significantly. There may be one, two and three phases (aqueous-hydrocarbon-gas) flow. Their presence depends on temperature, pressure, amounts and type of liquids and gases, constructive features of pipelines (diameter, bends, orientation, and presence of valves, tees, and branches). Thus, liquids may be ranged from a finely dispersed mist to separate stratified phases.

Considering flow regime in the overhead, it is important to emphasize that the mist (good dispersion and distribution) is the best condition for injection of neutralizer and corrosion inhibitor. Stratified flow prevents intimate mixing of liquids and thus slows down mass transport and affects the ability of corrosion inhibitors to reach metal surfaces. Turbulence flow can remove protective corrosion product layer or inhibitor film from the surface of the metal.

3.1.1.1 Chemical Compounds

Following contaminants in crude oil can influence low-temperature corrosion: water, air, salts, hydrogen sulfide and other sulfur inorganic compounds, S- and N-containing organic compounds. In addition, sour water, spent caustic, polythionic acids, organochlorine and organic chloride compounds, aluminum chloride and different alkaline solutions (including amines) make corrosion problems. Influence of salts $CaCl_2$ and $MgCl_2$ which hydrolyze with HCl evolution were described earlier (see Sect. 3.1 and Appendix C). Influence of other contaminants will be analyzed in this section.

Water. Water can be present in crude oils as an emulsion or a separate phase and it is impossible to remove it completely. Liquid water is formed during stripping of steam in atmospheric distillation column and is produced in hydrotreating operations. Water vapor can enter together with air, dissolve in hydrocarbons and then condense on metal surface causing corrosion.

Air. There are many ways to air to penetrate into equipment and processing streams: into vacuum system and into pumps if seals or connections are not tight, during transportation, storage in tanks, different operations (i.e. filling-emptying of tanks and other vessels), and during shutdowns. When air enters into atmospheric or vacuum distillation system, hydrogen sulfide can be oxidized to sulfate ions (SO_4^{2-}) which decrease pH values and increase corrosion of carbon steel. Evidence of this is that sometimes hundreds of ppm SO_4^{2-} are determined in atmospheric and vacuum water condensate (see Sect. 8.2.3). During shutdowns, it is impossible to isolate equipment from penetration of air (oxygen, water vapor, sometimes aggressive ions, such as chlorides). Thus, most equipment is exposed to aggressive air and uncontrolled atmospheric corrosion usually occurs during shutdowns. Corrosion products (mostly rust) can be wiped away during start-up in different locations of equipment, contaminate hydrocarbon streams and plugged heat exchangers, piping, and strainers. So, air can be detrimental not only in corrosion sense but also in fouling (see Chap. 7).

Hydrogen sulfide (see Appendix D). Hydrogen sulfide (H_2S) is present in original crude oils but also can be formed as a result of activity of sulfate reducing bacteria (SRB) during storage or transportation of crude and destruction of sulfur organic compounds at high temperatures (in furnaces and distillation columns). Hydrogen sulfide dissolves in water (forming weak hydrosulfuric acid H_2S, pH $\sim$ 4) but in hydrocarbons significantly better. Hydrogen sulfide is present in light fractions exiting from atmospheric distillation column, fractionation column in FCCU, in hydroprocessing (hydrocracker and hydrotreater) effluent streams, in stabilizer and splitter at Visbreaker unit, in sulfur recovery units (SRU), and in sour water strippers (SWS). Hydrogen sulfide is the main component of sour water.

We will describe corrosion in the presence of H_2S. Iron, copper, nickel, zinc and some their alloys are not resistant to H_2S:

$$Fe_{(s)} + H_2S_{(aq)} \rightarrow FeS_{(s)} + H_{2(g)} \quad (3.7)$$

The reaction (3.7) proceeds in two electrochemical processes (anodic and cathodic) in aqueous electrolytic phase (similar to reactions 3.1 and 3.2). This corrosion reaction occurs at $T < 100$ °C on inner surfaces of piping, air coolers and heat exchangers in the overhead of atmospheric distillation unit and after stabilizer at hydrocracker unit (see Sect. 4.1). We can call this attack by low-temperature sulfidic corrosion where electrochemical mechanism is realized (dissolved H_2S aqueous solution is electrolyte). This is opposed to high-temperature sulfidic corrosion where chemical mechanism takes place (H_2S gas is non-electrolyte) (see Sect. 4.1.1). Intensity of low-temperature sulfidic corrosion depends on conditions of forming of sulfide layers and their composition. Their structure and morphology effect corrosion of carbon steel. Corrosion products are different iron sulfides (ranged in increasing order of stability): iron(II) sulfide (FeS, amorphous); greigite (iron(II, III) sulfide, Fe_3S_4; analog to magnetite Fe_3O_4); pyrrhotite ($Fe_{1-x}S$, where x = 0–0.2, Fe_7S_8); troilite (FeS, crystallic); mackinawite ($Fe_{1+x}S$, where x = 0–0.1 or Fe_9S_8); marcasite (iron(II) disulfide FeS_2, orthorhombic); pyrite (iron(II) disulfide FeS_2, cubic). Sometimes mackinawite can alter to smythite (Fe_9S_{11}) [5–7]. Their relative content in layer depends on the chemical composition of alloy, H_2S concentration, temperature, and pH. At pH $\sim$4 and ambient temperature, mostly mackinawite is formed, while at pH $\sim$7 pyrrhotite is favored [8]. Elemental sulfur can coexist with various iron sulfides.

Iron sulfides are non dissolved in water. Like most things in the Universe, sulfides also have 'two faces'. These sulfide films can be protective or induce corrosion. Iron sulfide films are electrically conductive, their electric potential is 0.2–0.4 V more positive than that of iron and as a result they are cathodes to iron, namely, accelerate cathodic reaction and thus, corrosion [9]. When H_2S concentration is small, smooth thin protective film with thickness of 5–20 μm is formed (Fig. 3.1b).

This situation takes place in the overhead of distillation column and in amine treating units. These films are well adsorbed to carbon steel surface and are removed by diluted hydrochloric acid. When H_2S concentration is relatively large, thick fragile iron sulfide non-protective layers like powder (or easily broken) are formed. Thickness of these iron sulfide deposits is 50–100 μm (see Fig. 3.1c).

Nitrogen-containing compounds. Diverse organic compounds containing nitrogen can be present in crude oils. Usually these compounds are either inert or can inhibit (pyridine, some amides) corrosion of steel at ambient temperature. At high temperature, N-containing organic compounds can be destroyed with formation of ammonia (NH_3) and hydrogen cyanide (HCN) which contribute corrosion problems of some alloys. This chemical transformation usually takes place in catalytic cracking, hydrotreating, and hydrocracking where NH_3 and HCN, in combination with H_2S, become the major constituents of sour water that can be highly corrosive to carbon steel.

Sour water. *Sour* means 'having a sharp, bitter, unpleasant smell, or taste, like a lemon, and not sweet'. This is synonym to 'acidic' indicating the presence of H_2S in the medium, e.g., *sour crude*. *Sour water* is the wastewater produced during many refining processes. It contains not only H_2S, but also NH_3, HCN, mercaptans, phenols (carbolic acid), chlorides, fluorides, spent caustic, and suspended solids. Such components of sour water as NH_3, H_2S, mercaptans and phenols have a strongly unpleasant smell. The largest source of sour water comes from the steam stripping of the crude oil in the atmospheric and vacuum distillation columns. FCCU, catalytic reforming, HDS, crude desalting, hydrocrackers, isomerization, visbreaking, coking, SRU, wash systems and amine systems also generate substantial amounts of sour water (see Sect. 8.2.1.2). Ammonia and hydrogen sulfide are present in relatively large concentrations (3.5–4.8 wt%) in sour water and react to form ammonium bisulfide (NH_4HS) at pH ~ 9 which can cause corrosion (see Appendix D). Therefore, this type of corrosion is called *alkaline sour water* or *ammonium bisulfide corrosion*. Most problems of sour water corrosion exist in the vapor recovery section of catalytic cracking units and in reactor effluent and light ends sections of hydrotreating and hydrocracking units, and in the overhead of amine regenerators (strippers) of SRU. Depending on composition of sour water (especially NH_4HS concentration) and flow velocity, it can become extremely corrosive to carbon steel and copper alloys. Severe corrosion of carbon steel occurs when NH_4HS concentration exceeds 2 wt% and fluid velocity is greater than 6.8 m/s [10–12].

The H_2S partial pressure influences significantly corrosion rate of alloys, especially at higher NH_4HS concentrations, and even affects stainless steel 20Cb-3 and nickel alloys Incolloy 825, Inconel 600, and Inconel 625. When the H_2S partial pressure is 3.5 bar, corrosion rate of carbon steel is 0.2 mm/year at 55 °C and 6.1 m/s. When the H_2S partial pressure is 7 bar, corrosion rate of carbon steel is 1 mm/year under the same conditions [13].

Stainless steel is usually resistant to ammonium bisulfide corrosion. However, chlorides (even 6–9 ppm) can cause pitting and high fluid velocity (accompanied by turbulence) can result in erosion-corrosion of SS 316 trays in stripper [14, 15]. Titanium is resistant to these conditions (see Sect. 8.1.1). Cyanides accelerate ammonium bisulfide corrosion under alkaline conditions (see Appendix D). The presence of HCl in sour water can result in aqueous HCl corrosion, formation of NH_4Cl fouling, which also can be extremely corrosive because of its acidic character when hydrolyzed (see Sect. 3.1 and Appendix D). The initial water condensation point gives rise to formation of aqueous HCl and severe corrosion when the HCl is present in the effluent vapor. The HCl gas dissolves well in water, even at elevated temperatures (~100 °C). Ammonia present in the effluent vapor stream (together with HCl) also will be absorbed into the water condensate, but to a much lesser extent than HCl. Therefore, the first water to condense may be acidic and corrosive. Only in the downstream separator fully condensed aqueous phase will be alkali.

In addition to corrosion problems, NH_4HS and NH_4Cl salts can cause fouling problems (see Chap. 7). Solid NH_4Cl salt can be formed directly from NH_3 and HCl which are present in the effluent stream, depending on their concentrations and temperatures. These two salts can deposit at T > 100 °C. Dry NH_4Cl salt is not corrosive but can create significant fouling problems. 'Dry' is defined as solid salt in equilibrium with a vapor phase with less than 10 % relative humidity [16]. However, NH_4Cl salt is hygroscopic, and tend to absorb moisture from the process stream. Wet NH_4Cl salt can be very corrosive because it is acidic (see Appendix C). The following solutions are used in order to control corrosion and fouling occurring from NH_4Cl deposits.

(a) Determining the source of chlorides and attempting to lower their concentration or eliminate them from unit feed (hydrocarbons and hydrogen) streams. The hydrogen should be passed through an adsorbent (absorbent)-filled drum to scrub HCl.
(b) Maintaining the temperature above the salt deposition temperature.
(c) Intermittent or continuous water wash.

Sour water corrosion sometimes can be accompanied by hydrogen damages. Following anti-corrosion measures are recommended to control sour water corrosion [13, 17]:

(a) To maintain NH_4HS concentration less than 2 wt% and fluid velocity less than 6.8 m/s. Carbon steel can be used under these limit conditions. Copper alloys are not resistant to both NH_3 and H_2S, even Nickel alloy 400 (Monel) corrodes.
(b) Temperature in reflux should be above 40 °C to avoid sublimation of NH_4HS on trays.
(c) It is possible to reduce NH_4HS and NH_4Cl concentrations in the condensation area by injection of calculated amount of wash water. Water is injected to remove salts and scrub HCl gas. Quality of wash water is very important to prevent corrosion and fouling (Table 3.2). Water sources can be stripped (or recycled) sour water, steam condensate, or boiler feed water. Stripped sour water containing cyanides should be avoided. When water injection is not possible, the use of corrosion resistant alloys is required.
(d) The following alloys can be used in sour water: Ti Gr.2, SS 304 (in the absence of chlorides), duplex stainless steels, Incoloy 825, and aluminum. Following relative resistance of alloys to sour water corrosion is defined: Hastelloy C276 > AL-6XN > 2507 Duplex SS > Inconel 625 > Alloy 20Cb-3 > Incoloy 825 > Inconel 600 > Incoloy 800 > 2205 Duplex SS > SS 316 > SS 304 > SS 410> Ni Alloy 400 > carbon steel [17–19]. Chemical composition of alloys is given in Appendix F.
(e) Injection of corrosion inhibitors (water-soluble amines, ammonium or sodium polysulfide) into the effluent system to control both corrosion and hydrogen attack.

Table 3.2 Parameters of injected wash water [10]

Parameter	Unit	Maximum value	Desirable value
Oxygen[a]	ppb	50	15
pH		9.5	7.0–9.0
Total hardness	ppm (as $CaCO_3$)	2	<1
Dissolved iron[b]	ppm	1	0.1
Chlorides	ppm	100	5
H_2S	ppm	1000	<1000
NH_3	ppm	1000	<1000
Cyanide (CN^-)	ppm	0	0
Total suspended solids	ppm	0.2	0

[a]The presence of oxygen enhances the potential corrosion due to chloride pitting. Oxygen will react with the bisulfide ion to form elementary sulfur (see Appendix D). Stripped sour water is generally oxygen-free
[b]Iron concentration should be low because iron in the water will form insoluble iron sulfides and can deposit in the equipment. For instance, 11.3 m^3/h of wash water containing only 1 ppm of iron will form over 14 kg of iron sulfide in a year [10]

(f) Periodic or continuous blowdown of sour water to the sour water stripping unit to lessen the concentrations of corrosive components (mostly H_2S, NH_3, and HCN) in the overhead water condensate.
(g) Design of header boxes of reactor effluent air cooler systems should ensure good flow distribution through tube rows. It is not recommended to use U-tubes.
(h) Inspection plan should provide checking of all possible corrosion problems: general and localized corrosion, hydrogen damages (blistering, SSC, HIC, and SOHIC).

Phenols. Any extraneous oxygen present in the catalytic cracking zone (reactor riser) at the FCC unit can react to form a variety of oxygen compounds including phenols. Phenols are produced from naturally occurring oxygen containing compounds of the feed in which oxygen is covalently bonded to aromatic structures or by oxidation of some hydrocarbons by dissolved oxygen present in the feed (if it was stored under the atmosphere) or during air purging in the reactor riser. A lot of the phenol produced is washed out of the hydrocarbon phase with the water wash and leaves the FCC unit with the sour water effluent from the main column overhead receiver. The concentrations of phenols in the FCC sour water usually are 50–450 ppm. Phenols in the FCC unit do not cause problems in the unit itself, however sometimes high concentrations of phenols in sour water can kill microorganisms used for biotreatment. Two ways can be used for discharge of phenols. In the first technique, water containing phenols is injecting into the desalter, where phenols dissolve in the crude oil while the remaining water is decanted away. In the second procedure, sour water containing phenols is used as wash water in the crude unit where hydrocarbons absorb them. If the phenol enters the kerosene, it can produce color problems. Caustic treating of the kerosene removes phenols and other acidic compounds.

Spent caustic. Spent caustic (NaOH) is a waste caustic solution that has become exhausted and is no longer useful (or spent). Spent caustics are made of sodium hydroxide, water, and contaminants. The contaminants have consumed the majority of the sodium hydroxide and the caustic solution is spent. Thus, when H_2S gas is scrubbed by aqueous NaOH solution to form two salts NaHS and Na_2S, caustic soda is consumed. Spent caustic at refinery comes from the Merox processing: removing of mercaptans from LPG, light naphtha or gasoline, and kerosene/jet fuel. In these media, sulfides and organic acids are removed from the product streams into the aqueous caustic phase. The sodium hydroxide is consumed and resulting wastewaters (cresylic for gasoline; naphthenic for kerosene; sulfidic for LPG—spent caustics) are mixed and called *refinery spent caustic.* This solution is contaminated with sulfides, carbonates, and often with heavy organic acids.

Polythionic acids. Polythionic acids are oxoacids $H_2S_xO_6$ (see Appendix D). Polythionic acid SCC of austenitic stainless steel furnace tubes was first identified at the reforming units during shutdowns and can happen in several hours. Usually furnace and heat exchanger tubes, reactors, vessel linings, and thermowells made of austenitic stainless steels and nickel alloys in reforming, desulfurizing and hydrocracking units are susceptible to SCC by polythionic acids [20–22].

Not all S-containing acids are implicated in SCC by polythionic acids. Sulfuric acid may exacerbate the SCC and can result in intergranular corrosion, but does not cause SCC. Cracking requires relatively low tensile stresses for initiation and propagation. Sensitized austenitic stainless steels 304, 316 and nickel alloys Incoloy 800 and Inconel 600 are particularly susceptible to polythionic SCC (especially in weld HAZs) [23–26]. Low carbon (<0.03 wt% C: SS 304L, SS 316L, and SS 317L), the chemically stabilized SS 321 and SS 347 (especially after heat treatment) and the stabilized Incoloy 825 and Inconel 625, are resistant to polythionic SCC. However, these alloys also may become sensitized by prolonged exposure in the sensitized temperature range (see Appendix D, Table D.2).

Polythionic acid SCC cannot occur at the relative humidity ≤60 %. Resistance of stainless steels and nickel alloys to intergranular SCC in the presence of polythionic acids can be examined according to the standard [27]. The following preventive measures from polythionic acid SCC are in our arsenal [26]:

(a) Appropriate selection of alloys and fabrication practices. Low-carbon (SS 304L, SS 316L, SS 317L—with <0.03 % C) and chemically stabilized (SS 321, SS 347, SS 347 LN, SS 316Ti) stainless steels may be used. Low-carbon alloys contain so low amount of carbon that chromium carbides cannot form. Chemically stabilized alloys contain small additions of titanium or niobium, which have a stronger affinity to react with carbon and form carbides than does chromium at high temperatures.

(b) Purging with dry nitrogen (free of water vapor and oxygen) to exclude oxygen during shutdown. Ammonia gas (5000 ppm) may be added to the nitrogen.

(c) Washing of equipment surfaces with aqueous alkaline soda ash solution (2 wt% Na_2CO_3, pH > 9) to neutralize sulfides before shutdown. The chloride concentration in solution should be limited to 250 ppm. Corrosion inhibitor sodium nitrate (0.4 wt% $NaNO_3$) may be added to the alkaline wash solution to reduce the likelihood of chloride SCC. The equipment must be washed with the alkaline solution before any exposure to air. Alkaline solution should also be used for hydrotesting prior to returning equipment to service. SS TP 316Cb is resistant to sensitization up to 650 °C and therefore is resistant to polythionic acid intergranular corrosion and eliminates need for alkaline washing.
(d) Use of dry (dehumidified) air to prevent liquid water formation. Dry air is acceptable if the dew point temperature of the air entering the equipment is maintained a minimum of 22 °C lower that the internal surface metal temperature.

It is important to note that no of the above mentioned preventive measures does not guarantee 100 % protection from polythionic acid SCC.

Organochlorine and organic chloride compounds. *Organochlorine*, named also organochlorinated, or chlorinated hydrocarbons, are organic substances containing at least one covalently bonded atom of chlorine (R-Cl, namely, C-Cl). For example, chloroform ($CHCl_3$) and dichloromethane (CH_2Cl_2) are their representatives. They are relatively non-polar substances and, therefore, are immiscible with water and are effective solvents for removing paraffin and wax deposits and other degreasing operations and cleaning. They are oleophilic (having an affinity for oil) and cannot be extracted by water. *Organic chlorides* are salts, namely, ionic compounds, for instance, amine chloride $[R\text{-}NH_3^+]Cl^-$ salt, and dissolved in water. Both organochlorine and organic chloride substances can be present in original crude oil and can contaminate it from different processes. There are several sources of these compounds. First, organochlorine solvents are used for cleaning oil field tanks, equipment, pumps, and piping. Spent solvents are then added to crude oils. Second, halogenated biocides are injected in oil-bearing formation and pipelines. Chlorine dioxide (ClO_2) is used as a downhole biocide, can form organochlorine compounds and remain dissolved in hydrocarbons of crude oils.

Injection of amines ($R\text{-}NH_2$) as neutralizers and corrosion inhibitors in the overhead vapor line at the distillation columns can result in formation amine chloride salts which can enter into the slops and then are added to the crude oil. Slops in oil refineries represent emulsion of various hydrocarbons, chemicals (i.e. demulsifiers, catalysts, amines, etc.) and water coming from crude oil and sour tank bottoms, wastes, and spills from units. Sometimes inorganic chlorides can be encapsulated in high-melting-point waxes or asphaltenes. Crude oils can contain different amounts (ranged from 5 to 7000 ppm) of such organochlorine and organic chloride substances [3, 28, 29]. Such crude oils can cause corrosion and fouling in the overhead systems of crude distillation atmospheric and vacuum units, catalytic reformers (naphtha hydrotreating units), hydrocrackers and hydrotreaters that process gas oil, and desulfurizers.

Organic chlorides as ionic compounds can be removed by desalting but organochlorine compounds are not removed in desalter. Therefore, they received several titles, sometimes even unpleasant: 'nondesaltable chlorides', 'nonextractable (in water) chlorides', 'phantom chlorides', or 'rogue halogens' [29, 30]. These chlorides are not simple to detect. They are corrosively dangerous because the heating and processing in a crude distillation unit can cause hydrolysis (at $T \geq 160$ °C) or thermal breakdown to form HCl in the overhead system. This strong inorganic acid causes corrosion failure of the overhead condensers made from carbon steel after only 14 h service with corrosion rate 1000 mm/year [29]. The hydrolysis of organochlorine compounds is not complete and the remaining nonhydrolized chlorides usually flow into the naphtha. Following solutions exist to diminish the impact of organochlorine and organic chloride compounds:

(a) Source control: application of careful chemical analysis of organochlorine and organic chloride compounds. Sometimes this is impractical. It is recommended allowable chlorine concentration not more than 1 ppm. Some refineries flatly reject any crude oil containing these compounds [28].
(b) Blending with uncontaminated crude oils. This solution is similar to blending acidic with non-acidic crudes to lessen their corrosiveness.
(c) Prevention of charging contaminated crude oils.

Organochlorine compounds trichloroethylene and perchloroethane are used as sources of chlorine in catalytic reformers. The process stream contains a mixture of hydrocarbons and hydrogen gases. Upon being continuously injected into the hot process stream ahead of the reactor, these organochlorine compounds decompose into HCl and thus replace chlorides lost from the catalyst and maintain its activity. If injection equipment (tanks, piping, pumps, injection quills, valves, and meters) is made of SS 316L, in the presence of water (condensation during shutdown at ambient temperatures) chloride SCC (Cl SCC) can occur. Therefore, nickel alloy Inconel 600 (which is immune to Cl SCC) is recommended for use in this case.

Aluminum chloride ($AlCl_3$). This compound is used as a catalyst in several refining and petrochemical processes: ethylbenzene and polybutene production, butane isomerization [31]. When $AlCl_3$ is dry, it is not corrosive. Aluminum chloride has a strict affinity to water (see Appendix C). If traces of water are present in hydrocarbons, $AlCl_3$ hydrolyzes to form HCl, which is very corrosive not only to carbon steel, but also to stainless steel. Corrugated hoses made from stainless steel SS 304 and used for distribution the catalyst $AlCl_3$ should be always dry and not contact atmosphere. Otherwise, water vapor from the atmosphere can enter into the stainless steel hose with remainders of $AlCl_3$, and forming HCl will attack stainless steel hose up to its perforation. Equipment should be opened for the shortest possible time during shutdowns and outages. Upon closing, it should be dried with hot air (not containing even traces of water), followed by dry nitrogen blanketing.

Sulfuric acid. Sulfuric acid (H_2SO_4) is used in different applications in oil refining industry. Therefore, it is important to know the properties of H_2SO_4 and behavior of different materials in contact with it (see Appendix D). Aggressiveness of sulfuric acid depends on its concentration, temperature, and flow velocity. Concentrated H_2SO_4 (>90 wt%) does not influence corrosion of iron and its alloys as they are passivated in it. Therefore, it is possible to store and carry concentrated H_2SO_4 in carbon steel tanks. Concentrated H_2SO_4 vigorously absorbs moisture. Consequently, it is important to keep containers with concentrated H_2SO_4 hermetically closed from the atmosphere and any source of water. This is the reason why concentrated H_2SO_4 is used for drying of gases, but not H_2, because H_2SO_4 can be reduced by H_2. It is forbidden to wash with water carbon steel piping and tanks containing concentrated H_2SO_4, because it becomes diluted and very corrosive.

Most organic coatings are not resistant to concentrated H_2SO_4. However, some polymers (fluoro-organic, i.e. Teflon, Tefzel, Halar) possess excellent resistance to both concentrated and diluted H_2SO_4. Others, such as polypropylene, polyethylene and polyvinyl chloride, are resistant to sulfuric acid in restricted range of concentration (up to 60 wt% H_2SO_4).

Alkaline solutions. Sodium hydroxide (NaOH), sodium carbonate (Na_2CO_3), ammonium hydroxide (NH_4OH), and amines are used in refinery and petrochemical plants to neutralize acid substances (HCl, H_2SO_4, H_2S, CO_2, organic and other acids) in different processes and operations. For instance, dilute aqueous sodium caustic solution (3 wt%) is injected into crude oil after desalter to neutralize remaining HCl. Aqueous solution of ammonium hydroxide (25 wt%) is injected into the overhead vapor line of the distillation column to control acid dew point corrosion. Aqueous sodium carbonate solution of different concentrations is used in following applications: to neutralize acid gases in CCR unit; to neutralize deposits formed on outer surface of carbon steel and stainless steel tubes in furnaces to prevent under deposit acid corrosion and polythionic acid SCC; for degreasing metal surface of tubes, pipes, and other equipment. Therefore, we should know how alloys behave in contact with alkali solutions. Most metals (Fe, Cr, Al, Cu, Zn, Ni, Ti) and their alloys using in equipment and constructions of oil refinery and petrochemical plants, are amphoteric ones. This means that they corrode both in acid and alkali solutions. Type of alkali and conditions (i.e. temperature and contaminants) significantly influence corrosion resistance of metals, namely the boundary pH value. For instance, iron begins corrode in very strong soda caustic solution (pH > 13.5), zinc at pH > 12, aluminum at pH > 8.3 at 25 °C. At 90 °C, iron begins corrode at pH > 11 [32, 33].

All alkaline solutions play the role not only neutralizing agents but also corrosion inhibitors of carbon steel in water. In order to reach significant inhibitor efficiency, they (excepting amines) can be added in relatively large concentrations: NaOH—above 0.8 g/l; Na_2CO_3—above 5 g/l; NH_4OH—1.5 g/l [1]. Amines are injected in small concentrations (4–8 ppm) in the overhead vapor line of distillation column.

Sodium hydroxide (caustic soda or, simply, caustic, NaOH). Aqueous caustic solutions passivate carbon steel and it is resistant at pH < 13.5 at 25 °C. Caustic solutions can cause severe corrosion of carbon steel in two situations: when solution contains chlorides (general corrosion) and concentrated caustic can cause SCC at

T > 90 °C. Iron corrodes at pH > 13.5 at T > 90 °C with possible caustic embrittlement (type of SCC). This corrosion phenomenon can occur in any place of equipment or structure at T > 80 °C where NaOH can concentrate ('pockets') and pH exceeds 13.5. For instance, in places of injection of diluted soda caustic solution after desalter, or on upper trays of distillation columns at ethylene plants in contact with the wall. If there is no flow and good mixing of injected caustic solution with hydrocarbons, caustic can concentrate (~3 wt% NaOH) in particular places of the upper part of distillation column with increase pH above 13.5, and SCC can occur at ~90 °C. Cracks in the wall of the distillation column containing hot flammable hydrocarbons can cause many serious environmental problems.

In order to prevent caustic SCC, equipment and piping which contact caustic solutions must be stress relieved (postweld heat treatment, PWHT).

Another frequently encountered case of caustic corrosion occurs in boiler feed water (BFW) heat exchangers and steam generators. BFW is often treated with diluted alkali solutions to maintain pH = 9.3–9.5 and thus to minimize corrosion of carbon steel. Traces of caustic can become concentrated in some areas, i.e. steam generator tubes that alternate between wet and dry conditions (steam blanketing) because of overfiring. Heat exchangers for cracked gas in ethylene plants are also vulnerable if whitish deposits of caustic are accumulated on the bottom tube-sheet. Usually boiler tubes in contact with steam have protective magnetite (Fe_3O_4) layer which can be dissolved by concentrated alkali NaOH solution at high temperature (see Appendix E). As a result of caustic attack, elliptical hemispherical shallow pits are formed on carbon steel surface. Therefore, *caustic corrosion* also is named *caustic gouging* or *ductile gouging* [34]. Sometimes pits are filled by whitish (NaOH) and sparkling (Fe_3O_4) deposits.

Sodium hydroxide in no case should not be used in contact with aluminum alloys because of severe corrosion at pH > 8.3. Aqueous sodium hydroxide solutions should not be stored in glass vessels with glass stoppers because of 'freezing' of ground glass joints. It is possible to store aqueous sodium hydroxide solutions (up to 50 wt%) in polypropylene or polyethylene vessels.

Sodium carbonate (soda ash, washing soda, Na_2CO_3). Pure Na_2CO_3 is a white very hygroscopic powder and forms strong alkaline aqueous solution because of formation NaOH when dissolved in water (see Appendix E). Therefore, corrosiveness of aqueous Na_2CO_3 solutions are similar to NaOH. Pure aqueous soda ash solution (pH = 9–11.5) is not corrosive to carbon steel. However, if solution contains chlorides (~1000 ppm), severe corrosion of carbon steel occurs with rust formation and corrosion rate reaches ~1 mm/year. This can happen at the CCR unit where washing soda solution is used for absorption HCl.

Ammonia (NH_3) and ammonium hydroxide (NH_4OH). Ammonia is used as 25 wt% NH_4OH aqueous solution or compressed anhydrous gas NH_3. For ease of handling and reliable control during injection, aqueous ammonia solution is usually preferred. This solution is a corrosion inhibitor of carbon steel but causes corrosion and SCC of copper and its alloys (see Appendix E).

Amines. Amines are organic substances, derivatives of ammonia (NH_3) (see Appendix E). Amines are used as neutralizers of acids (HCl) and as corrosion inhibitors in the overhead vapor line of distillation columns, and as absorbers-neutralizers of acid gases (H_2S and CO_2) at sulfur recovery and amine treating units. Some amines in concentrated form can be corrosive to carbon steel. The stronger basicity of amine, the more corrosive it to iron, as the latter is an amphoteric metal. Primary amines (MEA, DGA) are the most corrosive, secondary amines (DEA) slightly less and tertiary amines (MDEA) exhibit the lowest risk of corrosion. Severe corrosion of carbon steel can occur if amine inhibitor is injected and is concentrated (because of bad atomization) in the overhead vapor line of the distillation column. In spite of alkali properties of amine solutions which would have to be protective to carbon steel, the majority of corrosion problems associated with amine applications occur at amine treating units (see Sect. 4.4).

3.2 High-Temperature Corrosion

High-temperature (T > 200 °C) corrosion pertains to the second large group of corrosion phenomena which occurs when exposed to non-electrolytes: sulfidic corrosion (oxidation by H_2S, S_8, and some other sulfur compounds), naphthenic acid corrosion, hydrogen attack, hot ash (fuel ash or salt deposit) corrosion, oxidation (by oxygen gas), carburization, and 'metal dusting'. Non-electrochemical mechanism is realized in these corrosion phenomena. Non-electrolytes are generally gases (i.e. H_2S, H_2, O_2, and hydrocarbons) or liquids (i.e. naphthenic acids or S_8) at high temperature. Sometimes these corrosive components are present together. In these cases, we should analyze sulfidic corrosion with the presence of H_2, and naphthenic acid corrosion in the presence of H_2S. Such situations complicate corrosion mechanism and under particular conditions can either induce or decrease corrosion. For example, sulfidic corrosion proceeds more intensively in the presence of H_2, and inhibition of naphthenic acid corrosion occurs by sulfide scale formed in the presence of H_2S. It should be noted that hot ash corrosion is caused by molten salts and oxides attached on tubes` surface in furnaces. These molten salts and oxides are liquid electrolytes at high temperature and conduct electricity. Certainly electrochemical mechanism is realized when hot ash corrosion takes place. Oxidation by air (oxygen gas) of outer surfaces of tubes also occurs in furnaces. Hydrogen attack can occur both at high (>200 °C) and at low temperature (<100 °C) (see Sect. 4.3).

Equipment and piping at most units subjected to high temperatures are under high pressures of hydrocarbons and corrosive compounds. These conditions lead to the situation, that corrosion risk of high-temperature corrosion is accompanied by danger of explosion and fire, followed by emissions of noxious substances. They can result in unpredicted destruction of equipment and structures around, but what is more significantly, injury to personnel and even loss of life, and severe damage to the environment. High-temperature corrosion phenomena are described in the next chapter.

References

1. Groysman A (2010) Corrosion for everybody. Springer, Dordrecht
2. Groysman A (2014) Corrosion in systems for storage and transportation of petroleum products and biofuels. Springer, Dordrecht
3. Gutzeit J, Merrick RD, Scharfstein LR (1987) Corrosion in petroleum refining and petrochemical operations. In: Corrosion, vol 13, ASM Handbook, ASM International, pp 1262–1287
4. Whitman GW, Russel RP, Altieri VJ (1924) Effect of hydrogen-ion concentration on the submerged corrosion of steel. Ind Eng Chem 16(7):665–670
5. Wikjord AG, Rummery TE, Doern FE, Owen DG (1980) Corrosion and deposition during the exposure of carbon steel to hydrogen sulfide-water solutions. Corr Sci 20(5):651–671
6. McNeil MB, Little BJ (1999) The use of mineralogical data in interpretation of long-term microbiological corrosion processes: sulfiding reactions. J Am Inst Conserv 38(2):186–199
7. Furukawa Y, Barnes HL (1996) Reactions forming smythite, Fe_9S_{11}. Geochim Cosmochim Acta 60(19):3581–3591
8. Smith SN, Brown B, Sun W (2011) Corrosion at higher H_2S concentrations and moderate temperatures. Paper 11081 presented at the NACE CORROSION 2011 conference, NACE International, Houston, TX, USA
9. Shreider AV, Shparber IS, Archakov YI (1976) Influence of hydrogen on oil and chemical equipment. Mashinostroyeniye, Moscow
10. Piehl RL (1976) Survey of corrosion in hydrocracker effluent air coolers. Mater Perform 15 (1):15–20
11. Damin DG, McCoy JD (1978) Prevention of corrosion in hydrodesulphurizer air coolers and condensers. Mater Perform 17(12):23–25
12. Scherrer C, Durrieu M, Jarno G (1980) Distillate and resid hydroprocessing: coping with corrosion with high concentrations of ammonium bisulphide in the process water. Mater Perform 19(11):25–31
13. API RP 932-B (2014) Design, materials, fabrication, operation, and inspection guidelines for corrosion control in hydroprocessing reactor effluent air cooler (REAC) systems, 2nd edn. API International, Washington
14. Al-Dhafiri S, Rawtal D, Reddy L (2010) Erosion—Corrosion of SS316L Trays in FCC's Sour Water Stripper—A Case Study. Paper 10273 presented at the NACE CORROSION 2010 conference, NACE International, Houston, TX, USA
15. API Publ 948 (1996) A study of variables that affect the corrosion of sour water strippers. API International, Washington
16. NACE 6A192/SSPC-TR3 (2000) Dehumidification and temperature control during surface preparation, application, and curing for coatings/linings of steel tanks, vessels, and other enclosed spaces. Joint Technical Committee Report, NACE International, Houston, TX, USA
17. Horvath RJ, Cayard MS, Kane RD (2006) Prediction and assessment of ammonium bisulfide corrosion under refinery sour water service conditions. Paper 06576 presented at the NACE CORROSION/2006 conference, NACE International, Houston, TX, USA
18. Cayard MS et al (2006) Prediction of ammonium bisulfide corrosion and validation with refinery plant experience. Paper 06577 presented at the NACE CORROSION 2010 conference, NACE International, Houston, TX, USA
19. Horvath RJ et al (2010) Prediction and assessment of ammonium bisulfide corrosion under refinery sour water service conditions—part 2. Paper 10349 presented at the NACE CORROSION 2010 conference, NACE International, Houston, TX, USA
20. Garveric L (ed) (1994) Corrosion in the petrochemical industry. ASM International
21. Nagashima E, Matsumoto K, Shibata K (1998) Effects of sensitization and service fluid chemistry on polythionic stress corrosion cracking of 18-8 stainless steel. Paper 592 presented at the NACE CORROSION 1998 conference, NACE International, Houston, TX, USA

22. Nishida H, Nakamura K, Takahashi H (1984) Intergranular stress corrosion cracking of sensitized 321 SS tube exposed to polythionic acid. Mater Perform 23(4):38–41
23. Tamaki K, Nakano S, Kimura M (1987) Application of CrNi stainless steel weld metals to polythionic acid environments. Mater Perform 26(8):9–13
24. Dove D, Messer B, Phillips T (2001) An austenitic stainless steel, resistant to high temperature creep and naphthenic acid attack. Paper 01523 presented at the NACE CORROSION 2001 conference, NACE International, Houston, TX, USA
25. Brophy AJ (1974) Stress corrosion cracking of austenitic stainless steels in refinery environments. Mater Perform 13(5):9–15
26. NACE Standard SP0170-2012 (2012) Protection of austenitic stainless steels and other austenitic alloys from polythionic acid stress corrosion cracking during a shutdown of refinery equipment. NACE International, Houston, TX, USA
27. ASTM G35-98 (2010) Standard practice for determining the susceptibility of stainless steels and related nickel-chromium-iron alloys to stress-corrosion cracking in polythionic acids. ASTM International, West Conshohocken, PA, USA
28. Gutzeit J (2000) Effect of organic chloride contamination of crude oil on refinery corrosion. Paper 00694 presented at the NACE CORROSION 2000 conference, NACE International, Houston, TX, USA
29. NACE Publication 34105 (2015) Effect nonextractable chlorides on refinery corrosion and fouling. NACE International, Houston, TX, USA
30. Veazey M (2002) Phantom chlorides create real problems for refiners. Mater Perform 41 (5):16–19
31. Treseder RS, Wachter A (1949) Corrosion in petroleum processes employing aluminum chloride. Corrosion 5(11):383–391
32. Townsend HE (1970) Potential-pH diagrams at elevated temperature for the system Fe-H_2O. Corr Sci 10(5):343–358
33. Ashworth V, Boden PJ (1970) Potential-pH diagrams at elevated temperatures. Corr Sci 10 (10):709–718
34. Port RD, Herro HM (1991) The Nalco Guide to Boiler Failure Analysis. McGraw-Hill, New York

Recommended Literature

35. Forsén O, Aromaa J, Hakonen T, Rintamäki (2007) Corrosion in the overhead system of a sour water stripper. In: Harston JD, Ropital F (eds) Corrosion in refineries, European Federation of Corrosion Publications Number 42, Woodhead Publishing Limited, Cambridge, pp 73–78
36. Price MD, Shargay CA, Jacobs GE (1999) Ammonium salt corrosion in hydrotreating unit stripper column overhead systems. Paper 99392 presented at the NACE CORROSION 1999 conference, NACE International, Houston, TX, USA
37. Rees AJR (1977) Problems with pressure vessels in sour gas service (case histories). Mater Perform 16(7):29–33

Chapter 4
Corrosion Problems and Solutions at Oil Refinery and Petrochemical Units

A pessimist sees difficulties in every opportunity; an optimist sees opportunities in every difficulty.
Winston Churchill (1874–1965), a British politician.

Abstract Specific corrosion problems of sulfur compounds, hydrogen gas, naphthenic acids, and amine solutions, physicochemical mechanisms, anti-corrosion and monitoring measures are described. Sulfidic corrosion and hydrogen attack can occur both at low and high temperatures. Naphthenic acid corrosion occurs only at high (190 °C < T < 360 °C) temperatures. Amine solutions are corrosive at ~ 100–130 °C. Case studies at refineries' units are analyzed and solutions are given.

Analysis of oil refining industry shows that whilst the number of refineries declined in the period between 1993 and 2007 the average capacity per refinery increased by nearly 30 % [1]. This increase is due in part to rationalization of smaller capacity and part to the addition of new, larger refineries. This tendency remains also in nowadays and brings corrosion problems related to "childhood diseases" of new refineries and "adult diseases" of existent plants.

There are corrosion phenomena which occur in any industry and in any place: corrosion in water, atmosphere, soil, at high temperatures, acid dew point corrosion, corrosion under thermal insulation, pitting, crevice, galvanic, intergranular corrosion, selective leaching, stress corrosion cracking (SCC), microbiologically induced corrosion (MIC), erosion, cavitation, and impingement. All these phenomena with different frequency occur in oil refining units. However, because of specific media and conditions in oil refining and petrochemical industries in addition to these corrosion phenomena there are particular corrosion problems intrinsic only to them. They are: corrosion in crude oils, petroleum products and fuels, sulfidic corrosion, hydrogen attack, naphthenic acid corrosion, sour water corrosion, corrosion in amine solutions, and polythionic acid corrosion. Specific corrosion problems at low temperatures and their solutions are partly described in Chap. 3. Specific corrosion problems of sulfur compounds, naphthenic acids, hydrogen gas, and amine solutions which we can face with at high temperature and their solutions are presented in this chapter.

A. Groysman, *Corrosion Problems and Solutions in Oil Refining and Petrochemical Industry*, Topics in Safety, Risk, Reliability and Quality 32,
DOI 10.1007/978-3-319-45256-2_4

4.1 Corrosion by Sulfur Compounds

Abstract 'Reactive or corrosive' and 'non-reactive or non-corrosive' sulfur compounds are described. Low- (T < 100–200 °C) and high- (T > 200 °C) temperature corrosion with the participation of sulfur compounds is analyzed. The mechanism of sulfidic corrosion, estimation of corrosion rates of alloys, influence of temperature and velocity, control measures against sulfidation in hydrocarbon streams containing only H_2S and mixture H_2S with H_2, inspection methods and safety actions are described. Case studies are also given.

Sulfur compounds play essential role in corrosion of the equipment at oil refining and petrochemical units. For our convenience, we will distinct low-temperature (T < 100–200 °C) and high-temperature (T > 200 °C) corrosion with the participation of sulfur compounds. Hydrogen sulfide can participate in low- and high-temperature corrosion. Low-temperature corrosion in the presence of dissolved H_2S in aqueous phase, corrosion by ammonium bisulfide, sulfuric acid and polythionic acids are discussed in Sect. 3.1.1 and Appendix D. Dissolved hydrogen sulfide influences significantly hydrogen absorption by metals and hydrogen damages (see Sect. 4.2). In this part, we will discuss corrosion by sulfur compounds at high temperatures, known as sulfidic corrosion.

4.1.1 Sulfidic Corrosion

Sulfidic corrosion is corrosion of metals and alloys by particular sulfur compounds at high temperatures, usually in the range 230–540 °C [2, 3]. This occurs in crude distillation (atmospheric and vacuum), hydroprocessing, catalytic reforming, and coking units, in fractionation and distillation facilities downstream from hydrotreaters and hydrocrackers. If there is no hydrogen (H_2) gas in hydrocarbon stream, sulfidic corrosion occurs alone. Not all sulfur compounds oxidize metals and alloys. Even some of them, thiophenes and sulfones, can inhibit corrosion of metals [4].

Elementary sulfur (S_8), hydrogen sulfide (H_2S), mercaptans (R-SH, thiols), organic sulfides (R-S-R), disulfides (R-S-S-R) and polysulfides (R-S_n-R) can react with metals under particular conditions, and usually corrosivity increases at elevated temperatures. Therefore, these compounds sometimes are called 'reactive or corrosive sulfur compounds'. Others, thiophenes and sulfones, are called 'non-reactive or non-corrosive sulfur compounds'. It is possible also to divide all sulfur compounds on highly corrosive (S_8, H_2S, R-SH), corrosive (R-S-R, R-S-S-R) and non-corrosive (alkyl thiophenes, alkyl benzothiophenes and dibenzothiophenes) [4]. Their corrosiveness is realized under appropriate concentrations and temperatures, and is differently manifested on various alloys, even steels.

When oil refineries purchase crude, they check only one sulfur parameter—total sulfur content, and nobody knows the distribution of reactive (corrosive) and non-reactive (non-corrosive) sulfur compounds in the crude. In spite of this situation, several rules were established [2, 5, 6]:

(a) Crude oils differ significantly in content of diverse sulfur compounds and total sulfur may varied in 0.1 wt% S (Ukraine, Kazakhstan) to 9.6 wt% S (Etzel, Germany).
(b) The relative amounts of corrosive and non-corrosive sulfur compounds differ from crude to crude and from petroleum distilled fraction to fraction of a crude. Usually 30–50 wt% of total sulfur compounds belong to corrosive substances.
(c) The content of total sulfur and corrosive sulfur compounds increases with increasing boiling point of distilled fractions.
(d) Usually sulfidic corrosion is manifested in crude units using crude oils containing above 0.6 wt% total sulfur (or above 0.2 wt% corrosive sulfur).

The mechanism of sulfidic corrosion. If there are so many different sulfur-organic compounds, how do they react with metals? All these compounds are not stable at high temperatures (T > 260 °C) and are decomposed to form H_2S. Different forms of sulfur-organic compounds play an important role in the kinetics of H_2S generation. Mercaptans and disulfides have been identified as being the most reactive sulfur species under thermal stress conditions, followed by aliphatic sulfides, thiophenic compounds, while benzothiophenic compounds appear to be the most stable organically bound sulfur species during crude oil thermal decomposition.

Use of heavy and extra-heavy crude oils brings particular risk of sulfidic corrosion because they do not only contain increased amounts of sulfur compounds but method of production and processing of these crudes leads to generation of huge quantities of H_2S (Table 4.1). Moreover, this quantity increases significantly at T > 235 °C [7–9].

Hydrogen sulfide may be formed during sulfate reduction in crude oil during thermal recovery. This reaction consists of the oxidation of some hydrocarbons in crude oil and reduction of aqueous sulfates (SO_4^{2-}) to form H_2S and CO_2 at T > 120 °C. Up to 95 % of the original sulfate in aqueous phase can be converted to H_2S at T ~ 300 °C [10]. Hydrogen sulfide is more stable than sulfur-organic compounds and can exist up to 650 °C [11]. This is explained by decrease polarity of bonds S-H and S-C in the row: H-S-H > R-S-H > R-S-S-R.

We are talking about corrosion processes at 230–540 °C. The mechanism of sulfidic corrosion is the direct chemical oxidation of metals by sulfur compounds (see Appendix D). As a result, solid metal sulfides are formed on the surface of the metal or alloy. Therefore, 'sulfidic corrosion' is also named 'high-temperature sulfidic corrosion' or 'sulfidation'. These corrosion reactions occur under 'dry' conditions, namely, without electrolyte, and represent 'pure' chemical (not electrochemical) reactions. Characteristic feature of this corrosion mechanism is the absence of electric potential and electric current on the metal surface during corrosion. These corrosion processes are similar to oxidation of metals by oxygen at

Table 4.1 Generation of H_2S from heavy crude oil with increasing steam temperature [9]

T °C	H_2S, ppm	Conditions
75	100	Field
100	500	
125	3000	
235	10,000	Laboratory
250	20,000	
300	30,000	

high temperature with formation of oxide scale on metal surface. Sulfide scale exhibits parabolic growth behavior, namely, the thickness of scale (corrosion gain) increases with time but the rate of its growth diminishes with time because corrosion rate is the differential of corrosion gain (Fig. 4.1). After removing scale by means of acid, we may define corrosion loss. Similar curves are observed for corrosion of iron in water, air, and soil [12]. Corrosion of iron in all these media slows down as corrosion products (iron sulfides or rust) are formed on iron surface and protect (but not fully) it from further corrosion. Thus, the corrosion rate is inversely proportional to scale thickness (see Fig. 4.1). We should relate not to the short-term corrosion rate measured during the first unstable period but to the steady non-changed corrosion rate value (plateau in the Fig. 4.1a). Sometimes the thickness of scale formed on steel is 3–4 times greater than the thickness of metal corroded away [6]. Sulfidation (oxidation by sulfur compounds) is considerably more destructive than oxidation by O_2. Sulfide scale is fragile and tends to crack and spall more readily than oxides.

The sulfidation mechanism is closely akin to air oxidation and the same alloying elements provide resistance to attack. Carbon steel, low-alloy Cr-Mo steels, nickel, copper, zinc and their alloys are attacked by sulfur compounds at elevated temperatures. High-nickel alloys are most susceptible to sulfidation. Decreasing nickel in nickel-base alloys usually diminishes susceptibility to sulfidation. Increasing chromium in iron-, nickel-, and cobalt-base alloys improves sulfidation resistance.

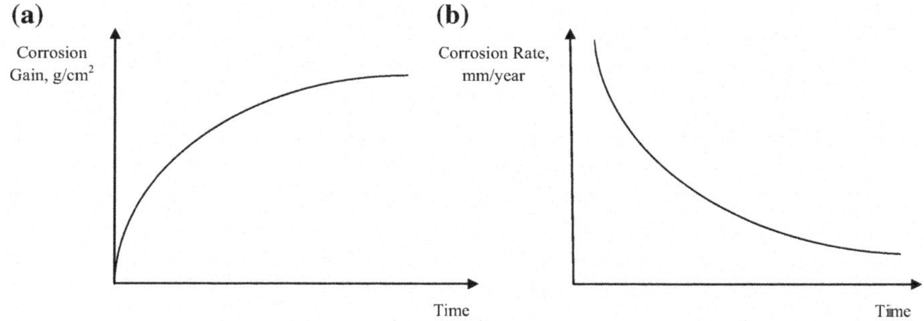

Fig. 4.1 **a** Kinetic curve for corrosion of iron in contact with H_2S (the mass of metal increases with time—corrosion gain). **b** Changes of corrosion rate of iron in contact with H_2S with time

Chromium and aluminum alloys are resistant to sulfidic corrosion. The greater chromium content in alloy, the better its resistance to sulfur compounds at elevated temperatures. Some authors point that chromium should be higher than 25 wt% [13]. When sufficient chromium is present, the relative quantity of nickel is of secondary importance. Other beneficial alloying elements include Co, Ti, Al, and Si.

Iron/chromium/molybdenum sulfide layers are formed on low-alloy Cr-Mo steels. Stability of sulfides can be arranged in the following sequence: V_2S_3 > MnS > TiS > CrS > TiS_2 > FeS > Co_4S_3 > NiS > FeS_2 [13]. For instance, steels alloyed with Cr have a two-layered scale: a mixed inner layer of $Fe_{1-x}S$ with $FeCr_2S_4$, and an outer layer of $Fe_{1-x}S$ [2]. The greater chromium content in steel, the more content of $FeCr_2S_4$ in layers which is more stable and more protective than $Fe_{1-x}S$. These sulfide layers, or scale, is tightly adhering to the metal surface and diminishes further corrosion. Scale stability and its degree of protection increase with rising chromium content in the alloy. Carbon steel forms less protective iron sulfide scale than mixed scale formed by low-alloy Cr-Mo steels. These complex scale is particularly resistant to turbulent flow of hydrocarbon streams. Chromium oxide (Cr_2O_3) layer is protective against sulfidation and turbulent flow. Destructive sulfidation attack takes place mainly at sites where the protective oxide has broken down. If sufficient sulfur enters the alloy, available chromium is converted to sulfides, and then less stable and less protective iron sulfides may form. It is these iron sulfides are responsible for the accelerated attack, because they grow much faster than the sulfides of chromium [14]. In some cases, Cr-Mo steels corrode at the same rates as carbon steels. Sometimes severe sulfidation even occurs when the total sulfur content in hydrocarbon feed (e.g., naphtha or LPG) is very low (10–20 ppm) [15]. As a result of sulfidation, large quantities of fragile metal sulfide (mostly FeS, CuS, and NiS) scale are formed (see Sect. 4.1.2). The volume of iron sulfide scale is greater than the volume of the corroded iron substrate. Compressive stresses appear inside of scale. Sometimes they are fell off and carried away with flow and result in fouling, namely, plug up heat exchanger tube bundles, tubes, trays, and filters. Iron sulfide can promote the adhesion of coke to the surface. In addition, iron sulfide can act as dehydrogenation catalyst to promote the formation of coke by asphaltenes [16, 17]. Thus, iron sulfide can interact with asphaltenes to form foulants (see Chap. 7).

Estimation of corrosion rates of alloys. Many scientists and engineers contributed in corrosion science in the fact that they created tables and curves which are named in honor of these peoples: Evans diagram, Pourbaix diagram, Tafel equation, to name a few. Even corrosion phenomena are called by their names, e.g., Galvanic corrosion was named in honor of an Italian physician Luigi Aloisio Galvani who discovered 'animal electricity'. Sulfidic corrosion in refinery units is distinguished by McConomy [18] and Couper–Gorman [19] curves. McConomy created curves in 1963 for hydrocarbon streams in the absence of hydrogen gas. Couper and Gorman created curves in 1971 for hydrocarbon streams containing mixture H_2S and H_2. McConomy curves allow to estimate corrosion rates of carbon steel, low-alloy steels (1–9 wt% Cr), SS 12Cr and austenitic SS 304 in hydrocarbon streams in the absence

Table 4.2 Maximum temperature for steels in hydrocarbons containing corrosive sulfur compounds [28]

Steel	Maximum metal temperature (°C) for	
	12.5-years life	25-years life
Carbon steel	290	270
2.25Cr-1Mo	310	290
5Cr-0.5Mo	330	310
9Cr-1Mo	400	360

of hydrogen as a function of total sulfur content (0.4–2.0 wt% S) and temperature (260–400 °C) [2–4, 6]. These curves can also be used to estimate maximum metal temperatures for different steels in hydrocarbon streams containing corrosive sulfur compounds to provide a specified service life time (Table 4.2).

Neither the McConomy nor the Couper–Gorman curves take into consideration the influence of flow regime (turbulence) and flow velocity. The McConomy curves showed higher corrosion rates of steels than that predicted by models considering corrosive sulfur content. They were modified and recommended for use [2, 3, 20]. The corrosion rates increase with the quantity of sulfur in hydrocarbons. A threshold allowable concentration of sulfur in hydrocarbons does not exist because sulfidic corrosion can occur even at 1 ppm H_2S [3]. The influence of sulfur content on the corrosion is less important than the effect of temperature.

Influence of temperature. Sulfidic corrosion intensifies with increase of temperature from 230 °C and reaches maximum at 425–450 °C [3, 6]. Then corrosion rates of sulfidic corrosion decrease with increase temperature till 540 °C. The rate of increase in the corrosion rates between 230 and 425 °C is faster as the temperature is increased [3]. At T > 425 °C, nearly all quantity of corrosive sulfur already reacted with metal surface (or corrosive sulfur compounds are decomposed) and formation of dense coke deposits can prevent further corrosion.

Influence of velocity. Intensity of sulfidic corrosion does not depend on liquid flow rate less than 6 m/s. Increase of fluid flow rate above this value intensifies sulfidic corrosion and at ~30 m/s can lead to severe corrosion. At vapor flow velocities above 30 m/s, when two-phase (liquid-vapor and vaporization) turbulent flow is present, erosion-corrosion with impingement attack of carbon steel bends, elbows, tees, headers, and circumferential welds can occur [6].

Control actions against sulfidation. The only realistic and practical way to control sulfidic corrosion is to use corrosion resistant alloys containing chromium. Carbon steel can be used only till 260 °C. Killed steel (containing above 0.10 wt% Si) possesses better corrosion resistance than usual carbon steel (see Appendix F). Above 260 °C, low-alloy 5Cr-0.5Mo steel is widely used up to ~343 °C (especially for piping where high flow velocities exist) [2]. Their resistance increases in direct proportion to chromium content. Another low-alloy 9Cr-1Mo steel is used in transfer lines and fired heater tubes up to 400 °C. Stainless steel containing 12 %Cr (12Cr SS), as a rule, is fully resistant to sulfidic corrosion. When temperature increases significantly above 260 °C, the resistance of alloys containing less than

12 % chromium diminishes and they become susceptible to sulfidic attack. Unfortunately even 12Cr SS has essential shortcoming—'475 °C Embrittlement'. It becomes brittle at 475 °C. In spite of curing by annealing above 575 °C, some refineries avoid the use of 12Cr SS for heater tubes, return bends, and headers [6].

Aluminum alloys and alonized carbon steel (steel coating with aluminum) can be used for vacuum column internals and packing. Austenitic stainless steels (SS 304, 316, 317, 321) are resistant to sulfidic corrosion.

Corrosion in hydrocarbon streams containing mixture H_2S and H_2. Hydrocarbons containing mixture H_2S and H_2 are processed in hydrotreating (catalytic reforming) and hydrocracking units. The presence of H_2 together with H_2S in hydrocarbon streams increases the severity of sulfidic corrosion but mechanism has not been established yet. Hydrogen converts sulfur organic compounds containing in hydrocarbon feedstocks (naphtha or gas oil) to H_2S, and corrosion intensity depends on H_2S concentration and temperature, but not H_2 partial pressure. Some authors assume that H_2 does not allow coke formation which can protect steel surface from sulfidic attack [2]. The composition and morphology of the iron sulfide scale are similar in both H_2S environments, in the absence and presence of H_2. However, 5 % Cr steel shows good performance in the absence of H_2, and behaves worse in $H_2S + H_2$ hydrocarbon environment. The cracks, fissures, and spalls containing in iron sulfide scale, also provide easy way for corrosive sulfur substances to directly reach the steel surface.

Control actions against sulfidation in the presence of H_2. The beneficial effect of Cr in the steel may be less effective in the presence of H_2. Resistance of some alloys to the mixture 16 $H_2S + H_2$ at 315–480 °C at 75 bar can be given in the following order of downgrade: Fe-20Cr-40Ni > austenitic stainless steels (18Cr-8Ni) > Fe-20Cr-65Ni ≈ Fe-20Cr > Fe-(12-16)Cr > 20Cr-80Ni > Fe-(0-9Cr) [13]. Adding of a slight to moderate amount (8–40 %) of nickel to Fe–Cr significantly improves the sulfidation resistance of alloys. Greater nickel addition diminishes sulfidation resistance. The alloy HR-160 (Fe-Ni-Co-28Cr-2.75Si) shows the highest resistance to sulfidation because of formation good protective oxide scale [13]. Usually, austenitic stainless steels (18Cr-8Ni) are used in hydrocarbons containing mixture $H_2S + H_2$ at $T > 260$ °C.

Inspection methods. Ultrasonic thickness measurements, radiography, pulsed eddy current, and visual inspection during shutdown are used for detection sulfidic corrosion.

Safety. Iron sulfides deposits are pyrophoric and can ignite when become dry and exposed to air. Pyrophoricity is a spontaneous combustion of a material upon exposure to air (atmospheric oxygen). Pyrophoric means 'fire-bearing' from Greek. Pyrophoric substances ignite spontaneously in air at $T \leq 55$ °C. Most materials are not pyrophoric unless they are in a very finely divided state: dust, powder, fine particles, thin sections, fine turnings, thin slices, or small chips. Examples are iron sulfides, some reactive metals (Ti, Mg, Al, even Fe), hydrocarbon sludge (wastes formed at oil refineries and petrochemical plants). Iron sulfides formed on carbon steel surface as a result of corrosion by H_2S can be oxidized exothermically (release of heat energy) and eventually be ignited. This occurs when equipment is taken out of service and exposed to air. Namely, when the tube bundle in heat exchanger is taken out but removal of iron sulfides is delayed. In this case, iron sulfides smolders and then ignites. The empty heat exchanger shells on each sides and interconnected piping act like a chimney, increasing the severity. Precautions are to keep the equipment wet enough so that oxidation heat is removed by evaporation and ignition cannot occur. Another problem may arise during cleaning by acid—evolving very dangerous H_2S (see Sect. 7.6 and Appendix D).

4.1.2 *Case Studies*

4.1.2.1 Corrosion of Heat Exchanger Tubes (Vacuum Distillation Unit)

Conditions. The heat exchanger is intended to cool the bottom residue of the vacuum column at the crude distillation unit. The bottom residue entered into the tubes at 340 °C and 25 bars and exited at 310–320 °C. Crude oil flowed in the shell side. Crude oils Ural (Russia), Maya (Mexico), and Ras-Garib (Egypt) were used. The bottom residue contained 4.2–5.6 wt% sulfur. TAN values of crude oils ranged in 0.4–2.4 mg KOH/g. TAN values of HVGO ranged in 0.4–0.7 mg KOH/g.

Material of construction. Tubes—A199 T5 (5 wt% Cr) with the wall thickness 2.41 mm.

Service period before the failure: 8 years. The design life of the heat exchanger is 15 years.

Visual examination and findings. Thick black scale and deposits were formed inside and outside tubes' surface (Fig. 4.2).

Outside scale contained iron sulfide (86 wt%), chlorine (13 wt%), and chromium (1 wt%). Deposits inside the tubes contained mostly iron sulfide (97.1 wt%), chromium (2.4 wt%), and chlorine (0.5 wt%). Severe corrosion was found inside of the heat exchanger tubes. The thickness of the tube walls from inner side and tube-sheets (especially in the entrance of the bottom residue) decreased

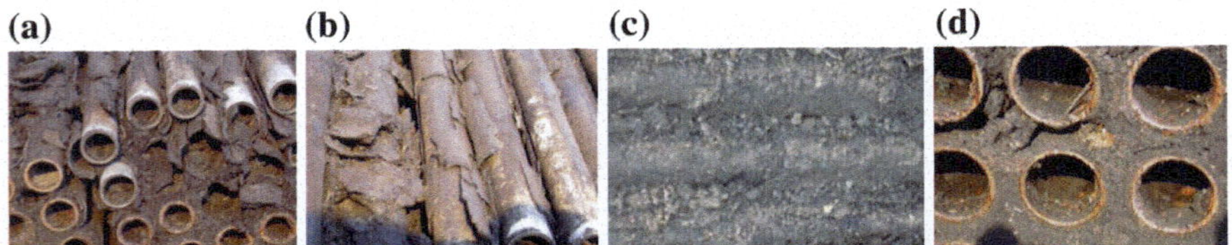

Fig. 4.2 **a–c** Deposits and scale between tubes (crude oil side). **d** Tubes turned into thin "paper" (HVGO side)

significantly, nearly to nil. Corrosion rate was 0.3 mm/year when allowable value is 0.11 mm/year for the tube material.

Failure phenomenon: *High-temperature sulfidic corrosion.*

Cause of failure and its explanation. Corrosion occurred from both sides of tubes. When using crude oil and distillates containing sulfur more than 1 wt% in the temperature range >290 °C, low-alloy steel 5 % Cr must be used. However, if hydrocarbons contain naphthenic acids (or other organic acids) with the TAN values greater than 0.5 mg KOH/g in a vacuum unit above 230 °C regardless of sulfur content, SS 316L should be used.

Solution and recommendations. Use SS 316L tubes.

4.1.2.2 Corrosion of Tubes in Heat Exchanger (Visbreaker)

Conditions. The heat exchanger is intended for cooling cracked gas oil. Cracked gas oil entered into the shell of the heat exchanger at 340 °C and 4 bar and exited at 100–120 °C.

Material of construction. Tubes—carbon steel ASTM A179.

Service period before the failure: 6 years.

Visual examination and findings. Severe corrosion of outer surfaces of tubes with many holes was found in the area of the impingement plate (Fig. 4.3). This plate consisted of carbon steel 'protective' strips from impingement attack and not as one whole part. Black deposits and thick black scale were found on outer surface of tubes. Chemical analysis showed that deposits and scale consisted of iron sulfides.

Failure phenomenon: *High-temperature sulfidic corrosion-erosion.*

Cause of failure and its explanation. H_2S containing in cracked gas oil reacted with iron at high temperature (~340 °C). H_2S is especially aggressive to carbon steel at T > 290 °C. Cracked gas oil entered into the shell at 340 °C and 4 bar. Turbulent flow of hot (~340 °C) gas oil in the area of carbon steel impingement strips (not plate!) caused erosion (impingement attack) in addition to sulfidic corrosion.

Fig. 4.3 **a** General view of failed heat exchanger. **b** The tubes with iron sulfide deposits at the top of the bundle in the area of 'protective strips' from impingement attack. **c** The remains of corroded tubes in the area of 'protective strips'

Solutions and recommendations.

(a) To install the impingement plate made from SS 304 (whole part and not in the form of strips). The area of the impingement plate should be 30 % larger than the cross sectional area of the gas oil inlet pipe.
(b) To replace the material of tubes (carbon steel) with 5 % Cr which recommended to use in contact with hydrocarbons containing >1 wt% S at T > 290 °C.

4.1.2.3 Corrosion of Pipes in Furnaces (HDS Kerosene)

Conditions. Kerosene flowed with velocity 0.5 m/s at 300 °C and 6 bar inside the pipes located in the furnace. Skin temperature was ~400 °C. The furnace contains 104 pipes. Fuel oil has been used during the first 3 years of service for the burning in the furnaces with the aim of heating kerosene inside the pipes. Then natural gas has been used instead fuel oil for the burning during the last 3 years before the failure.

Kerosene contained hydrogen gas, 600 ppm sulfur, 10 ppm mercaptans, 130–150 ppm H_2O, and its TAN = 0.0028–0.0092 mg KOH/g.

Material of construction: 5 % Cr (required). The pipes of two sizes: Sch 40 (original thickness 7.1 mm) and Sch 80 (original thickness 11 mm). Really the material of pipes was carbon steel ASTM A106 Gr. B.

Service period before the failure: 6 years.

Visual examination and findings. Grey uniform deposits were found on the outer surface of the pipes. Black layer of deposits with thickness 0.6–1.2 mm was found on the inner surface of the pipes (Fig. 4.4). Wall thickness of pipes in convection area near the furnace wall diminished on 3.6–4.8 mm. Corrosion rate of pipes was 0.6–0.8 mm/year (after 6 years of service).

Chemical analysis of deposits. Outer surface. Deposits consisted of corrosion products of iron (mostly ferric sulfates) and elements (S, V, Ni, and Si) typical after the burning of fuel oil. When these deposits were dissolved in distilled water, pH = 2.3–2.5.

Inner surface. Iron sulfides were formed inside the pipes.

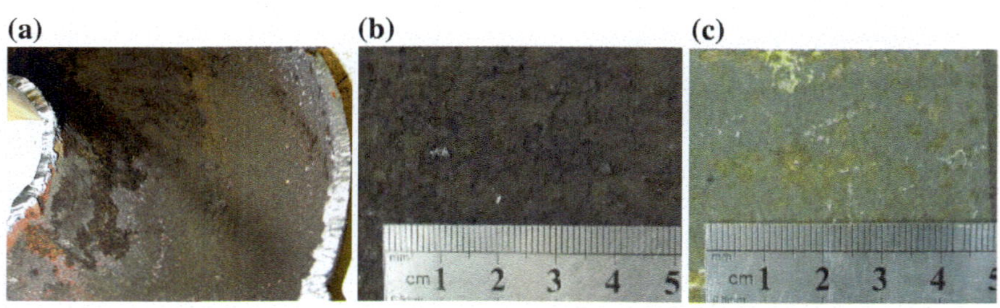

Fig. 4.4 **a** Inner surface of corroded pipe (black deposits—iron sulfides). **b** Magnification of the figure **a**. **c** Outer surface of corroded pipe

Chemical analysis of pipe steel. Carbon steel ASTM A106 Gr. B (0.03–0.11 wt % Cr) when 5 % Cr was required.

Cause of failure and its explanation. Chemical analysis of pipe material showed that it was carbon steel and not 5 % Cr as had to be. Corrosion occurred on both sides of carbon steel pipes.

Corrosion of inner surface. Sulfur compounds (mostly H_2S and mercaptans) in kerosene at ~300 °C resulted in severe corrosion of inner surface of carbon steel pipes. Dissolved hydrogen gas in kerosene converted S-organic compounds into H_2S, which was a major reason of corrosion of inner surface of carbon steel pipes. The Couper–Gorman curves show that the corrosion rate of steels depends on the content of chromium in steel, concentration of H_2S in kerosene, and temperature [19, 21, 22]. According to these curves, if kerosene contains 600 ppm H_2S, corrosion rate of carbon steel is 0.4 mm/year at 300 °C and 1.75 mm/year at 400 °C [22]. If to take into account that iron sulfide deposits were inside pipes and skin temperature was ~400 °C, it coincides with the calculated corrosion rates 0.4–0.8 mm/year (after 6 years of service of pipes in the furnace) in kerosene at 300 °C.

Metal temperatures of furnace pipes' material are usually 85–110 °C higher than the temperature of hydrocarbon stream passing through the pipes [21]. Furnace pipes corrode normally at a higher rate on the hot side (flame area) than on the cold side (wall area).

Corrosion of outer surface. Low values pH = 2.3–2.5 of aqueous solution containing dissolved deposits (ferric sulfates) forming on outer surface of the pipe shows that while shutdown of furnaces deposits absorbed water vapor from the atmosphere at ambient temperature creating an acid environment on the exterior surface of the pipes which in its turn caused corrosion of outer surface of carbon steel pipe (see Eqs. 2.3.1 and 2.3.2).

Failure phenomenon: *High-temperature sulfidic corrosion* (inside the pipes); *acid under deposit corrosion* (outer surface of pipes during shutdowns).

Solutions and recommendations.

(a) Carbon steel and even 5 % Cr are not recommended for use in kerosene containing 600 ppm S at 300 °C [22]. Increase of concentration of chromium in steel results in a significant increase of the resistance of steel to sulfidic corrosion at high temperature. It is recommended to replace carbon steel with SS 321 pipes in contact with kerosene containing $H_2S > 0.01$ mol% at $T > 230$ °C.

(b) To examine the conformity of material type of pipes to requirements before they have to be used.

(c) If fuel oil is used for burning in furnaces, at the beginning of shutdown, immediately after cooling, to remove acid deposits (ferric sulfate salts) from outer surfaces of pipes by cleaning with aqueous solution of soda ash (3 wt%).

4.2 Naphthenic Acid Corrosion

Abstract Naphthenic acids (NA), their place among organic acids and physico-chemical corrosion mechanism are described. Acidity of crude oils and distillate fractions is discussed. Problems caused by NA (formation of naphthenates, stable emulsions and fouling, appearance of nonextractable chlorides, environmental problems, diminishing inhibitor efficiency, and corrosiveness of NA) are presented. Mechanism and factors influencing naphthenic acid corrosion (NAC), combined influence of NA and sulfur compounds on corrosion of carbon steel, NAC control and monitoring are analyzed. A unique case study is given.

The oil refining industry has seen a trend toward refining more highly acid crude oils since the early 1990s. This tendency also remains nowadays. The global supply of high acid crudes will increase fourfold since 2000–2020 [23, 24]. There are about 100 acidic crude oils (TAN > 0.5 mg KOH/g oil), and their geography is very diverse: Romania, Russia, Azerbaijan, USA (Texas, California, the Gulf Coast), Canada, Venezuela, Brazil, Mexico, North Sea, Africa (Congo, Nigeria, Sudan, Chad, Equatorial Guinea, Angola, Cameroon), India, China, Far East, Australia, and Indonesia [25–27]. Oil sands from Canada (Athabasca) and Albania, some Syncrudes (Canada oil shale) are also acidic. Naphthenic acids (NA) are responsible for acidity of these crude oils and oil sands. Nearly all crudes can contain some NA. Usually heavy crudes (i.e. crudes of high density and asphaltene content) have more NA than paraffinic crudes. Use of these acidic crude oils increased corrosion problems.

NA are the generic name wrongly used for all the organic acids present in crude oils.

According to IUPAC (International Union of Pure and Applied Chemistry), the term *organic acids* applies to a broad range of organic compounds which contain the organic acid group

–COOH:

- aliphatic (fatty) acids RCOOH, where R is a straight $CH_3(CH_2)_n$ or branched chain;
- aromatic acids ArCOOH, where Ar is a benzene ring or substituted benzene rings;
- naphthenic acids $XR(CH_2)_nCOOH$ ($n \geq 0$), where X is a cycloparaffinic ring (Fig. 4.5).

All of these three groups can be present in crude oils. It is very difficult to isolate NA separately from the acyclic aliphatic acids and aromatic acids. The Russian chemist Markovnikov (1838–1904) was the first who isolated NA from the Azerbaijan crude oil and studied their physical and chemical properties. He determined a new class of organic substances—*naphthenes*. *Naphthenes* are cycloalkanes—saturated monocyclic hydrocarbons (cyclopentane, cyclohexane and their alkyl derivatives). NA are cycloaliphatic mono-carboxylic acids considering as derivatives of *naphthenes* and can be mono-, bi-, tri-, or polycyclic. Therefore, the

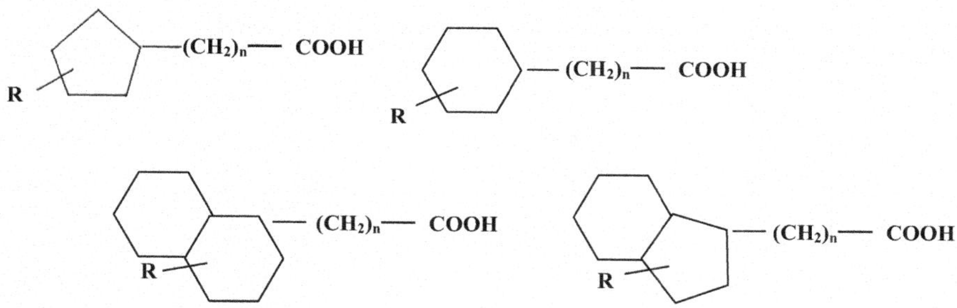

Fig. 4.5 Naphthenic acids. R represents alkyl group (e.g., $-CH_3$, $-C_2H_5$, etc.)

acids received the name *naphthenic*. The carboxylic group (–COOH) is attached to a side chain (see Fig. 4.5). There are hundreds of different NA (134 have yet been detected in acid crude oils) [28]. The terms *naphthenic* and *naphthenes* seem to be obsolete, except in the petroleum and petrochemical industries. NA are natural constituents in crude oils and oil sands originated from long-time oxidation of naphthenes (cycloalkanes) or from either aerobic or anaerobic biodegradation of hydrocarbons in fossil deposits at suitable temperature, pressure, water pH, salinity, and nutrients [29, 30]. NA can be used as biochemical markers of crude oil origin and maturation.

NA are liquids at ambient temperatures with a wide boiling point range (216–350 °C). Most of pure NA are completely soluble in organic solvents and oils but insoluble in water. Low molecular weight NA can slightly dissolve in water. Chemically, NA behave like typical carboxylic acids with an acid strength similar to those of the large molecular weight aliphatic acids such as heptanoic ($C_6H_{13}COOH$) and nonanoic ($C_8H_{17}COOH$), but are slightly weaker than low molecular weight carboxylic acids such as formic (HCOOH) and acetic (CH_3COOH).

Di- and tri-cyclic NA can undergo thermal decomposition to produce lower molecular weight naphthenic and aliphatic acids. Large molecular weight aliphatic acids in its turn can also be thermally cracked into smaller formic, acetic, propionic and butyric acids which are more corrosive than large molecular weight organic acids.

NA containing in acid crudes (see Appendix B) can cause diverse problems in oil refining industry. Therefore, interest to NA is increasing significantly since 2000. There were a few (1–3) publications each year between 1920 and 1960. Then there was a small increase in the number of publications (7–10) per year in the late 1960s, which coincides with the beginning of the Athabasca acid oil sands refining in Canada [31]. The rapid increase in the number of publications (20–90) per year after 2000 accompanies improvements of analytical methods of detection of NA and use of opportunity acid crudes.

4.2.1 Acidity of Crude Oils and Distillate Fractions

Total Acid Number (*TAN*) values (also called *neutralization numbers*) are used to evaluate the corrosiveness of crude oils and their distillate fractions [32]. TAN is a measure of the numbers of milligrams of potassium hydroxide (KOH) needed to neutralize 1 g of crude oil or its distillate fraction. Because crudes and petroleum distillates can contain in addition to NA other acid compounds such as inorganic acids (H_2S, HCl, $H_xS_yO_z$, CO_2), mercaptans, phenols, aliphatic and aromatic acids, hydrolysable esters, acid nitrogen compounds, certain additives (including corrosion inhibitors) used during crude production and other operations, TAN does not reflect real content of NA and thus their actual threaten and participation in corrosion. Therefore, NAN (naphthenic acid number) was suggested which is defined from extraction by chromatography method and then analysis by FTIR [33, 34]. Sometimes NAN values may be two times lower than TAN. Unfortunately NAN determination needs more complicated equipment and more skilled personnel. In any case, TAN and NAN values give rough estimation of corrosiviness of crudes and distillate fractions.

TAN for different crudes ranges in 0.1–6 mg KOH/g. TAN values for most pure NA range in 220–486 mg KOH/g. When $0.5 < TAN < 1.0$ mg KOH/g, crude is named *acid*; when $TAN > 1.0$ mg KOH/g, the crude is named *high acid.* As a rule there is no correlation between TAN of crude oils/distillates and their corrosiveness. For the same TAN value, molecular size and structure of the acid have an important influence. Some NA are very corrosive, others are relatively inert, and high molecular weight NA can inhibit corrosion of carbon steel under certain conditions [35, 36]. In spite of this, some rules of thumb exist for definition of crude oil or distillate corrosion threaten. It was suggested using a threshold of 0.5 mg KOH/g of crude oil and of 1.5 mg KOH/g of distillate. NA distribution can be used to determine likely areas of possible corrosion threatens. Such assay (library) exist [37]. Huizhou refinery (China) is the first large-scale refinery designed for the full conversion of high TAN crude (3.6 mg KOH/g) and has been operated since 2009 [38].

4.2.2 Problems Caused by Naphthenic Acids

A. Formation of naphthenates, stable emulsions and deposits (fouling). Naphthenates are the salts of NA. They are formed when NA contact metal ions in the produced water under suitable pH and temperature. Produced water in the oil industry is water that is produced as a byproduct along with the oil and gas. Naphthenates of sodium and potassium and NA themselves have nonpolar (hydrocarbon hydrophobic) and polar (hydrophilic, COO^- or Na^+) groups. They can form stable emulsions (like surfactants) during desalting, thick intermediate layer oil-water, and lead to foaming problems. The presence of

NA and naphthenates in desalter reduces crude oil desalting efficiency, which in its turn can result in corrosion problems in the overhead of distillation column. Thus, naphthenates play havoc with oil-water separation (dehydration of the crude oil) and de-oiling of the produced water. Naphthenates of calcium, magnesium and iron can form deposits (fouling), which accumulate in oil/water separators, desalters, pipelines, heat exchangers, valves and filters, and cause operational problems because of plugging leading to unexpected shutdowns.

B. Appearance of nonextractable chlorides. NA increase hydrolysis of $CaCl_2$ and $MgCl_2$, promote the formation of HCl from asphaltene-encapsulated chloride salts, and cause fouling by destabilizing asphaltenes [39]. These salts are not removed by desalting (see Sect. 3.1.1.1).

C. Environmental problems. NA and other organic acids can enter into wastewaters and present severe environmental problems. NA are toxic components in wastewaters, however are susceptible to biodegradation [40–43].

D. Diminishing inhibitor efficiency. NA may also interact with injected corrosion inhibitors, and thus negatively impact inhibitor efficiency.

E. Corrosiveness of naphthenic acids.

NA are not corrosive at ambient temperatures and their aggressiveness is materialized at 190–400 °C. Therefore NA not pose a corrosion risk during crude oil transportation in pipelines and its storage in tanks at ambient temperatures. NA are not corrosive at T > 400 °C because they can break down or coke deposits can form and physically (as barrier coating) protect from NA attack. NA can be very corrosive to carbon steel, low-alloy Cr-Mo steels, 12Cr SS, copper, copper-nickel and other alloys under particular conditions at refinery units. Influence of NA on corrosion of metals, case study and preventive actions are described below.

4.2.3 *Mechanism and Factors Influencing Naphthenic Acid Corrosion*

Naphthenic acid corrosion (NAC) intensity is defined by the following factors: crude oil type, kind of acids, TAN (in some cases), naphthenic acid content, their distribution in petroleum distillates, S- and N-containing substances, temperature, number of phases (presence of liquid or liquid/vapor phases, boiling-condensing phenomenon), flow velocity and flow regime, type of alloy and its surface condition (presence of passive layers), and duration. Different boiling points of NA result in their appropriate distribution in various petroleum distillate fractions, and as a result, their corrosiveness is usually also different. As a rule, NA are concentrated in heavy gas oils (boiling temperatures 220–320 °C), but sometimes the kerosene fraction of narrow boiling temperatures (190–210 °C) can contain also NA and be highly corrosive (Fig. 4.6) [44–46].

Different NA are concentrated in specific areas in distillation columns. Distillation towers' internals (including walls, trays and bubble caps), piping, heat

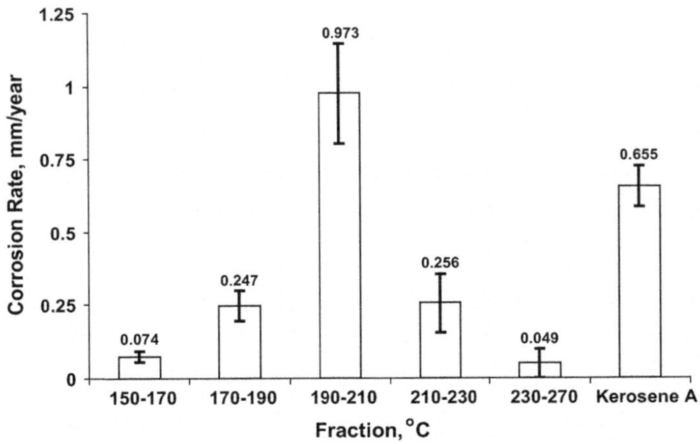

Fig. 4.6 Corrosion rate of carbon steel in various fractions of kerosene "A" (150–270 °C) [46]. Acid kerosene fraction was distilled from acid crude oil 'Azeri'

exchangers, collection headers, check valves, which contact kerosene, light and heavy atmospheric gas oil, atmospheric residue, light and heavy vacuum gas oil and vacuum residue operating at temperatures 190–400 °C are possible areas for naphthenic acid attack.

In spite of NAC being known since 1920s and systematic study begun since 1950s, the corrosion mechanism as well as the chemical reactions describing it are not fully known because of the many factors influencing the corrosion and also the complex interaction between these factors. Usually corrosion between NA and iron is written by the reaction (4.2.1):

$$2XRCOOH_{(l)} + Fe_{(s)} \rightarrow Fe(XRCOO)_{2(l)} + H_{2(g)} \quad (4.2.1)$$

This mechanism suggests formation of iron naphthenate [$Fe(XRCOO)_{2(l)}$] and hydrogen gas (H_2), which allows controlling NAC via the evolving hydrogen [47]. Iron naphthenates dissolve in hydrocarbon phase, corrosion products do not remain on the metal surface, which will be continuously corroded by NA. The chemical structure type of NA defines their corrosiveness. For instance, the cyclic structures were more corrosive than the acyclic [48].

Appearance of NAC. NAC may appear in different forms of localized attack on refinery units:

- Shallow, deep, pin-head, large and round pits because of boiling and condensation of NA on metal surface inside the atmospheric and vacuum distillation columns (wall shell, trays and bubble caps). NAC occurs primarily in liquid and mixed liquid/vapor streams.
- Metal thinning and uniform etching of stainless steel cladding on the vacuum distillation column shell [49].

- Grooves, striations and deep localized attack (droplet impingement) due to velocity effects in transfer lines and heater (furnace) tubes [49]. Severe NAC occurs when flow is turbulent and in bends, elbows, tees, and pumps.
- General corrosion of carbon steel trays and bubble caps if constant change of acid (low sulfur) and non-acid (high sulfur) crudes are used in refinery (see Sect. 4.2.7).

Combined influence of NA and sulfur compounds on corrosion of carbon steel. Most oil refineries run alternately different kinds of crudes and crude blends. Usually high sulfur (sour) crudes have low acidity and opposite, low sulfur (sweet) crudes have high acidity. Only some types of crude oils have high acidity and low sulfur. For example, crudes from California and China (having TAN up to 6 mg KOH/g and relatively low concentrations of sulfur compounds) are highly corrosive because insufficient iron sulfide scale is produced to counteract NAC [49]. In contrast, certain crudes from Venezuela (having intermediate TAN up to 3 mg KOH/g and relatively high concentrations of sulfur compounds) are less corrosive because sufficient sulfide scale is produced to partly counteract NAC. Probably Piehl [50] was the first who noted that iron sulfide scale had a protective role against NAC, and competition between the sulfidic and NAC occurred.

Constant alternation of acid sweet crudes and low acid sour crudes in distillation units leads to complicated influence of corrosive components (sulfur compounds and organic acids, including naphthenic) in crudes on corrosion of carbon steel and low alloy steels. For instance, when TAN = 3.5 mg KOH/g (T = 343 °C and flow rate 60 m/s), at low H_2S concentrations (<0.16 wt%) corrosion rate of carbon steel is high. Then at H_2S concentrations 0.16–2.0 wt% corrosion rate is very low. When H_2S more than 2.0 wt% corrosion rate is high again [51, 52]. Thus, NAC can be successfully inhibited on carbon steel and low-alloy steels (e.g., 5Cr and 9Cr) by the moderate concentrations of H_2S in hydrocarbon media but can also be aggravated with higher H_2S concentrations [53]. High concentrations of sulfur compounds increase high-temperature sulfur corrosion in hydrocarbon media. This situation is similar to passive region for dependence electric current–potential for some metals in definite electrolyte solution. Thus, the firmly established fact exists that H_2S can inhibit NAC at particular conditions depending on TAN, type of NA, temperature, flow rate, flow regime, and duration [51–54]. It is not simply to define this passive region (based on H_2S and NA concentrations) in real conditions at crude units because of many influenced factors. Reaction of H_2S with iron results in formation of a protective layer composed of iron sulfides; however, high turbulence of stream can shear away this film to expose new surfaces for corrosion [55].

Iron naphthenate (particular case of iron carboxylates) forming in reaction (4.2.1) can be decomposed into iron oxide in (4.2.2) [56–58]:

$$\mathrm{Fe(XRCOO)_{2(l)}} \rightarrow \mathrm{FeO_{(s)}} + \mathrm{XRCORX_{(l)}} + \mathrm{CO_{2(g)}} \qquad (4.2.2)$$

Then iron oxide can react with S-containing compounds with iron sulfide formation (4.2.3, 4.2.4):

$$FeO_{(s)} + H_2S_{(g)} \rightarrow FeS_{(s)} + H_2O_{(l)} \quad (4.2.3)$$

$$FeO_{(s)} + R_1{-}S{-}R_{2(l)} \rightarrow FeS_{(s)} + R_1{-}O{-}R_{2(l)} \quad (4.2.4)$$

When there are no enough S-containing compounds, it is possible to detect mixture of iron oxide (red-brown rust) and iron sulfide (black) on carbon steel surface.

Both NAC and sulfidic corrosion occur at high temperatures (T > 200 °C), and we need to distinguish between them. When high temperature sulfidic corrosion occurs protective sulfide scale is formed on steel surface, corrosion intensity diminishes with time, and increase of chromium in steel slows significantly corrosion by sulfur species. In contrast, there is no scale during NAC because only soluble corrosion products are formed, corrosion intensity does not decrease with time, and carbon steels, low-alloy (5Cr, 9Cr) and 12Cr SS are usually attacked by NA.

Thus, high sulfur crudes having low acidity when attacking carbon steel surface at the beginning iron sulfide protective scale is formed. This passive film usually protects carbon steel surface from further corrosion if acids are not present (or present in small amounts) in streams. This is the case, for instance, in amine treating units, where H_2S absorbing by aqueous amine solution reacts with carbon steel surface, forming passive iron sulfide layer and protects inner surface of constructions from the corrosion. This is also the case in the overhead distillation column when H_2S creates tenacious uniform protective iron sulfide film with thickness 5–10 μm (see Fig. 3.1). Acid crudes usually are low sulfur, for instance, 'Azeri'. When low sulfur acid crudes alternate with high sulfur low acid crudes, organic acids (including NA) can dissolve iron sulfide protective film and attack carbon steel surface. This mechanism concerns both crudes and their distillate fractions. The Sect. 4.2.7 described below confirms such approach to the impact of the combined influence of sulfur content and acidity in crude oil and its fractions on their corrosiveness.

4.2.4 Naphthenic Acid Corrosion Control

NAC is understood in many cases and appearances, and as a result prevention actions are developed. There are no reliable data available that can be used to predict where, when, and to what extent NAC is likely to occur. In any case, it is recommended using following anti-corrosion actions if crudes with TAN > 0.4 mg KOH/g are used at crude distillation units.

A. The use of corrosion resistant alloys (materials selection).
B. The blending of acid with non-acid crude oils (or distillate fractions).
C. Neutralization and removing, naphthenic acid extraction for their further application in industry.

D. The use of corrosion inhibitors.
E. The application of metalizing coatings (e.g., aluminized carbon steel).
F. Limiting fluid flow velocity.

A. *Materials selection*. The use of corrosion resistant alloys is the only practical way of NAC prevention. The greater the content of molybdenum (Mo) in stainless steels, the better their resistance to NAC. Minimum content of molybdenum must be 3 wt% (SS 317L). Therefore, even SS 316L (containing 2.1 wt% Mo) can be attacked by NA. Stainless steels containing molybdenum are more perfect and they have significantly less defects than stainless steels without molybdenum. Stainless steels posses anti-corrosion properties due to chromium oxide (Cr_2O_3) protective passive film. However, this film is 'metastable' which means 'quasi-stable', not absolutely stable, namely, stable at its present condition and can exit this state at any moment as a result of changes of outer conditions or inner properties of the metal-environment system. Iron and chromium exit earlier (the first), nickel and molybdenum later. When molybdenum is present in stainless steel, the surface is enriched by molybdenum which in its turn increases stability of protective passive film.

Copper and copper-nickel alloys, SS 304 and SS 321 should never been used against NAC. Aluminum, titanium, their alloys, SS 317LM (4–5 wt% Mo), 904L (4 wt% Mo) and AL6XN (6 wt% Mo), Ni-based alloys Hastelloy C-276, Incoloy 825 and Inconel 625 are resistant to NA attack.

B. *The blending (dilution) of acid with non-acid crude oils (or distilled fractions).* The aim of the blending is to reach TAN < 0.5 mg KOH/g oil. We will show how blending of acid with non-acid kerosene can solve the problem of corrosiveness of kerosene fraction because of its high acidity (TAN = 0.57 mg KOH/g kerosene) [44–46]. Corrosion rate of carbon steel in this acid kerosene was 0.4 mm/year. The non-acid kerosene that was used for blending had a TAN < 0.01 mg KOH/g. In blending, various volume ratios of acid and non-acid kerosene were examined on their corrosiveness to carbon steel (Fig. 4.7). One can observe the existence of a critical acidity of 0.464 mg KOH/g kerosene at which the corrosion rate of carbon steel drastically decreases with increase of dilution. This dilution found for the particular acid kerosene at a ratio 3 to 1 (three volumes of acid kerosene and one volume of non-acid kerosene). This method was successfully used at crude distillation unit [46]. Disadvantage of dilution is that part of finished product (kerosene) must be returned for re-distillation.

C. *Neutralization* and *removing* NA from crudes [59–62]. Three methods are used for this: extraction, destruction, and adsorption.

Extraction. Simple alkaline treatment by soda solution of crude rarely works because naphthenate salts are well soluble in the crude, stable emulsions can be formed and then fouling can occur. Other compounds, tetraalkylammonium hydroxide and alkali metal trialkylsilanolates, can be used for this purpose [63, 64]. Extraction with polar organic solvents (e.g., methanol and ionic liquids) also can be used [65]. NA are typically recovered by caustic extraction of crude oil distillates

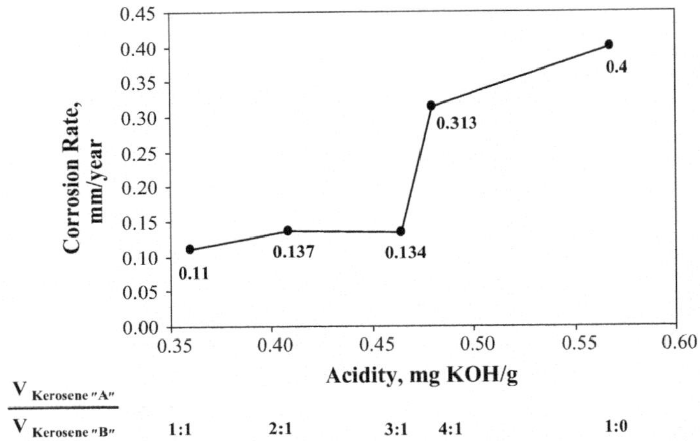

Fig. 4.7 Corrosion rate (mm/year) of carbon steel versus total acidity of kerosene (blended samples). Kerosene "A" is acidic (TAN = 0.57 mg KOH/g) and kerosene "B" is non-acidic (TAN < 0.01 mg KOH/g). T = 185–200 °C; P = 1 bar; Duration—3 days [46]

boiling in the 200–370 °C range. Therefore, corrosion problems remained during distillation and are solved by application of corrosion resistance alloys.

Naphthenic acid neutralization and removal. Removing NA from crude oil downstream using basic compounds (alkali salts or amines) has been carried out for many years to avoid corrosion and other problems bearing by NA [66]. Catalytic hydrotreatment under mild conditions was developed for selective removing of NA [67]. Oxidation of NA by peroxodisulphate ($S_2O_8^{2-}$) was efficient too [68].

Destruction. Decarboxylation (with CO_2 formation); catalytic decarboxylation [69]; mild catalytic hydrotreating; heating to 400 °C; mix with lime (CaO), heating to 260 °C and separation from reacted lime [65].

Adsorption. Nickel oxide (for a kerosene stream), strong base ion exchange resins (for lube oils), ionic silica based hybrid materials can be used for this purpose [65, 70].

D. *The use of corrosion inhibitors*. Traditional N-contained filming corrosion inhibitors, widely used for the protection of the overhead in the atmospheric and vacuum distillation columns, are not effective at temperatures above 140 °C. Thus, they cannot be used for the NAC control, and other compounds were found which are effective against NAC at temperatures 175–400 °C at concentrations from 1 to 5000 ppm [24, 36, 71–89]. The active ingredients of organic inhibitors contain one or more functional groups with one or more hetero atoms (N, O, S, or P), through which the inhibitors anchor on to the metal surface. The inhibition efficiency of the homologous series of organic substances differing only in the heteroatom is usually in the following sequence: P > S > N > O. The electronegativity, that is, the ability to attract electrons, increases in the reverse order [90]. As a result, phosphorus containing inhibitors ("P") are more effective than sulfur containing inhibitors ("S") (Fig. 4.8).

The efficiency of inhibitor "P" (phosphate ester) is higher (94–99 %) than that of "S" (sulfur containing) inhibitor (80–85 %) at 50–75 ppm. The former inhibitor at concentrations above 50 ppm protects carbon steel at nearly 100 % efficiency in acidic kerosene forming a dense protective passive layer. The inhibitor "P" was examined and used in the distillation unit during a run of acid crude oil 'Azeri' (Fig. 4.9) [44–56].

Over 40 successful applications to reduce NAC using phosphorus based inhibitor have been reported [24, 25, 37]. It is preferred to add the inhibitor "P" at a relatively high initial dosage of about 100 ppm, and to maintain this level for a relatively short period (about 24 h) until the presence of the inhibitor induces the build-up of an anti-corrosion protective layer on the metal surface (pretreatment, or pre-passivation stage). Once the protective, tenacious, persistent, passive layer is established, the dosage may be reduced to about 10 to 25 ppm. This film can persist more than 14 days, but this period certainly depends on flow rate, turbulence, and temperature. Phosphorus based inhibitors' concentrations of 20–30 ppm are recommended for pre-passivation and then reduced to 5–15 ppm for normal control [25]. Their main drawback is that they can poison catalysts in downstream units and deteriorate the kerosene quality by lack of control of the content of phosphorous.

E. *The application of metalizing coatings*: aluminized carbon steel, niobium and niobium oxide thermal spray coating.

F. *Limiting fluid flow*. The critical flow velocity for carbon steel is 25 m/s and for SS 316 is 120 m/s without turbulent flow [38].

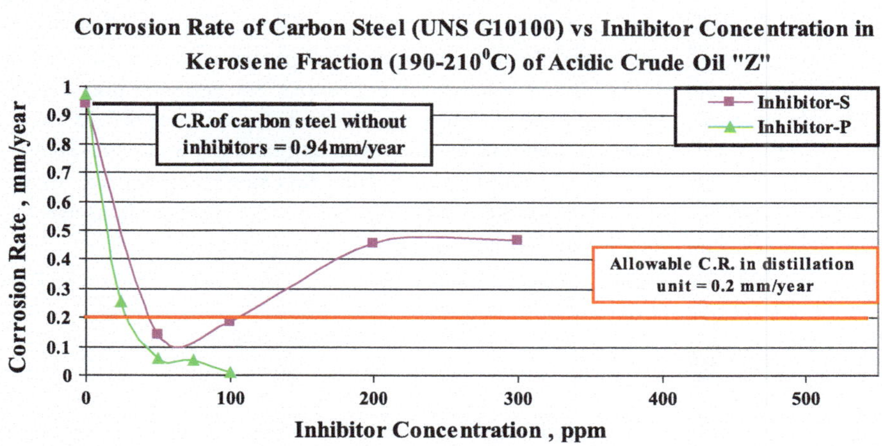

Fig. 4.8 Corrosion rate of carbon steel versus phosphorus (P)- and sulfur (S)-based inhibitor concentration in acidic kerosene at 195 °C [91]

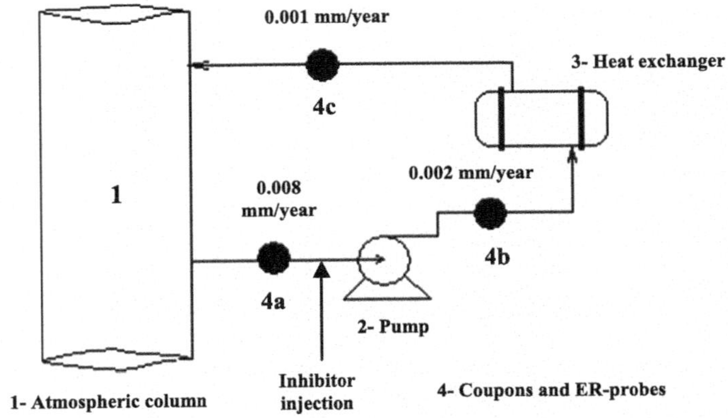

Fig. 4.9 Naphthenic acid control by injection of phosphate ester inhibitor in acid kerosene 'Azeri' at the distillation unit. Corrosion rates of carbon steel coupons are average during 3 months run

4.2.5 Monitoring of Naphthenic Acid Corrosion

Physical and analytical methods are used for monitoring of NAC. Analytical methods include measuring of Fe, Cr and Ni in process streams and measuring hydrogen flux [47, 92, 93]. It is important to measure trend of concentrations of metallic ions in media and hydrogen flux with time and follow alternation of crudes with different TAN. Any sudden increase of concentrations of metallic ions in process streams and hydrogen flux will show possible NA attack.

Physical methods include ER-probes, WL coupons (with retractable holders), ultrasonic measurements, and visual examination. If coupons from carbon steel, low-alloy steel (5Cr, 9Cr), and 12Cr SS corrode with similar rate this means that possible NA attack occurs. Allowable corrosion rates for carbon steel components are 0.1 mm/year for heat exchanger tubes (when original thickness of tubes is 2.333 mm) and 0.2 mm/year for pipes and column's walls (original thickness 6 mm and more). However, ER-probes and WL coupons can be installed not in all hot locations. Ultrasonic wall thickness measurements and visual examination during shutdown are good inspection methods for detection of NAC.

4.2.6 Conclusion

There is a sizeable increasing global supply of high acid crudes and this situation causes special attention to corrosion and other problems at oil refinery units. High acid crudes may offer value relative to other grades of crudes and this situation is defined by cost of mitigation anti-corrosion actions.

(a) Corrosiveness of high acid crudes is defined by TAN values, naphthenic acid content, their distribution, temperature, the number of phases (presence of liquid or liquid/vapor phases including boiling-condensing phenomena), flow regime and velocity, and duration of contact of NA with equipment.
(b) In spite of the firmly established fact that there is no correlation between TAN values and corrosiveness of crude oils and distillate fractions, no one wants to endanger his oil refinery because there is no reliable method of determining of corrosiveness of crude oils.
(c) Corrosion mechanisms of high-temperature (above 190 °C) corrosion in refinery units include naphthenic acid corrosion, sulfidic corrosion and interactions which can either accelerate or inhibit corrosion. Hydrogen sulfide can react with iron to form protective (against NAC) iron sulfide layer under particular conditions (crude type, particular temperature range, and TAN values). It is impossible to predict the region of this passivity and only real experiment at the unit can define the passive area.
(d) In spite of insufficient understanding of corrosion mechanism occurring in the presence of naphthenic acids, mitigation measures against NAC are developed which are successfully proved and used.
(e) The use of corrosion resistant alloys is the only practical way of NAC prevention. The greater the content of molybdenum (Mo) in stainless steels, the better their resistance to NAC. Minimum content of molybdenum must be 2.5–3 wt% (SS 316L, 317L). Stainless steels 317LM, 904L and AL6XN, Ni-based alloys Hastelloy C-276, Incoloy 825 and Inconel 625, aluminum, titanium, their alloys, are resistant to NA attack.

Interesting about naphthenic acids. NA are treacherous in the oil refining industry because of their high, unpredictable corrosiveness, and many facet appearance. However, NA and their derivatives (naphthenates, esters and amides) are used as wood preservatives (replacing creosote), corrosion inhibitors, surfactants (emulsifiers), fuel and lubricant additives (e.g., defoamers in jet fuel), paint and ink driers, catalyst of rubber vulcanization and production of alkyl and polyester resins, tire cord adhesives, heat/UV stabilizers, flame retardants in fabrics, improver water resistance and adhesion of concrete, insecticides, fungicides, acaricides, vinyl resins stabilizers, and regulators of plant growth. Such wide application of NA and their derivatives impart importance their extraction from acid crude oils and thus prevent NAC. There are chemical plants and units in some countries which extract NA for further application in industry.

NA have negative sense not only in corrosion of refinery equipment but also because of 'napalm'—a flammable liquid developed in the USA in 1942 and used during the World War II and other warfare. The word 'napalm' is derived from the words naphthenic acid and palmitic acid as it is produced by their combination.

4.2.7 *Case Study*

Conditions. Various blends of crude oils have been used during 5 years and acid crude 'Azeri' has often been used during the last 2 years before the detection of severe corrosion of trays in the atmospheric crude distillation column (Fig. 4.10).

Recirculation LAGO enters to the tray number 24 in the distillation column (the trays have downward numbering). Data about distillate fractions, temperatures, trays' materials and numbers in the crude distillation column are shown in Table 4.3.

Material of construction: Trays numbers 24 and 25—carbon steel; the tray number 33—SS 410 (12Cr).

Service period before the failure: 5 years.

Visual examination and findings. Severe *general corrosion* of trays (especially numbers 24, 25) and diminishing in 1 mm of the wall shell thickness of distillation column was found in the area of trays numbers 24–33 (see Fig. 4.10). The thicknesses of the trays numbers 24, 25 diminished to zero with the formation of the holes. Black (mostly), white and reddish (a little) deposits were found on these trays. Most deposits (62–90 wt%) consisted from corrosion products of iron and chromium sulfides. Severe corrosion of the tray number 35 and bubble caps made from SS 410 (12Cr) was also detected.

Analyzing PI (Plant Information) data at the crude distillation unit, namely, by plotting on-line corroded thickness (yellow line) obtained by ER-probe data against the sulfur concentrations in crudes, one can reveal periods of running low sulfur crude 'Azeri' (high corrosion trends!), changing by the periods of running high sulfur crudes (low corrosion trends!) (Fig. 4.11).

Examination of corrosiveness of acid crude 'Azeri'. The physical and chemical properties of crude oil 'Azeri' are shown in Table 4.4.

The TAN of crude oil 'Azeri' is 0.4 mg KOH/g. This is close to the threshold value 0.5 mg KOH/g. Therefore, it is hard to predict whether or not crude oil 'Azeri' will be corrosive. Sulfur is 0.15 wt%, thus, this is a sweet crude oil (<0.5 wt% S). The salt and water contents of the crude oil are low and would not cause its corrosiveness.

Corrosion tests (estimation of the corrosiveness of the crude oil 'Azeri') were carried out in an autoclave and of the distillate fractions in a special device [46].

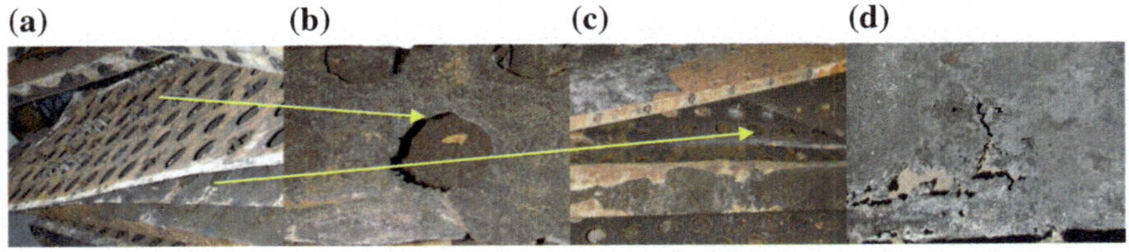

Fig. 4.10 Corroded carbon steel trays (original thickness—3.5 mm; LAGO area) in crude distillation tower: **a–c** tray number 24 (236–243 °C). **d** tray number 25 (244–250 °C)

Table 4.3 Data about trays and distillate fractions in the atmospheric crude distillation column

Tray number	Material	Original thickness, mm	Distillate fraction	T (°C)
24	Carbon steel	3.50	LAGO	236–243
25	Carbon steel	3.50	LAGO	244–250
33	SS 410 (12Cr)	1.88	HAGO	260

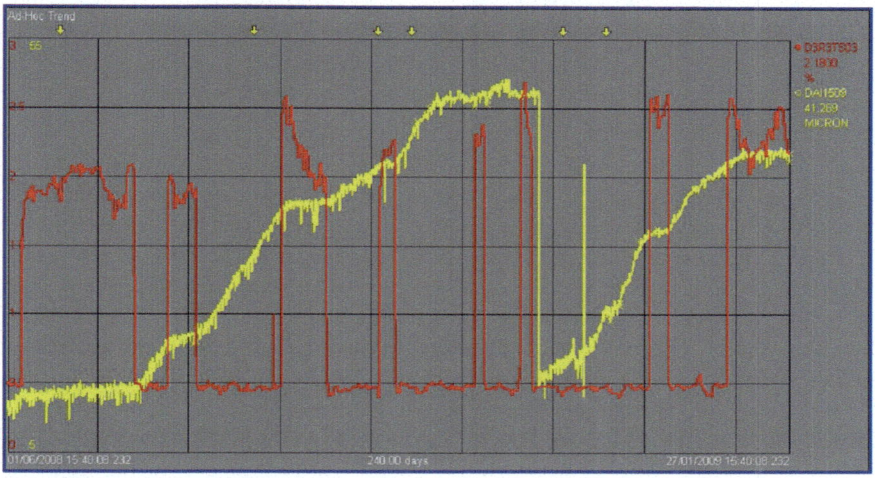

Fig. 4.11 Trend in atmospheric distillation column (PI data, 240 days): ER-probe readings (*yellow*) and sulfur content (*red*) in crudes. When corrosion rate (calculated as the differential of *yellow line*) is low there is no increase of *yellow line* with time (the corroded thickness of ER-sensor in microns is constant—no corrosion). When corrosion rate is high, the corroded thickness of ER-sensor increases with time and yellow line also increases (there is a constant slope of *yellow line*—corrosion rate is high)

Table 4.4 Physical and chemical properties of the crude oil 'Azeri' [46]

Parameter	Unit	Value
TAN	mg KOH/g	0.4
Sulfur	wt%	0.15
Density, 15 °C	kg/L	0.8515
Viscosity, 20 °C	cSt	15.62
Pour point	°C	−3
Salt	Lb/1000 bbl	4.2
Water	% volume	<0.05

The carbon steel coupons were immersed in petroleum distillate fractions in 1 L flask equipped with a reflux condenser. The operating conditions for the crude oil were 300 °C and 10 bar; for the fractions tested distilled from 150 until 370 °C the pressure was atmospheric. The results of the determinations of corrosiveness of crude oil 'Azeri' and its distillate fractions are shown in Table 4.5.

Table 4.5 Corrosiveness of crude oil 'Azeri' and distillate fractions [28]

Crude oil 'Azeri' or its fraction, °C	TAN, mg KOH/g	S, wt %	Corrosion rate of carbon steel, mm/year
Crude oil 'Azeri', 300 °C	0.40	0.15	0.02–0.1
150–270	**0.72**	**0.025**	**0.26–0.66**
230–260	**0.76**	**0.032**	**~1**
270–300	0.25	0.07	0.048
300–322	0.25	0.08	0.019
322–342	0.30	0.12	0.019
342–370	0.35	0.16	0.052

Note Laboratory data, 3 days

Cause of failure and its explanation. Examination of corrosiveness of low sulfur relatively acid crude oil 'Azeri' shows that corrosion rate of carbon steel varies from 0.02 to 0.1 mm/year at 300 °C (see Table 4.5). However, the most corrosive kerosene fraction boils at 150–270 °C with corrosion rate ~0.66 mm/year for carbon steel, and even can reach 1 mm/year for the fraction boiling at 230–260 °C. General corrosion rate of trays nos. 24 and 25 was 0.7 mm/year in the operating distillation column (during 5 years). This value (0.7 mm/year) coincides with corrosion rates (0.66–1 mm/year) obtained under laboratory conditions. Crude oil 'Azeri' probably contains light naphthenic acids, cyclopentane and cyclohexane carboxylic acids, boiling at 216 and 232 °C respectively. Therefore, they appear in the kerosene fraction boiling from 150 to 270 °C and give it a high acidity (~0.76 mg KOH/g) and consequently a high corrosiveness. This fact confirms high corrosiveness of kerosene fraction distilled at 150–270 °C. Thus, the presence of light naphthenic acids accelerated significantly corrosion of carbon steel and SS 410 (martensitic stainless steel containing 12 % Cr but does not contain Mo). SS 410 should be resistant to H_2S but does not resist to naphthenic (and other organic) acids.

Thus, the crude oil 'Azeri' was pinpointed as being the most critical in creating increased corrosion risk during the processing at the crude distillation unit. Alteration of acid low sulfur crude oil 'Azeri' and non-acid high sulfur crudes at the distillation unit during 5 years lead to rare situation (see Fig. 4.10). The presence of H_2S in distillate fractions in this case caused severe corrosion of carbon steel pure surface with formation of iron and chromium sulfides. Alteration of high sulfur non-acid crudes on acid low sulfur crude 'Azeri' lead to dissolution and removing of iron and chromium sulfide layers and refreshing steel surface for further corrosive attack by hot organic acids (naphthenic acids among them) present in the crude 'Azeri'. Further change on non-acid high sulfur crude caused severe corrosion of carbon steel by H_2S with formation new scale, etc.,… to the complete failure.

Solutions and recommendations.

(a) When acid low sulfur crude oils are interchanged with non-acid high sulfur crudes, alternate NAC and high sulfur corrosion occurs. In these cases, trays and bubble caps in distillation column in areas with T > 200 °C must be made from SS 316L or even better SS 317L.

(b) It is possible to inject corrosion inhibitor (Tannic Acid + IPA + Glycerine; 5–25 ppm) into LAGO circulated stream.

4.3 Hydrogen Damages

Abstract Hydrogen attack and prevention at the oil refineries' units are described and analyzed. Hydrogen damage failures are differentiated into two groups based on two mechanisms: electrochemical processes (mainly at low temperature, up to 100 °C) arising from acid corrosion or cathodic protection, and high temperature (between 200 and 900 °C) arising from the presence of hydrogen gas at high pressures. Theory of hydrogen attack is described. Examples of hydrogen failures, causes, analysis, monitoring and remedies are given for the various units at oil refineries.

About 60 % of the hydrogen produced today is used in numerous processes for obtaining high octane gasoline. Use of hydrogen was increased significantly during the 1980–2015 period as desulfurization and hydrocracking processes increased quantity of crude processed.

Hydrogen molecules (H_2) are indifferent to metals but atoms (H) are not apathetic to metals. Like very small bacteria and viruses can infect and cause illnesses and death of plants, animals and people, tiny hydrogen atoms can cause very dangerous corrosion phenomena under the general term of cracking which relates to their penetration into metals and alloys. When hydrogen gas is released into the atmosphere as a result of crack in metallic equipment it reacts with oxygen gas of the atmosphere producing a flame hot enough to melt such refractory materials such platinum, fire brick, porcelain, and corundum (crystals Al_2O_3 with $T_{melt} = 2044$ °C). Hydrogen failures can result in explosions, emissions of toxic substances, fires, injuries, and even death of people.

Hydrogen damages have a wide spread effect in oil refining and petrochemical industry. One of the most important and spread hydrogen damages is hydrogen embrittlement. The latter is the result of corrosion processes because of penetration of hydrogen atoms inside of metals or alloys when they become brittle and appear as a fracture (crack or fissure).

Hydrogen gas occupies an essential place in the *hydro*processing in the oil refining industry: *hydro*treating, *hydro*desulfurizers, *hydro*crackers, and catalytic reformers (see Chap. 1). In addition to these processes proceeding at T > 200 °C and high pressure, there are sources of hydrogen gas arising from electrochemical corrosion of carbon steel equipment in contact with aqueous solutions of acids, such as H_2S, HCl, H_2SO_4, HCN, and HF at low temperatures (20–100 °C) and corrosion by organic acids (including naphthenic acids) at 100–360 °C.

Hydrogen damages can be divided into two groups based on two mechanisms: electrochemical processes (mainly at low temperature, up to ~100 °C) arising from acid corrosion (including chemical cleaning and etching—“useful corrosion”) and cathodic protection (when water molecules can be reduced to hydrogen atoms at certain negative electric potentials), and high temperature (between 200 and 900 °C) arising from the presence of hydrogen gas at high pressures (4–30 MPa). It was suggested differentiating special case of appearing of hydrogen gas on carbon steel surface because of corrosion by organic acids (naphthenic acids among them) at 100–360 °C [94, 95].

Low-temperature (<100 °C) hydrogen damages take place with the participation of hydrogen ions (H^+) which are reduced to hydrogen atoms (H) on the surface of steel in electrolyte solution and then penetrate in its interior. This type of hydrogen damages can occur in the overhead of atmospheric and vacuum distillation columns, amine treating, isomerization, fluid catalytic cracking, and alkylation units. In cathodic protection at potentials below −0.72 V versus Saturated Calomel Electrode (SCE) electrode (pH = 7), water molecules can be reduced with release of hydrogen gas which can penetrate into steel in the form of hydrogen atoms and cause hydrogen embrittlement. This can occur under cathodic overprotection—when protective potentials are maintained significantly less than −0.72 V versus SCE electrode. Hydrogen gas in the form of molecules H_2 cannot penetrate into metals.

Corrosion by organic acids (naphthenic acids among them) which are present in crude oils and petroleum distillates results in formation of hydrogen gas on carbon steel surface and penetration inside in the form of hydrogen atoms. This type of hydrogen damage can occur in distillation units: rectification columns, piping, and heat exchangers.

High-temperature hydrogen damage named *high temperature hydrogen attack* (HTHA) takes place as a result of thermal dissociation of hydrogen molecules with the formation of hydrogen atoms, and their further penetration into metal or alloy. Hydrogen atoms can react with carbides (i.e. cementite Fe_3C) and carbon inside carbon steel forming methane gas (*decarburization*). As a result of this hydrogen damage, steel irreversibly loses mechanical properties, such as strength and toughness. HTHA can occur at *hydro*desulfurizers, *hydro*crackers, and catalytic reformer units. General in low- and high-temperature hydrogen corrosion is that hydrogen atoms when forming on the surface of metal do not retain a memory of its origin, and damage often occurs beneath the surface and not always possible to detect it until failure occurs. We will describe the mechanism of low- and high-temperature hydrogen damages, their forms, monitoring, preventive actions, and case studies.

Preliminary note. It would be wrong to think that hydrogen attack begins from the contact with aggressive hydrogen media. From the very beginning we do not use original metallic equipment and structures free and pure from hydrogen inside. Iron and steel which were not in contact with aggressive hydrogen environment can contain ‘metallurgical’ hydrogen in concentration 10–90 cm^3/kg at 25 °C [96]. This hydrogen appears as a result of contact of melting metal with furnace atmosphere, different additives, passes from cast iron and scrap-metal both as hydrogen is dissolved in scrap and in the form of rust containing hydrates of iron oxides, during smelting. Concentration of hydrogen in steel depends on particular features of metallurgical processes and can reach 1080 cm^3/kg in seams and heat affected zones (HAZ) after welding as a result of contact of metals with water containing in welding electrode coatings, flux and rust [96]. Thus, we get equipment for hydrogen service with some initial virgin amounts of dissolved hydrogen within the alloy. Nobody checks this quantity of hydrogen inside the alloy but even at this early stage it is possible to take preliminary measures to get rid of it.

4.3.1 Low-Temperature Hydrogen Attack (LTHA)

Most hydrogen damages of equipment at oil refinery and petrochemical units occur in the presence of "the eternal companion" of many refining processes—H_2S. This compound in aqueous media is the strongest promoter of both corrosion and an absorption of hydrogen atoms by steel. Electrochemical reaction between H_2S and iron results in the formation of nascent atomic hydrogen (H) on carbon steel surface on cathodic areas (regions with higher electric potentials than neighboring anodic zones) (see Sect. 3.1. and Appendixes D, G). Cathodic protection of steel can result in hydrogen forming on the surface of a bare or under coated construction as a result of the decomposition of water under proper electrode potentials (below −0.72 V vs. SCE electrode at pH = 7) (see Appendix G) [97]. Nascent hydrogen atoms forming in these two cathodic processes are very mobile and active, which can diffuse into metals. 'Nascent state' originates from 'in statu nascendi' (in Latin), which means 'in the state of being born'. This monoatomic short-lived nascent form of hydrogen ($H^•$) is really a 'radical' and as all radicals in chemistry is most reactive than its other forms. When hydrogen radicals ($H^•$) are formed on the metal surface, they find themselves at a crossroads: recombination into H_2 molecules or absorption and diffusion into metallic lattice (see Appendix G). Recombination of hydrogen atoms (radicals) into H_2 molecules occur much faster on metal surface than in the gas phase. Because of large dimensions of hydrogen molecules their intrusion into metals is impossible (see Appendix G). Some chemical substances (mostly H_2S, sulfides, sulfur S_8, cyanide ions CN^-) named cathodic poisons prevent the recombination of hydrogen atoms to form molecular hydrogen and enhance the absorption of nascent atomic hydrogen by the steel [98–100]. Sometimes protective layer of iron sulfide (FeS) is formed on carbon steel surface. Cyanide ions (CN^-) can destroy this protective layer (even at pH > 8) and stimulate dissolution of hydrogen atoms in steel [101]. Proportionality between hydrogen absorption rate by steel and corrosion rate of steel in acids not always occurs.

Content and concentration of electrolytes in aqueous solution have significant and complicated influence on hydrogen absorption by steel. Salts $CaCl_2$ and $MgCl_2$ are hydrolysed in aqueous phase of crude mixture, pH decreases to ~1 and promote hydrogen attack. Usually decrease of pH causes increase of hydrogen absorption by metals. Absorption of hydrogen by carbon steel cannot occur at pH ≥ 4 in HCl and H_2SO_4; pH ≥ 6 in H_2CO_3 and C_6H_5OH. The amount of hydrogen absorbed by steel during its corrosion in nitric acid (HNO_3) is much lower than in HCl, H_2SO_4 and H_3PO_4 because reduced hydrogen atoms are oxidized by nitrates (NO_3^-) [102].

Dissolved oxygen can react with H_2S with formation of dispersed sulfur (see Appendix D). Further oxidation leads to formation of two acids H_2SO_3 and H_2SO_4 which also can contribute to hydrogen absorption. When pH of aqueous solutions containing H_2S increases from 1 to 7 the portion of hydrogen that diffuses into steel increases up to pH = 4 and then decreases. Increasing of hydrogen diffusion is explained by rising of S^{2-} and HS^- concentration as a result of exacerbation of dissociation of H_2S in aqueous solution. Further decreasing of the portion of hydrogen relates probably to diminishing of iron sulfide solubility with rising of pH. The iron sulfide film can be passive under particular conditions and slows down both hydrogen penetration and general corrosion of steel (see Sect. 3.1.1.1). Increasing of H_2S concentration in gaseous hydrocarbon phase (above aqueous media) and diminishing of temperature stimulate hydrogen absorption by steel.

Traces of arsenic, antimony, mercury, selenium and tellurium in solutions, and precipitation of phosphorus and sulfur on a carbon steel surface, promote the hydrogen absorption by steel [103].

Intensity of hydrogen absorption by steel is defined by ratio of rates of cathodic reduction of H^+ and formation of H_2. Absorption of hydrogen by metal is greater when the rate of cathodic reduction of H^+ is higher and the recombination rate is lower. The function of H_2S is that S^{2-} and HS^- ions accelerate cathodic reduction of H^+ and inhibit recombination [96]. The most other poisons act by similar way. Minimal critical concentration exists for poisons when they loss activity of increasing hydrogen absorption by steel. For instance, for arsenic, this value is 1 ppm As in solution and 0.25 wt% As in alloy [96].

Only small part of nascent hydrogen atoms dissolves and enters into solid metal. The question is how hydrogen atoms can dissolve and penetrate into dense crystal structure of metal? It is not easy to imagine how small hydrogen atoms dissolve in solid material and then cross inside through it. Iron and carbon steel resemble a vast "empty" space with islands of ions (positive charges) which are arranged into crystal lattice and floating in a sea of electrons (negative charges). The ions form strong crystal structure in the metal.

In this point, we could compare penetration of hydrogen atoms through solid metal with small mosquitoes which penetrate through window grid. Not all but some insects possess energy which allows penetrate through tiny grid. Another comparison is dissolution of non-polar gas O_2 in polar liquid H_2O molecules where oxygen is arranged inside of "empty" space and even can move inside of liquid water. An appropriate comparison is tunnel effect (tunneling) in physics (quantum-mechanics) when a particle succeeds pass through a potential energy barrier even though it does not have sufficient energy to surmount the barrier. The Alfa decay of radioactive nuclei and the emission of electrons from the metals or semiconductors occur through tunneling.

Hydrogen atoms absorbing and penetrating into metal volume can be named 'tunneling hydrogen' [104]. This phenomenon can be due to high temperature, high pressure, and, of course, high gradient hydrogen concentrations on the metal surface and inside. Two scenarios or fates are expected for this ‘tunneling hydrogen’. The fate of some hydrogen atoms is the collection in voids and recombination with the formation of H_2 molecules inside. The fate of others is the diffusion through full metallic thickness and their appearance on the opposite side of the metal (in the case when only one side of metal contacts electrolyte, for instance, pipe, tank, or heat exchanger shell). There is opinion that hydrogen crosses metal surface also in the form of hydrons H^+ [98, 105]. Because of small size and high mobility H^+ ions can diffuse in steel faster than other ions and atoms. Some part of hydrogen in the form of H^+ can remain in solid solution inside the metal. For real equipment, penetration rate of hydrogen is limited by its diffusion. Significant difference between dimensions of hydron (H^+) and parameters of iron crystal lattice determines the diffusion by moving of hydrons through the interstices (see Appendix G). Some part of H atoms are polarized with formation of negative H^- ions. Gaseous molecular hydrogen H_2 and very small amounts of H atoms are present in equilibrium inside steel: in micro-voids, blowholes, pores, non-metallic inclusions, cracks, accumulation of dislocations, and grain boundaries. Such presentation of hydrogen in steel agrees with dimensions of hydrons (H^+), hydrogen atoms (H), hydrides (H^-), molecules (H_2), interatomic distances in H_2, and parameter of crystal lattice of iron (see Appendix G). Due to the accumulation of hydrogen inside of collectors-traps its total amount in steel can be three-four times more than the solubility of hydrogen in iron crystal lattice.

Hydrogen permeability into steel increases with increase of carbon amount to 0.9 wt%. Further increase of carbon amount in steel causes slowdown of hydrogen absorption. Alloying elements have relatively low effect on solubility and diffusion of hydrogen in steel in the case if their addition does not influence structural transformation. Adding of nickel, chromium, molybdenum, silicon and manganese have insignificant effect on hydrogen absorption by steel [96]. Structure of steel has greater influence on absorption of hydrogen. Much greater hydrogen solubility occurs in austenite steel than in ferrite. For diffusion, opposite dependence takes place (see Appendix G, Table G.3). The rate of hydrogen permeability is greater in perlite structure than in martensite.

Physico-mechanical properties of steel (plasticity, ductility, toughness and tensile strength) are changed significantly when hydrogen is absorbed by steel. Hydrogen atoms, recombined in H_2 molecules in traps, collect as gas in structural defects up to the pressures, capable to strain surrounding metal volumes. The concentrating of such traps in subsurface layer of steel complicates diffusion of hydrogen atoms in deep-seated metal layers as the arisen field of mechanical stresses in metal acts against a flow of diffusing hydrogen atoms, and meeting on the way the collectors with hydrogen atoms. They leave on their internal surface, where the recombination of H atoms into H_2 molecules develops. Original properties of steel influence significantly on character and degree of change of its mechanical properties as a result of hydrogen absorption. Elongation of mild steels and tensile strength of high strength steels are drastically diminished after hydrogen absorption. Significant deterioration of mechanical properties of steels as a result of hydrogen absorption results in cracks.

Any sour (H_2S containing) environments (crude oil, aqueous solutions, and gaseous hydrocarbons) can result in hydrogen damages. When adsorbed hydrogen atoms are formed it does not matter their origin. Their further fate and behavior in metals are similar. All types of hydrogen damages occur in four steps:

(a) Formation of hydrogen atoms and their adsorption on metal surface.
(b) Absorption—a process of uptake (intake) of hydrogen atoms by bulk solid metal.
(c) Diffusion—a penetration of hydrogen atoms into and through metallic lattice.
(d) Accumulation of hydrogen atoms inside of metals' defects, leading to increased internal pressure, and thus to blistering or cracks.

The question is how hydrogen species cause cracks of dense crystal structure of metals?

Under normal conditions, metals can undergo substantial plastic deformation when subjected to strain. This plasticity stems from the ability of nano-cracks to generate 'dislocations' within the metal—movements of atoms that serve to relieve stress in the material. Dislocations can be considered as vehicles to carry plastic deformation, while the nano-cracks can be viewed as centers (concentrators) to transmit those vehicles. Some important mechanical properties of metals, such as ductility and toughness, rely on the centers functioning well. Unfortunately those concentrators also attract hydrogen atoms which can gather. The way hydrogen atoms embrittle metals is by causing a kind of traffic jam: they collect around the concentrator and block all possible routes for vehicle (dislocations) pass on. This situation eventually leads to the breaking down of material.

4.3.1.1 Types of Hydrogen Damages

Hydrogen damage is close to SCC, but is caused only by hydrogen atoms and molecules and has the following types (Fig. 4.12): *hydrogen blistering*, *hydrogen-induced cracking* (*HIC*), *stress-oriented hydrogen-induced cracking* (*SOHIC*), and *sulphide stress cracking* (*SSC*). Cracking occurring because of hydrogen is also called *hydrogen embrittlement* or *hydrogen cracking*.

Destruction occurs under stresses which can be both static and cyclic. The latter case is called *hydrogen fatigue*. Hydrogen damages occur in the presence of tensile stresses which may be residual or applied (as in SOHIC and SSC), or internal stresses resulting from the interior pressure of hydrogen gas (hydrogen blistering and HIC).

Hydrogen blistering is the phenomenon of accumulation of hydrogen molecules H_2 inside metals and the formation of blisters because of large hydrogen pressure (10–20 MPa). Several weeks to several years (4–6) is needed for formation of blisters. Dimensions of blisters can range from several mm to dozens cm. HIC, also

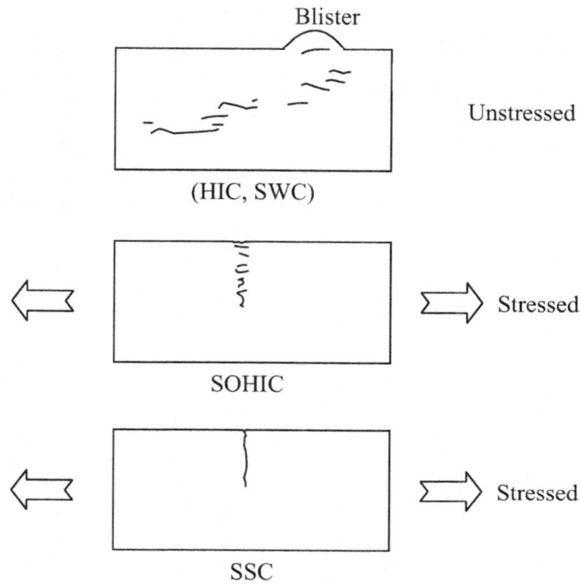

Fig. 4.12 Typical hydrogen damage in metals and alloys. *HIC* hydrogen-induced cracking, *SWC* stepwise cracking, *SOHIC* stress-oriented hydrogen-induced cracking, *SSC* sulfide stress cracking [106]

known as stepwise cracking (SWC), is a form of blistering in which laminating-type fissures (exfoliation) parallel to the metal surface link in the through-surface direction. HIC can occur without external stress (unlike to SCC) and mainly in low-strength steels. SOHIC is a variation of HIC, where the laminations are arranged in parallel arrays perpendicular to the surface. SOHIC usually evolves from areas near the surface where hydrogen blisters are the largest. Cracks slowly grow under applied stresses. SSC is a form of hydrogen embrittlement that occurs in the presence of hydrogen sulfide and water in high-strength steels (quenched-and-tempered and precipitation-hardened steels) and on localized hard zones in welds of susceptible alloys. SSC usually occurs at $T < 90$ °C. Unlike SCC, cracks caused by SSC usually do not branch, and the fracture mode is often a brittle intergranular fracture.

Kinetics of crack development includes the following four stages: incubation period; crack conceiving; its slow growth; and its quick passage through the intact metal cross section.

Steel structure influences significantly on tendency of steel to hydrogen cracking. Mild steels are susceptible to hydrogen blistering, whereas steels with relatively large tensile strength are prone to cracking. This tendency is very high in steels with martensite structure. Ferrite structures have the greatest incubation period and maximum inner stresses when cracking is possible. Resistance of martensitic steels to hydrogen cracking does not depend on their chemical composition [96]. Additions of nickel (>1 wt%) and manganese increase the susceptibility of steel to hydrogen cracking [107]. Addition of silicon increases resistance to hydrogen embrittlement. These additions influence on vulnerability to hydrogen cracking but not on the tendency to hydrogen absorption. The susceptibility to hydrogen cracking increases with increasing steel hardness, applied stresses, intensity of plastic deformation, and decreasing temperature [108]. Usually high-strength low-alloying steels are more susceptible to hydrogen embrittlement than low-strength steels.

Similar to carbon steel, stainless steels (ferritic, martensitic, austenitic, and duplex) are susceptible to hydrogen embrittlement [104, 109–111]. It is impossible to repair equipment exposed hydrogen blistering and delamination. Severe hazard risk is created because of sudden destruction of such equipment especially working at high pressures.

4.3.1.2 Preventive Actions Against LTHA

The following preventive actions of hydrogen failures exist partly described in standards [112–114]:

Metallurgical measures: selection of materials resistant to hydrogen damages; use of steels containing chromium and molybdenum; use of HIC resistant or killed steels (see Appendix F); decrease of carbon content in steel. Steels are not subjected to cracking at hardness values ≤22 HRC [114]. Monel, Hastelloy, Nimonic and Stellite alloys are resistant to hydrogen cracking [96].

Correct fabrication technique: optimal heat treatment and proper welding. To minimize detrimental martensitic formation at the weld by controlled cooling from 800 to 500 °C [104]; to use induction heating stress improvement, dehydrogenation heat treatment, and PWHT; to lessen cold working. Steel welded joints are characterized by increased susceptibility to hydrogen cracking due to large inner stresses in welds and HAZ. Preheating steel sheets to 100–150 °C and decreasing cooling rate after welding can diminish this susceptibility [96].

Change of environmental conditions: removing sulfides, cyanides, and arsenic compounds; neutralization of acids; injection of inhibitors of acid corrosion and hydrogen penetration. Practically some organic inhibitors of hydrogen absorption can turn into stimulators under changes of certain conditions [96]. Therefore, it is needed to examine any potential inhibitor in industrial conditions.

Use of organic, inorganic, and metallic coatings.

4.3.2 High Temperature Hydrogen Attack (HTHA)

If there were no hydrogen gas it would be difficult for our society to survive. Hydrogen saved many people from famine by producing ammonia from hydrogen and nitrogen gases (since 1920s) which was hereinafter the source of ammonium fertilizers used in agriculture. The second large use of hydrogen gas was in *hydro*processing in oil refineries (since 1940s) which allowed to obtain gasoline of high quality. Both processes are carried out at high temperatures (230–500 °C) and high hydrogen pressures (46–250 bar). Such harsh conditions gave itself felt. The first cracks in carbon steel vessel having a wall thickness of 30 mm and used in ammonia synthesis were detected after only 80 h of service in 1933 [115]. Ten years have passed and similar failures were found in *hydro*processing units. On the one hand, hydrogen is a 'positive' gas (in the sense of producing high quality gasoline, ammonia synthesis, and hydrogenation processes in chemical industry). On the other hand, hydrogen brought severe corrosion problems related to cracks and then explosions, fires, injuries, in short, to catastrophes [116]. We will describe influence of hydrogen on alloys (mostly carbon steel) at high temperature and pressure, how we can control and prevent its deleterious attack, monitoring, and analyze case study of the HTHA.

4.3.2.1 Mechanism of HTHA

How molecular hydrogen can cause cracking of carbon steel at high temperature and pressure?

Molecular hydrogen (H_2) at T > 232 °C can dissociate on steel surface into atomic hydrogen (H) which absorbs and diffuses into steel (see Appendix G). Steel becomes permeable to hydrogen atoms under high pressure (>7 bar). These hydrogen atoms react with carbon (C) dissolved in the steel and with cementite (Fe_3C). As a result, methane gas (CH_4) is formed at the grain boundaries, dislocations, inclusions, in the voids and various imperfections which mostly are present in welds. The attack starts with the nucleation and growth of numerous small bubbles on a minority of the ferrite/perlite or ferrite/ferrite boundaries [117]. In addition hydrogen atoms (H) can recombine into molecules (H_2) inside steel. These bubbles grow together to form tears, primarily in the rolling plane, leading to more rapid expansion normal to this plane than parallel to it. Methane (CH_4) and hydrogen (H_2) molecules cannot exit the volume of steel and as a result large internal pressures are formed causing fissuring and cracks in steel. At T > 300 °C, dissolved carbon diffuses to outer surface of steel, react with hydrogen gas, and methane is formed. Carbon is a major strengthening constituent in steel and by removing carbon, the strength of the steel is significantly decreased. Thus, general result of HTHA are a loss of carbon inside pearlite (ferrite α-Fe + cementite Fe_3C), named decarburization or selective leaching of carbon out the steel. As a result, structure of such decarburized steel consists only of mild ferrite (losing hard cementite), and steel losses important mechanical properties such as tensile strength and toughness. Any stresses in such steel which lost cementite cause fissuring and cracks. Surface decarburization results in a decrease in hardness and increase in ductility of the steel near the surface.

The main factors influencing HTHA are the temperature of the steel, the hydrogen partial pressure, the steel type, and the duration of the exposure [115, 118]. Damage and catastrophic failure usually occur after incubation period, which can vary from several hours to many years, depending on the environment and the composition of alloy. The latter has essential influence on the resistance to HTHA.

Such elements as Cr, Mo and V which form stable carbides improve significantly the resistance to HTHA. Usually steels with more than 5 wt% Cr and austenitic stainless steels are not susceptible to HTHA under conditions of hydroprocessing at oil refinery units.

4.3.2.2 Margins of Use of Steels in Hydrogen Service at High Temperature and Pressure (Nelson Diagram)

It was very important to define the margins of use of carbon steel and low alloy steels in hydrogen service at high temperatures and high hydrogen pressures. Probably Schuyten J. was the first who investigated these two parameters in 1947 [119]. American engineer George Nelson collected a large number of experimental observations on different steels in hydrogen service in 1949 and created diagram which received his name [120–122]. Nelson G.A. is also famous by corrosion charts which show regions of resistance of different metals and alloys as a function of temperature and concentration of various chemicals [123]. The Nelson diagram shows operating limits (temperature and hydrogen partial pressure) for carbon steel and low alloy steels in hydrogen service to avoid HTHA (decarburization, fissuring, and hydrogen embrittlement) [121]. Carbon steel is represented by low curve and the most susceptible to HTHA when compared to low alloy steels [121]. At temperatures above each curve, HTHA is possible for that material of construction, and at temperatures below the curve, the prediction is that HTHA will not occur for that material. Based on the Nelson curves, the standard [121] was published in 1970 and widely used for design and fabrication pressure vessels in hydrogen service in the oil refining and petrochemical industry.

Shortcomings of Nelson curves.

Nelson curves are empirical ones based on practical statistics, and define for specific steels the temperature and hydrogen pressure regime, where operation is supposed to be safe. There is no scientific basis explaining and underlying the positions and shapes of the curves. Nelson curves have disadvantages: there are no kinetics data and they are not relevant to the newer steels, i.e. modified 2.25Cr-1Mo steels (2.25Cr-1Mo-V and 3Cr-1Mo-V). Analytical models were developed to predict the kinetics of HTHA [124, 125]. The risk of HTHA is a function of more than solely the three variables described by the Nelson curves: temperature, hydrogen partial pressure and materials of construction. Other factors can also influence the potential for HTHA occurrence: stresses, type of weld, grain size, carbide stability, and time of operation [126–128]. Welding performed on steel parts creates residual stresses which represent additional risk of HTHA. Thus, welds are the most vulnerable parts, and this problem is not reflected in Nelson curves. In spite of disadvantages of Nelson curves they are proven and can be used in most applications. Understanding of HTHA mechanism allowed to develop measures preventing HTHA.

4.3.2.3 Preventive Actions Against HTHA

Steels may be more resistant to hydrogen attack at high temperature by following actions:

Proper and careful choice of alloy and correct design are the most reliable method to prevent HTHA. The basis for this selection must be the standard API RP 941 [121]. Austenitic stainless steels (300 series), as well as 12Cr, 9Cr and 5Cr are resistant to HTHA at conditions normally seen at oil refinery units.

Lowering the carbon content in the alloy and binding the remaining carbon into stable, dispersed carbides (addition of Cr, Mo, and V).

Compliance with technological regime: strictly control designing processing parameters (temperature and pressure).

Post-weld heat treatment (PWHT). This procedure can reduce residual stresses in steel welds which were generated from the welding. PWHT consists of a sequence of controlled heating and cooling applied to the welded structure [129]. The process gives the alloy enough time to modify to its original state and removes residual stresses.

Carrying out the adequate process hazard analysis including identification, evaluation, and control the hazards of processes with participation of hydrogen gas at high temperature and pressure (see Sect. 9.3). This integrative way is connected to monitoring methods.

4.3.3 Monitoring Methods of Low- and High-Temperature Hydrogen Attack

It is very difficult to inspect occurrence of HTHA because the damage can be microscopic and may be present only in small localized areas of the equipment. Different monitoring methods were developed for detection of possible hydrogen damages. We can differentiate them into two groups: detection of hydrogen gas and detection of physical discontinuities (fissures, cracks, blisters) in alloys resulting in changes of their mechanical properties.

Detection of hydrogen gas. Hydrogen gas can be detected either in intrusive or non-intrusive devices called hydrogen probes. Hydrogen that penetrates through a metallic wall can be detected by manometric (hydrogen pressure) or vacuum method, electrochemically (hydrogen ionization from H atoms into H^+ ions), heat conduction (gas chromatography), vacuum extraction at 400 °C, or hydrogen effusion [106, 130].

Detection of physical discontinuities. Ultrasonic methods are the most effective in early detection of HTHA of the microfissure stage for the alloy. The combined use of an automated ultrasonic backscatter method with the velocity ratio, attenuation, and/or spectral analysis is also used [121, 131, 132]. Angle-beam techniques based on backscatter and spectral analysis is used for weld and HAZ inspection. In situ metallography (replicas or field metallographic replication) is used for detection of the early stages of HTHA at the surface of the steel (see Sect. 4.3.4.3 and Appendix H). Wet Fluorescent Magnetic Particle Testing (WFMT), Magnetic particle examination (MT), Time of Flight Diffraction (TOFD) are also used for detecting and sizing HTHA damage below the surface.

There are significant limitations with relying solely on inspection and monitoring methods to identify and to prevent equipment failure from possible HTHA. For example, equipment must already be damaged by hydrogen attack for the inspection and identification. HTHA in the form of inner fissures and cracks is especially difficult to identify. Usually identification of large crack happens when likelihood of equipment failure is high.

In spite of knowledge of low-temperature and HTHA mechanism, preventive actions, monitoring methods, extensive experience in oil refining industry, vast literature including standards, hydrogen attack takes place and in most cases is unpredictable. Evidence of three case studies described in the next section.

4.3.4 Case Studies of Hydrogen Damages

Three examples of hydrogen damages (hydrogen embrittlement) are analyzed: low temperature (at 76–78 °C) in the isomerization unit (two examples), and high temperature (at 260–320 °C) in the HDS unit. Reasons of these failures were determined, preventions of hydrogen failures are reviewed and recommendations are given.

4.3.4.1 Hydrogen Induced Cracking of the Heat Exchanger Shell

Introduction to the failure. Naphtha was the feed for the stabilizer in an isomerization unit. Isomerization of naphtha is a process in which low octane number hydrocarbons (such as n-butane, n-pentane and n-hexane) are converted to a higher octane number products (iso-butane, etc.) over a catalyst (see Chap. 1).

LPG flowed from the stabilizer at 76–78 °C and pressure 24 bar into the air cooler and heat exchanger. The shell of the heat exchanger was made of carbon steel A516 Gr. 70. Internal crack (parallel to the surface) in the form of delaminating was detected inside the shell (the mid, or the center part) of the heat exchanger after 3 years of service (Fig. 4.13a).

It is very interesting to emphasize that management usually asks may be this shell might be manufactured of two steel layers, and delaminating was formed as a result of simple delamination. Metallographic examination gives a clear picture of the source of such delaminating.

Metallographic examination. Microscopic cross-section of the failed shell was prepared 10 cm from the crack (Fig. 4.13b). A branched crack developed parallel to the surface in the mid part of the shell. This crack was formed as a result of hydrogen attack called by the general name hydrogen embrittlement, or in this case hydrogen induced cracking (HIC), also known as stepwise cracking (SWC—see Fig. 4.12) [106, 133].

The pipes and valves around this heat exchanger were also made from carbon steel. Tubes of the bundle inside of the heat exchanger were made of copper-nickel 90/10 alloy. After 3 years of service, black iron sulfide deposits were detected inside of all pipes and valves made of carbon steel around the failed heat exchanger (Fig. 4.13c). Tubes in the heat exchanger were covered with black deposits of copper and nickel sulfides. This means that naphtha feed into the stabilizer and in its turn LPG effluent from the stabilizer contained hydrogen sulfide and water vapor. Low temperature (78 °C) at high pressure (24 bar) of flowing gases resulted in condensation of water vapor and formation of aqueous solution of hydrogen sulfide which attacked iron, copper and nickel.

Analysis of failure and its causes. Failure analysis showed the following mechanism of hydrogen attack. Aqueous solution of H_2S was formed in the lower part of the shell in the heat exchanger. Mechanism of formation of atomic hydrogen and its diffusion is described in Sect. 4.3.1. Inner surfaces of carbon steel pipelines, valves and pumps located before and after the failed heat exchanger were severely corroded by H_2S (see Fig. 4.13c). Hydrogen atoms diffused through grain boundaries and formed hydrogen molecules (H_2) in voids, laminations, on the surface of impurities (such as MnS) and imperfections existing inside the carbon steel shell. Hydrogen molecules were large enough and could not move through the grains of carbon steel. They collected inside (on the grain boundaries, imperfections, or inclusions) and large pressures were formed. These enormous pressures of hydrogen gas caused blistering, fissuring, and cracks called hydrogen embrittlement, mostly prevalent at ambient temperature. Thus, grain boundaries were the favored sites for embrittlement.

Among the different types of damage mechanisms which can be found in the presence of H_2S, the most understood phenomenon is HIC. Cracks of this type are more or less parallel to the plate surface with some steps. Mostly they are located in mid thickness position of the plate (see Fig. 4.13b). Hydrogen embrittlement differs

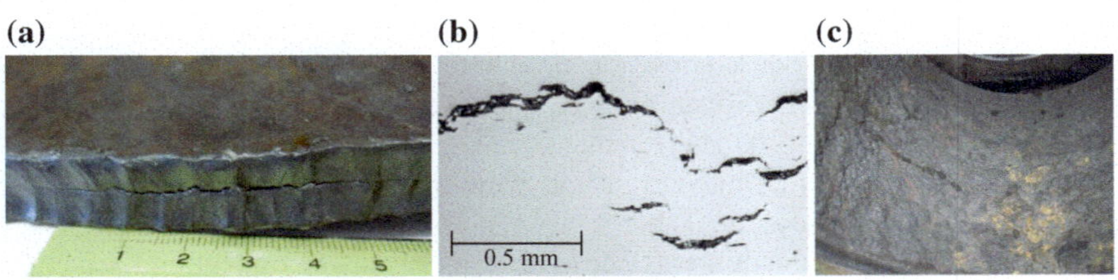

Fig. 4.13 **a** Cross-section through the shell of a heat exchanger, with crack. **b** Metallographic cross-section (×60) through the shell wall of the heat exchanger with delamination (typical stepwise cracking—see Fig. 4.12). **c** Corroded inner surface of the valve (close to failed heat exchanger) covered with iron sulfide

from SSC. The latter always propagates from the surface into the material (see Fig. 4.12) whilst hydrogen embrittled regions are often located inside the material, isolated from the surface (see Fig. 4.13b).

Conclusion. Hydrogen embrittlement as a particular case of HIC occurred in the shell of the heat exchanger as a result of presence of H_2S in aqueous solution, electrochemical corrosion and hydrogen penetration into the carbon steel.

Recommendations. Injection of organic corrosion inhibitor in the overhead of the stabilizer. Usually corrosion inhibitors prevent or diminish cathodic reduction of hydrogen ions and thus lessen the amount of hydrogen charged into the steel.

Replacing the material of the shell from usual carbon steel A516 Gr. 70 with A516 Gr.70 HIC (steel resistant to HIC), or alternatively to killed steel (see Appendix F). The HIC resistant steels are carbon steels characterized with ultra low sulfur concentration, vacuum degassing, normalizing heat treatment, calcium (Ca) addition to cause sulfide shape control, and fully killed. The HIC resistant steels should have been used in wet H_2S environment with H_2S partial pressure ≥ 10 psi and presence of cyanide ions (≥ 20 ppm in aqueous phase). The HIC resistant steels must be tested in accordance with NACE Standards [113, 134]. One should remember that even if HIC resistant steels are used, they are resistant but not immune to cracking [135, 136]. The refinery should continue to keep them on the list of equipment that should be inspected for cracking at some frequency, depending on service and corrosion risk.

Use of austenitic stainless steel SS 304 clad in heat exchangers shell made of carbon steel.

Carrying out an ultrasonic test of the upper part of stabilizer and the air cooler after the stabilizer.

Examination of water in the boot after the heat exchanger for the presence of cyanides (CN^-) and other cathodic poisons.

4.3.4.2 Sulfide Stress Cracking of Stud Bolts in Heat Exchanger

Introduction to the failure. The shell of the heat exchanger failed as a result of HIC (see Sect. 4.3.4.1). Several stud bolts from the failed heat exchanger were found broken after its opening (Fig. 4.14). The stud bolts were made from low-alloy steel according to ASTM A193 Grade B7 and were in service 3 years. The failed stud bolts were covered by black corrosion products of iron sulfides, and then were transformed into rust after opening the heat exchanger and immersion in the atmosphere (see Fig. 4.14c). The fracture surface was also covered by rust.

Mechanical tests of the failed stud bolt were done according to ASTM A370 [137] and results are shown in Table 4.6.

Metallographic examination was carried out by means of optical and scanning electron microscopes. Longitudinal metallographic cross-section was made through the bolt (Fig. 4.15). A crack initiated at the fracture surface and propagated perpendicularly to it. Corrosion products were detected in all cracks.

EDS analysis of the surface of the failed bolt and inside the crack showed large quantities of sulfur.

Analysis of failure and its causes.

The reason of the failure is SSC. SSC is a form of hydrogen embrittlement that occurs in the presence of H_2S and water in high-strength steels (quenched-and-tempered). SSC usually occurs at $T < 90$ °C [113, 138, 139]. Unlike SCC, cracks caused by SSC usually do not branch (see Fig. 4.12). Really, at first, H_2S and water were present in LPG. The second, the bolts were quenched-and-tempered during the manufacturing which made them more susceptible to cracking in the presence of H_2S. Mechanical properties of the bolt material are incompatible with the requirements of the standard ASTM A193 Grade B7. Yield strength is much greater than the minimum allowable value (150 ksi against 105 ksi min) and is close to the tensile strength (153 ksi), indicating that the material of the bolt was quenched-and-tempered. This heat treatment increased significantly yield strength and thus raised brittleness and correspondingly sensitivity to SSC. The difference between the tensile strength and yield strength should be at least 20–25 ksi and not 3 ksi as in this case. When yield strength of steel is above 145 ksi, even small concentrations of H_2S (lower than 10 ppm in a gas phase and 1 ppm of total sulfide in the aqueous phase) can cause SSC [133, 138].

There is the meaning of the difference between real elongation (14 %) which is lower by 2 % than the minimum required value (16 %), but even less than typical elongation for the bolt material. Thus, mechanical properties show that the bolts manufactured in the plant did not meet requirements of the standard. Overloading during installation also contributed in the occurrence of SSC.

Conclusion. SSC as a specific case of hydrogen embrittlement of stud bolts occurred as a result of the presence of H_2S in aqueous solution, electrochemical corrosion reaction with cathodic reduction of H^+ ions, and hydrogen penetration into the material of low-alloy steel stud bolts. The mechanical properties of these stud bolts did not comply with the requirements of the standard ASTM A193 Grade B7.

Recommendations. Acceptable bolting material resistant to SSC is carbon steel ASTM A193 Grade B7M [113, 138, 139]. The hardness of this material must not exceed 22 HRC. There are significant differences in both the mechanical properties

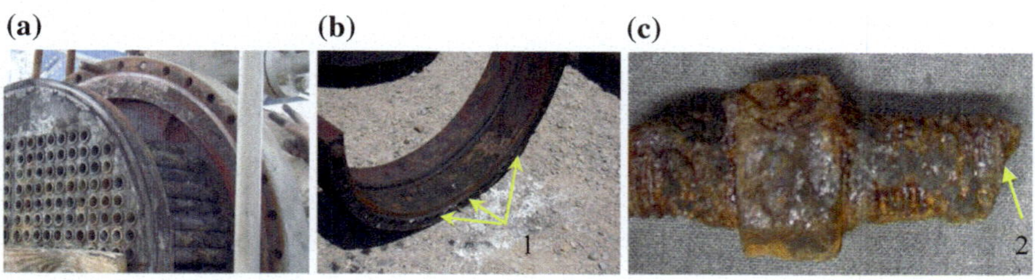

Fig. 4.14 **a**, **b** The heat exchanger parts. 1 Places of failed stud bolts. **c** The failed stud bolt. Rust was formed after the opening of heat exchanger during immersion on the air. 2 Fracture surface

Table 4.6 Mechanical properties of failed stud bolts

Property	Yield strength $\sigma_{0.2}$, ksi	Ultimate tensile strength σ_{UTS}, ksi	Elongation ε, %	Reduction of area, %
Result of the examination	150	153	14	53
Requirement according to ASTM A193 Grade B7	≥105	≥125	≥16	≥50

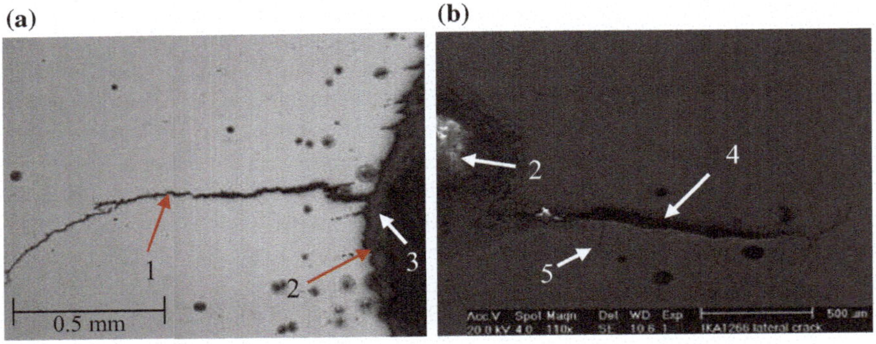

Fig. 4.15 **a** Longitudinal metallographic cross-section through the failed stud bolt: 1 Crack; 2 Corrosion products; 3 Fracture surface. **b** SEM image of the longitudinal metallographic cross-section through the failed stud bolt: 4 A crack parallel to the fracture surface; 5 A secondary crack

and the testing requirements of A193 Grade B7M and A193 Grade B7 (failed). The maximum hardness of A193 Grade B7M is 22 HRC, while the maximum hardness of A193 Grade B7 is 35 HRC. When providing A193 Grade B7M bolts, every individual stud bolt in the lot must be tested on hardness. We have to take into consideration that materials included in the standard [113] are resistant to, but not necessarily immune to SSC.

4.3.4.3 High-Temperature Hydrogen Attack of Piping Welds (HDS Unit)

Introduction to the failure. Heavy vacuum gas oil (HVGO) containing hydrogen gas flowed into the furnace at the HDS unit. A crack was formed in a seamless tube-elbow weldment (diameter 100 mm) after 16 years of the hydrogen service (Fig. 4.16).

The tube-elbow catastrophically ruptured. Highly flammable hydrogen and gas oil at ~280 °C were released from the ruptured tube and ignited causing an explosion and intense fire. As a result, the nearby equipment (mostly heat exchangers) was damaged.

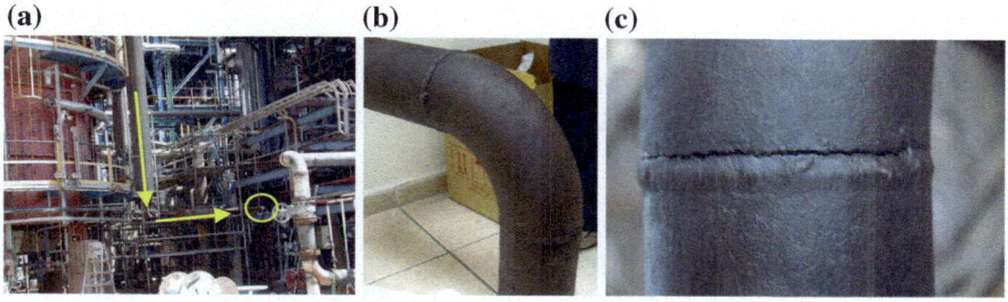

Fig. 4.16 **a** General view of the failed pipe with the elbow (*yellow circle*) to the furnace at the HDS unit. **b**, **c** Crack in the heat affected zone (HAZ). Seamless carbon steel pipe (diameter 100 mm) after 16 years of the hydrogen service. T = 280 °C (with peaks to 320–380 °C), P_{H2} = 46 Bar

The tube was made from carbon steel A106 Gr.B Sch.80, and the elbow from carbon steel A234 WPB 90 LR BW Sch.80. The design temperature of the hydrogen had to be held at 240 °C and pressure 46 bar (657 psi). Analysis of the Plant Information System data (PI) showed that the real temperature of hydrogen gas in the pipeline was above 240 °C: most of the time ~280 °C and sometimes with peaks of 320 to 380 °C (Fig. 4.17).

Fractographic examination of the fracture surface after crack opening was carried out by means of SEM. Characterization of the rupture shows intergranular fracture, as a result of decarburization.

Macroscopic examination was performed in sections through the area of the crack at the welding area and through the same weldment at a location with no visible crack (Fig. 4.18a). In the two sections, a smooth and a dark heat affected

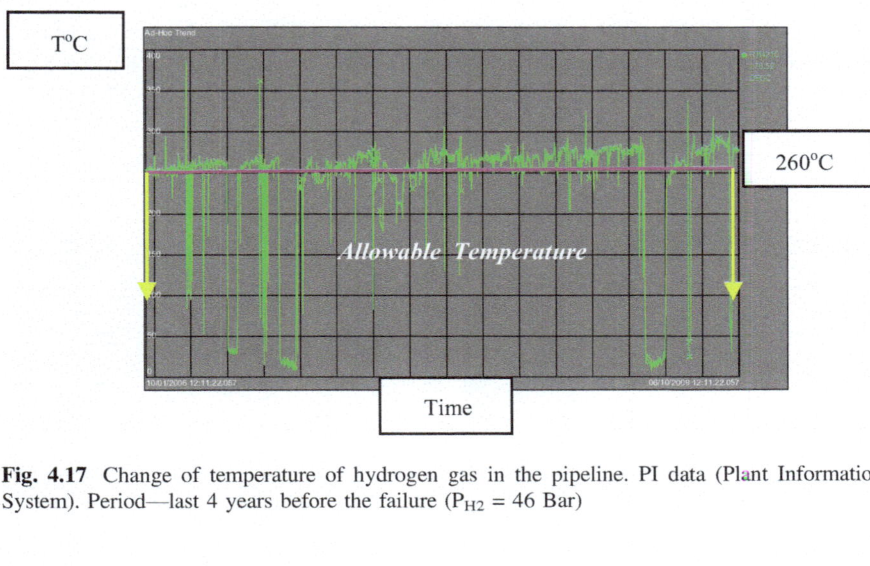

Fig. 4.17 Change of temperature of hydrogen gas in the pipeline. PI data (Plant Information System). Period—last 4 years before the failure (P_{H2} = 46 Bar)

zones were seen. A crack propagated through all of the wall thickness at the boundary between the weld metal and the HAZ can be seen at the failure area.

Microscopic examination by means of optical microscopy was performed in a section through the failure area after etching (Fig. 4.18b). A branched crack initiated at the inner surface of the tube and propagated through all of the wall thickness at the boundary between the weld metal and the HAZ. The microstructure of the HAZ contains only ferrite grains (there are no carbides), indicating decarburization in this area (white area). Weld metal contains typical perlite (mixture of ferrite α-Fe and carbides Fe_3C). SEM image (×1000) reveals brittle crack (Fig. 4.18c).

A microhardness Vickers test was performed in the HAZ at the failure area, in two locations: one with a microstructure containing only ferrite grains and a second with a typical microstructure of heat affected zone. The hardness at the first location (ferrite grains) was 120–127 $HV_{0.5}$, and the hardness at the second location (HAZ) found to be higher: 160–163 $HV_{0.5}$.

Conclusion and mechanism of failure.

What happens at temperatures above 260 °C? Molecular hydrogen at high temperatures can dissociate on a steel surface into atomic hydrogen which is easily absorbed and diffuses into the steel. The amount of tunneling hydrogen through carbon steel drastically increases at T > 260 °C (Fig. 4.19a). Absorption of hydrogen in carbon steel increases with increase of temperature as this is an endothermic process.

Hydrogen embrittlement with general name HTHA occurred (see Sect. 4.3.2). As a result carbon steel lost mechanical properties and became brittle like glass. Margin temperature for carbon steel use is 260 °C at hydrogen pressure 46 bar according to the Nelson diagram [121]. Usually hydrogen attack begins at the regions where large tensions exist as hydrogen diffuses to such regions. Therefore, HAZ are the first to be damaged by hydrogen. Nelson's diagram shows that carbon steel can be used up to 260 °C at 46 bar [121, 140]. Above this temperature, steels ASTM A182 F11 containing chromium (1.25 wt%) and molybdenum (0.5 wt%) are recommended. The question is how much time is needed for HTHA to occur? Usually HTHA needs initiation period, and at 280 °C and 46 bar the time for incipient attack equals 10,000 h (1.14 years); at 290 °C and 46 bar this time equals

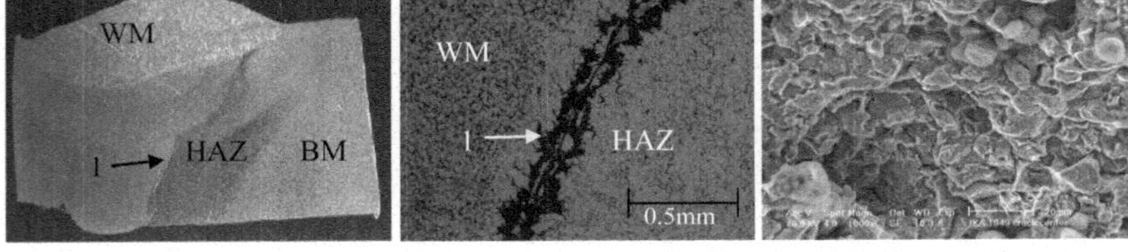

Fig. 4.18 **a** Macroscopic section through the failure area. 1 Crack in HAZ; WM Weld metal; BM Base metal. **b** Microscopic section through the failure area. **c** SEM photo of brittle crack (×1000)

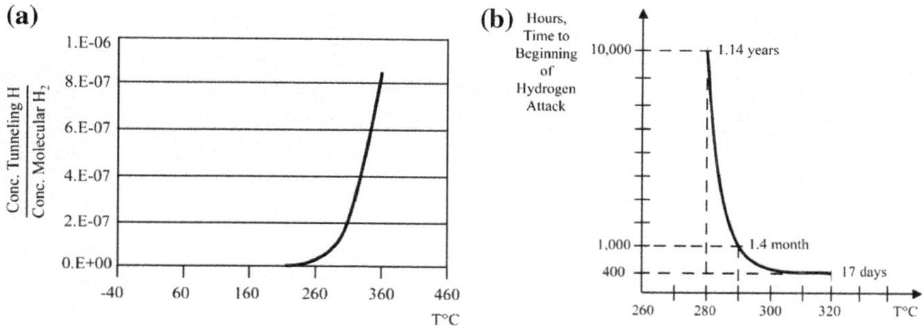

Fig. 4.19 **a** Available tunneling hydrogen to molecular hydrogen ratio versus temperature [104]. **b** Time for incipient attack of carbon steel versus Temperature (P_{H2} = 46 bar). Calculated on the basis of API 941 data [121]

1000 h (1.4 months); and at 320 °C and 46 bar the initial time is 400 h (17 days) [121, 140]. This assessment points out that it was really enough time (16 years!) to HTHA to occur (Fig. 4.19b).

Recommendations. Analysis of temperatures of hydrogen gas inside the pipeline (PI data) showed that it was in danger of hydrogen embrittlement which could occur in a period between 1 month and a year. Therefore, it was recommended to reduce temperature below 260 °C and to work under temperatures not more than 240 °C.

The feature of failure had such a form that usual nondestructive testing (excluding replicas) did not allow detecting decarburization, fissuring, and cracks. In order to define if decarburization occurred, metallurgical (replicas and impact test) examinations were needed.

Use of low-alloy steels containing Cr above 1.25 wt% instead of carbon steel was recommended for prevention of HTHA.

It was suggested that other welds, pipes and heat exchangers could be damaged from HTHA and fire. Therefore, additional welds (mainly HAZ) and different parts of the heat exchanger and pipes were examined in situ. These recommendations were carried out, field metallographic replication was applied, and results of replicas examination of the heat exchanger are described in the Appendix H.

Conclusions

Hydrogen damages are differentiated into two groups based on two mechanisms: electrochemical processes (mainly at low temperatures, up to ~100 °C) arising from acid corrosion and cathodic protection, and HTHA arising from the presence of hydrogen gas at high temperatures (200–900 °C) and high pressures (4–30 MPa).

Pure petroleum distillates (not containing aqueous phase and H_2S) are not corrosive towards carbon steel at T < 100 °C in refinery units [106, 141, 142]. Only the presence of aqueous phase and H_2S (wet sour condition) are needed for proceeding of electrochemical corrosion with reduction of H^+ ions to nascent hydrogen atoms and further penetration into steel. This process occurs not only in acidic and slightly acidic solutions (pH ≈ 4–5), but also in alkali aqueous solutions at pH ≈ 8–12 [96].

It is important to emphasize that H_2S stimulates hydrogen absorption by steel significantly greater than general corrosion of steel in electrolytes. Therefore, the main danger for steel equipment processing crude oil containing H_2S represents not general corrosion but hydrogen absorption and further hydrogen damages of steel.

Hydrogen atoms do not retain a memory of their origin whether from gaseous or aqueous solutions and are the same once absorbed, penetrated and moving inside the alloy.

In spite of existing of different methods of hydrogen monitoring, there is no reliable method allowable predicting where and when hydrogen embrittlement will happen.

4.4 Corrosion by Amine Solutions

Aqueous amine solutions have been used since 1930s to remove acid gas contaminants (H_2S or CO_2) by absorption from a process stream [143]. Following four amines are used: monoethanolamine (MEA), diethanolamine (DEA), methyldiethanolamine (MDEA), and diisopropanol amine (DIPA) (see Appendix E). In oil refineries, as a rule only H_2S should be absorbed. Corrosion problems at amine treating units usually happen because of faulty plant design, poor operating practices, and amine solution contamination [144].

The four amines themselves are not corrosive to carbon steel, even can play the role of corrosion inhibitors under certain conditions. However, they can degrade during long circulation at high temperature (if steam temperature exceeds 150 °C), and destructed products of amines together with H_2S and other contaminants can cause corrosion problems.

Corrosion in amine treating units depends on amine type, acid gas removal service and molar ratio H_2S/amine (called also loading), and type and concentration of amine degradation products [145, 146]. Corrosion in amine units using MEA is more severe than in those using DEA, MDEA, and DIPA, because the former is more prone to degradation. MDEA is considered to cause the less corrosion problems compared to MEA, DEA and DIPA. Amine degradation products bring severe corrosion, erosion, and operational (fouling) problems. In addition to H_2S, such contaminants as formic acid, acetic acid, glycolic acid, propionic acid, oxalic acid, chlorides, sulfates, thiosulfate, and thiocyanate can be present or produced by oxidation of hydrocarbons in the amine solution. These contaminants react with amine producing salts formate, acetate, glycolate, propionate, oxalate, and also chloride, sulfate, thiosulfate, and thiocyanate (see Appendix E). They got the name *heat stable amine salts* (HSAS) or simply *heat stable salts* (HSS) because it is impossible to destroy them under normal regenerator conditions, namely, they cannot be recovered by heating. As a result, HSAS are accumulated in the amine solution which becomes more and more corrosive with time. The presence of HSAS also diminishes the efficiency of the absorbing ability of amine solution by reducing the amount of pure amine. Therefore, maximum allowable concentration is 1 wt% HSAS in amine solution. When its quantity more than this value, unfavorable HSAS accumulating in amine solution give rise to severe corrosion, erosion, precipitation of solids, causing fouling and under deposit corrosion. Corrosion is most harsh at locations where H_2S is removed from rich (by H_2S) amine solution, namely, in reboilers. Really, temperatures, appearance of two-phase turbulence stream and flow velocity are the highest in these locations. Corrosion can be a problem on the rich-amine side of the heat exchangers, in reclaimers, and pumps. Hydrogen damages (blistering because of H_2S presence) can occur in the bottom of the absorber, overhead condensers, and reflux drums. Aqueous amine solutions can cause SCC of carbon steel welds, therefore, post-weld heat treatment at 620 °C should be applied. The following solutions to lessen and prevent these detrimental corrosion-erosion and fouling problems exist:

A. Molar ratio H_2S/amine is very important. For instance, when MDEA is used and 12 wt% formate HSAS is formed, at loading 0.002 mol H_2S/mole MDEA, corrosion rate of carbon steel is 0.40 mm/year; at loading 0.01 mol H_2S/mole MDEA, corrosion rate of carbon steel decreases to 0.14 mm/year; and at loading 0.05 mol H_2S/mole MDEA, corrosion rate of carbon steel decreases to 0.07 mm/year [146]. Increasing of H_2S concentration (loading) decreases significantly corrosion rate to allowable values. Certainly these values depend on amine type, concentration of HSAS and types of contaminants. The mechanism of protection is that H_2S reacts with iron and protective iron sulfide layer is formed on carbon steel surface. Similar protective iron sulfide thin film can be formed in the overhead of atmospheric distillation column (see Fig. 3.1). This protective iron sulfide film has passive character which means that it is metastable and can be disturbed under changes of the composition of media and environmental conditions. The presence of acid components (even when amine solution is alkaline) and mechanical attack by hard particles can remove protective iron sulfide layer. Some HSAS (i.e. formates) can be slightly regenerated to the undissociated acid under deposits and in crevices.

Operating practice requires to maintain the following parameters in amine treating units: MEA concentration should not exceed 20 wt%; acid-gas loading should be 0.35 mol H_2S/mol MEA and 0.05 mol H_2S/mole MDEA; and reboiler steam temperature less than 150 °C.

B. Neutralization with proprietary neutralizers. Neutralization can significantly improve operational problems, mostly decrease fouling in heat exchangers, absorber trays, and filters. Usually fouling results in the necessity of frequent replacing of filters (sometimes every 3–5 days), washing heat exchangers and absorber trays. Sometimes it is recommended to inject caustic soda solution to neutralize acids that appear in amine solutions. However, soda can cause caustic embrittlement of carbon steel at high temperature ($\sim$120 °C) of the reboiler. Therefore, caustic soda solution should not be injected. Injection of corrosion inhibitors is not always effective.

C. Clean-up the amine solution: change out, purge and make up fresh amine solution, vacuum thermal distillation, ion exchange, and electrodialysis [146–148]. The latter method found wide application. It is desirable to neutralize HSAS prior to electrodialysis. This process can diminish concentration of HSAS to any required level.

D. It is recommended to replace carbon steel with SS 304 if corrosion problems cannot be solved by improving operating conditions and purification of amine solution.

Five case studies are analyzed in this chapter.

4.4.1 Case Studies

4.4.1.1 Failure of the Reboiler at the Amine Treating Unit

Conditions. The reboiler is intended for returning of the amine solution to the stripper. Aqueous solution of diethanolamine (DEA) (20–22 wt%) enters at one inlet at the shell bottom, flows between the tubes and exits through two outlets at the shell top at 120 °C (Fig. 4.20). Steam at 150 °C flowed inside the tubes.

Materials of construction: Shell (wall thickness is 10 mm)—carbon steel ASTM A516 Gr.70. U-tubes—SS 304.

Service period before the failure: 12 years.

Visual examination and findings. The holes (~3 mm in diameter) were detected inside the top of the shell at the distance 60 cm from the amine solution exit from the reboiler (Fig. 4.20).

Inner surface (30 %) of the upper part of the reboiler shell was severely corroded. The wall thickness of this part of the shell (about 50 cm from the amine solution exit from the reboiler) diminished drastically, nearly to zero. Measurements of the shell thickness after 10.5 years of the service (1.5 years before

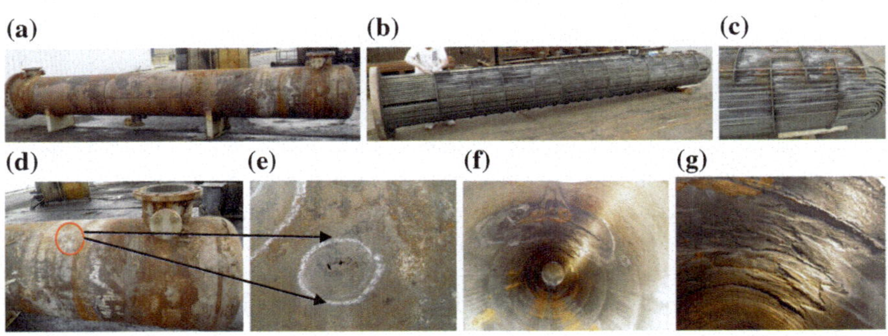

Fig. 4.20 **a** Failed shell of the reboiler. **b** U-tube bundle of the reboiler. **c** White-grey-black deposits on the tube bundle. **d**, **e** Holes in the shell. **f**, **g** Corroded inner surface of the shell

the failure) showed that there were no significant changes in the thicknesses. This means that corrosion occurred during the last 1.5 years of service with the rate 6.7 mm/year.

White-grey-black deposits and grey-white scale were detected on the top of the SS tubes (see Fig. 4.20c). After removing these deposits and scale, SS tubes showed good appearance.

Chemical analysis of DEA aqueous solution showed the presence of following corrosive substances to carbon steel: H_2S (0.1 wt%); HSAS (0.23–0.76 wt%); chlorides (11–53 ppm); sulfates (120–570 ppm); thiosulfates (120–890 ppm); thiocyanates (160–1250 ppm); acetate (790–1100 ppm); formates (210–270 ppm); oxalates (100–185 ppm). Electrical conductance of the solution ranged in 3160–4440 μs/cm, and pH = 10.3–10.5.

Four years before the failure, concentration of HSAS increased to 3 wt% (when allowable concentration was <1 wt%), and DEA solution was purified from HSAS. After the purification, concentrations of corrosive substances were in allowable limits.

Molar ratio H_2S/DEA (loading) must be >0.04 in reboiler, however really this ratio was 0.01–0.02. Excessive purification of DEA solution could give rise to H_2S deficiency and as a result to the lack of sufficiently good protective iron sulfide film on carbon steel surface. Excessive purification of DEA solution also could cause greater temperature in the reboiler (120 °C instead recommended 110 °C). Temperature increase could result in acceleration of corrosion processes and consumption of excess energy.

Chemical analysis of deposits showed that they contained black corrosion product of iron sulfides and probably calcium carbonate and calcium sulfate scale (white-grey) (Table 4.7).

Morphology of corroded inner surface was examined by SEM. Signs of erosion damage and hard particles were defined on the surface near the holes. EDS analysis of hard particles showed that they consist of Si, Fe, O and S (probably, silicon oxides, iron oxides, and iron sulfides). Macroscopic cross-section through the failed area also showed erosion of inner surface.

Failure phenomenon: *Corrosion-erosion* of inner surface of the upper part of the reboiler shell.

Cause of failure and its explanation. Aqueous DEA solution containing corrosive substances to carbon steel entered into the lower part of the reboiler, was heated to 120 °C and exited from the upper part of the reboiler. The top of the reboiler was the hottest place, and solution boiled. Actually two-phase system, liquid solution

Table 4.7 Chemical composition of deposits formed in the reboiler (XRF results)

Deposit	Element, wt%					
	Fe	S	Ca	Si	Cr	Zn	P
Black	75.60	18.65	4.75	0.07		0.94	0.03
White-Grey	9.90	1.20	87.40	1.35			0.16
Scale on SS tubes	70.8	22.9	3.93	0.75	0.69	0.49	0.50

containing aggressive ions and vapor, flew in the reboiler top. In ideal case, H_2S reacts with iron, and tenacious passive iron sulfide layer is formed which protects carbon steel from corrosion. Boiling of solution which contained organic acids (formic, acetic and oxalic) and mechanical attack by the flow of two-phase solution containing corrosive aggressive ions to carbon steel removed iron sulfide passive layer located on the inner surface of the reboiler top shell. The light organic acids, formic and acetic, are mostly corrosive at their boiling points 100 and 118 °C respectively. Thus, the passive protective iron sulfide film was removed and corrosion occurred at the top of the inner surface of the shell of the reboiler. Due to the flow of liquid-vapor containing hard particles (corrosion products), the damage occurred in the form of erosion-corrosion. In addition, the level of liquid dropped in the reboiler (as a result of intensive boiling) and deposits were formed on the top of the tube bundle.

Conclusions, solutions, and recommendations.

(a) Choice of SS 304 material of tubes was correct. Stainless steel is resistant to the environment and conditions in the reboiler.
(b) SS 304 is also recommended for the shell, or alternatively for cladding of inner surface of carbon steel shell.
(c) A temporary solution may be the fabrication of the shell with increased wall thickness (12.5 mm instead 10 mm) from carbon steel A516 Gr. 70.
(d) Periodical proper purification and filtration of DEA solution from corrosive substances should be carried out.

4.4.2 *Failure of the Reboiler at the HDS Unit*

Conditions. Aqueous solution of rich amine (20–25 wt% MDEA) at 120 °C entered into the bottom of the reboiler, flowed between the tubes, H_2S was desorbed from rich amine solution during this flow and amine became lean. Then this solution flowed down to the separation plate from the upper part (the top of the tube bundle) and exited behind the separation plate in the bottom of the reboiler (Fig. 4.21). Vapor (mostly H_2S) at 130 °C exited from the upper part of the reboiler. Steam flowed inside the tubes at 140–160 °C and heated rich amine solution. Five months before the failure, severe corrosion (48 mm/year) was defined in the bottom of the floating head where lean amine exited.

Materials of construction. The shell and separation plate–carbon steel ASTM A516 Gr. 70 HIC. The thickness of the separation plate is 16 mm. Tubes—SS TP304L. The flange in the shell–carbon steel (ASTM A105).

Service period before the failure: 4 months.

Visual examination and findings. The hole was found at the bottom of the reboiler near the separation plate and flange (see Fig. 4.21). Pits were detected on the inner surface of the shell at the top (exit of lean amine solution) of the reboiler.

Chemical analysis of aqueous solution of rich amine (20–25 wt% MDEA) showed that it contained thiosulfates $S_2O_3^{2-}$ (3.1–4.8 %, when 1.0 % max allowable), sulfates SO_4^{2-} (70–1920 ppm, when 500 ppm max allowable), oxalates (610–815 ppm, when 250 ppm max allowable).

Failure phenomenon: *Erosion-corrosion.*

Cause of failure and its explanation. Among four amines (DEA, MEA, MDEA, DIPA) used in oil refineries for H_2S and CO_2 absorption, MDEA causes the lowest corrosion problems. During continuous prolonged use of amine solutions HSAS could be formed. The formation of HSAS is directly related to chemical compounds at feed facility (in this particular case, SO_2 gas could reach MDEA solution or air could penetrate into the amine storage tank or through untightened seals in pumps). It is recommended that the concentration of the HSAS does not rise above 1 %, otherwise corrosion problems can start.

Corrosion rate of carbon steel in MDEA solution (containing 1 % oxalic acid) at 125 °C equals 4 mm/year. Thiosulfates in MDEA give rise to lower corrosion rate. Therefore, it is impossible to explain real corrosion rate 48 mm/year only by the influence of these compounds in MDEA. Experiments with pure organic acids (formic, acetic, propionic, and butanoic) gave corrosion rates of carbon steel in the range 11–27 mm/year. Even if these acids were in MDEA, they might be present in concentrations not so large to cause high corrosion rate of 48 mm/year. Therefore, additional reason, namely, flow-assisted corrosion resulted in severe corrosion and then to the hole in carbon steel separation plate in the reboiler (see Fig. 4.21d). Rich

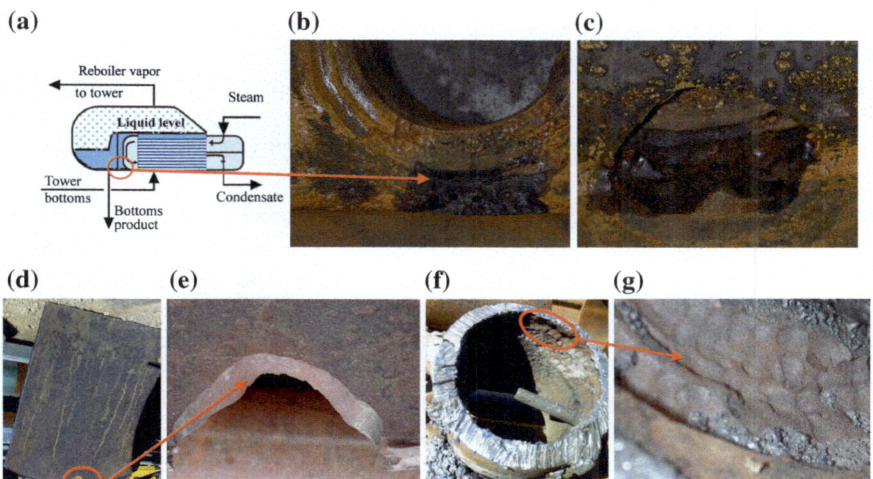

Fig. 4.21 **a** Reboiler (a scheme). **b** Corroded failed area (lean amine exit). **c** Corrosion hole in the separation plate (tube bundle side). **d** Corrosion hole at the bottom of the separation plate of the reboiler (near the lean amine exit, shell side). **e** Magnification of the picture **d** (the view from the lean amine exit, shell side). **f** The flange (lean amine exit). **g** Magnification of the picture **f** (the surface morphology is typical to erosion)

amine solution contains H_2S which reacts with carbon steel surface to form iron sulfide layer. These corrosion products do not dissolve in MDEA solution, but can exfoliate (when uneven thick layers are formed in the presence of large quantities of H_2S) and enter as solid particles into the reboiler. Flowing regime of lean amine inside the bottom of the reboiler resulted in erosion behind the separation plate. Severe corrosion (48 mm/year) of the bottom of the floating head 5 months before the discussing failure verifies this version of corrosion mechanism. In addition, the morphology of the corroded surface proposed that the phenomenon was erosion by particles and aggressive substances, and HSAS among them. Anions of these salts are good complexants of iron ions and accelerate corrosion. Thus, *erosion-corrosion* occurred in the bottom of the reboiler.

The MDEA could be decomposed under overheating with the formation of corrosive substances. When the hole was formed in the bottom of the separation plate, the level of MDEA solution decreased, and in the upper region of the tubes which were uncovered with solution, overheating took place, which in its turn forced the amine solution to boil and caused pitting corrosion on the inner surface of the shell of the reboiler.

Solutions and recommendations.

(a) To change the construction of the reboiler: rich amine should enter into the reboiler in several places (through two nozzles at least) for creation uniform flow between the tubes and not in one place (one nozzle). The distance between the separation plate and the tube bundle must be greater than 150 mm.
(b) It is possible to solve corrosion problems arising from HSAS, corrosion products (FeS) and suspended solids by means of their removing by filtration, ion exchange technique and electrodialysis.
(c) To increase concentration of MDEA to 45–50 %.
(d) To replace the material of construction of the shell and the separation plate with SS 316L. This can be done only in the case if concentration of chlorides is lower than 500 ppm. Such values will not cause Cl SCC or pitting corrosion of SS 316L.
(e) To check regularly concentrations of HSAS, formates, acetates, oxalates, sulfates, thiosulfates, cyanides, chlorides and iron ions in MDEA solution. Recommended maximum values of these corrosive contaminants are shown in Table 4.8. It is important to follow the trend of concentration changes.
(f) Lean amine solution should contain less than 10 ppm suspended solids (mostly iron sulfide). If the concentration of suspended solids is higher it is needed to filtrate solution. Large amounts of suspended solids not only lead to erosion but also cause fouling. High concentration of suspended solids (particularly those less than 5 μm) will also increase the foaming characteristics of the circulating amine solution. In addition to suspended solids, hydrocarbons and acids also increase foaming tendency of amine solution. This foam may reduce the ability to regenerate the MDEA adequately. MDEA has a much greater affinity for hydrocarbons than MEA and DEA, and foam stability is

greatly increased by the presence of liquid hydrocarbons. This is particularly important in such a low pressure system.

(g) To keep steam temperature less than 140 °C, in order to prevent MDEA decomposition.

(h) Injection of corrosion inhibitor into amine solution can protect areas where H_2S concentration is low and corrosion of carbon steel by organic acids and amine degradation products is most severe. Ideally this would be injected in the line from the base of the regenerator to the reboiler. It is important to emphasize that if there is high concentration of suspended solids (>10 ppm) in amine solution, corrosion inhibitor is unable to protect against erosion.

(i) To install ER-probes in relevant locations of the reboiler in order to keep tracking corrosion situation.

4.4.3 Corrosion of Inner Surface of Stripping Tower (HDS Unit)

Conditions. Original thicknesses of the wall are 16 mm at the lower part and 12 mm at the upper part of the stripping tower. Aqueous amine (MDEA) solution (20–40 wt%) at 112–120 °C flowed inside the stripper. The required MDEA concentration is 45–50 wt%.

Material of construction. Carbon steel.

Service period before the failure: 2 years.

Visual examination and findings. Severe corrosion of inner surface of walls in the stripping tower was found during shutdown (Fig. 4.22). Shallow pits 3–5 mm in deepness were detected. The concentration of HSAS was 2.5–3 wt% during 2 years of service (when recommended concentration is less than 1 wt%).

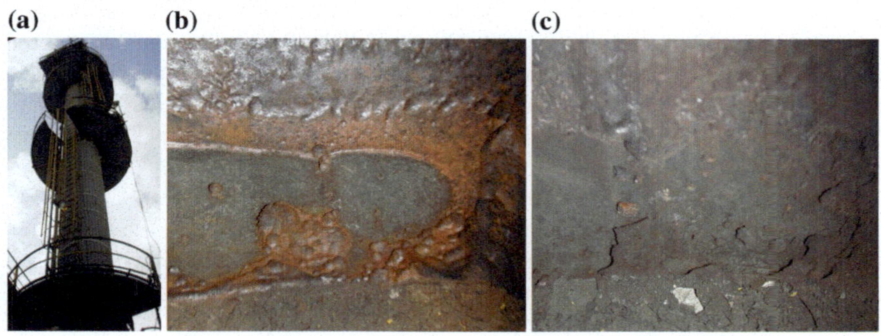

Fig. 4.22 **a** General view of the stripping tower. **b**, **c** Corroded inner surface of wall in the stripper tower

Table 4.8 Maximum recommended concentrations of corrosive compounds in Heat Stable Amine Salts (HSAS)

Component	Concentration, ppm
HSAS	10,000
Thiosulfate ($S_2O_3^{2-}$)	10,000
Thiocyanate (SCN^-)	1000
Acetate (CH_3COO^-)	1000
Formiate, glycolate, malonate	500
Chloride (Cl^-)	500
Sulfate (SO_4^{2-}), sulfite (SO_3^{2-})	500
Oxalate	250
Iron ions	2
Suspended solids	10

Cause of failure and its explanation. Heat stable amine salts at 112–120 °C resulted in severe general and localized corrosion of inner surface of the carbon steel stripping tower.

Failure phenomenon: *Pitting and general corrosion by HSAS.*

Solutions and recommendations.

(a) To fill shallow pits by welding.
(b) To carry out SS 304L clad on the inner surface of the stripping tower.
(c) To maintain correct operating conditions in the stripper in order to reduce the decomposition of MDEA and formation HSAS. To keep 45–50 wt% MDEA and not 20–40 wt%.
(d) To carry out periodical purification of amine solution from HSAS.

4.4.4 *Impingement Attack in Heat Exchanger with Amine Solution*

Conditions. The heat exchanger is intended to cool the residue of the distillation vacuum column. Amine solution (20–25 wt% MDEA aqueous solution) entered vertically in the upper part of the heat exchanger (Fig. 4.23). The impingement plate of dimensions 12.5 × 12.5 cm^2 and 9 mm in thickness was installed in order to protect upper tubes in the bundle from impact (impingement attack) by amine solution flow.

Material of construction. Carbon steel (ASTM A179).

Service period before the failure: 4 years.

Visual examination and findings. Large holes and severe corrosion were detected in nine upper tubes close to the impingement plate (see Fig. 4.23).

Failure phenomenon: *Impingement attack.*

Cause of failure and its explanation. The thickness of impingement plate was less than required value. The area of impingement plate was significantly less than the

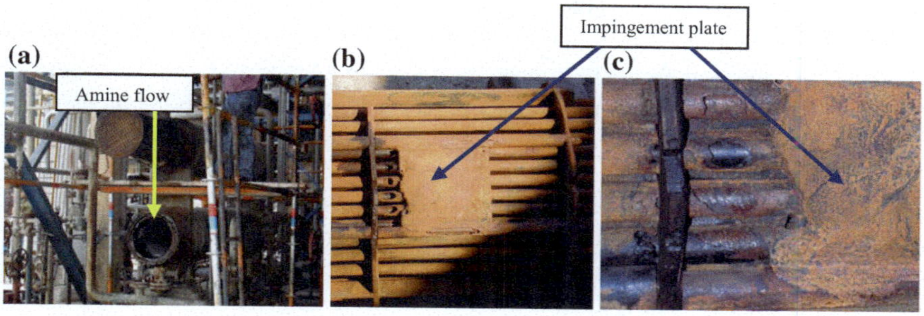

Fig. 4.23 **a** General view of the heat exchanger with failed tubes. **b**, **c** The impingement plate and holes formed as a result of impingement attack

cross sectional area of inlet pipe with amine solution, and this situation gave rise to severe impingement attack of the surfaces of upper tubes in the bundle.

Solutions and recommendations.

(a) To increase the area of impingement plate so that to have it 30 % more than the cross sectional area of inlet pipe with amine solution.
(b) To increase the thickness of impingement plate to 15 mm.
(c) It is not recommended to change the material of heat exchanger tubes.

4.4.5 *Corrosion of Pipe for the Transporting of Gaseous Mixture from the Absorber (HDS Unit)*

Conditions. Gaseous mixture containing N_2 (94–95 vol.%), H_2S (5–6 vol.%), and H_2 (traces) at 40 °C entered into the absorber (tower) for washing by aqueous solution of MDEA (had to be 50 wt%), namely, for absorption H_2S. At the exit from the tower (after the washing), the gases contained remains of H_2S (~0.2 vol. %) and flowed into the incinerator. The temperature in the tower was 50–60 °C and sometimes even more.

Material of construction. Pipe—carbon steel. Diameter—610 mm.

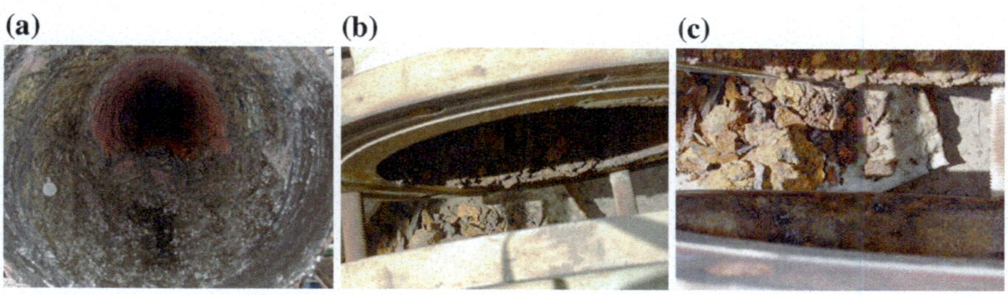

Fig. 4.24 **a** Corrosion products (iron sulfides) in the pipe in the exit from the absorber. **b**, **c** Corrosion products (rust) in the pipe exit from the absorber

Service period before the failure: 6 years.

Visual examination and findings. Large amounts of corrosion deposits were found in the pipe exit from the tower during shutdown (Fig. 4.24). Iron sulfides were detected inside the pipe outlet end and rust far from the inlet to the incinerator.

Failure phenomenon: *Low-temperature sulfidic corrosion (general corrosion by wetted H_2S).*

Cause of failure and its explanation. According to PI data, concentration of MDEA aqueous solution was 25–40 wt% (when recommended 50 wt%) and temperature in the absorber was 50–60 °C (when recommended 40 °C for better absorption H_2S). The temperature of the gases exiting from the tower was 60–70 °C (even up to ~100 °C), when recommended 40 °C. Operating (processing) discrepancy was the reason of severe corrosion in the pipe for gas transporting. Corrosion of inner surface of exit pipe from the absorber occurred as a result of improper absorption of H_2S by MDEA solution. Gases from the absorber were saturated with water and contained remains of non-absorbed H_2S which attacked inner surface of carbon steel pipe.

Solutions and recommendations.

(a) To maintain the temperature in the absorber not more than 40 °C and concentration of MDEA ~50 wt%.
(b) To replace carbon steel pipe with SS 304(if operating improvement will not overcome the corrosion situation).

References

1. Study on oil refining and oil markets (2008) Prepared for European Commission by Purvin & Gertz
2. NACE International Publication 43104 (2014) Overview of sulfidation (sulfidic) corrosion in petroleum refining hydroprocessing units. NACE International, Houston, TX, USA
3. API RP 939-C (2008) Guidelines for avoiding sulfidation (sulfidic) corrosion failures in oil refineries. American Petroleum Institute, Washington
4. Gutzeit J, Merrick RD, Scharfstein LR (1987) Corrosion in petroleum refining and petrochemical operations. In: Corrosion, vol 13, ASM Handbook, ASM International, pp 1262–1287
5. Groysman A (2014) Corrosion in systems for storage and transportation of petroleum products and biofuels. Springer, Dordrecht
6. Gutzeit J (2006) Crude unit corrosion guide. PCC Process Corrosion Consultants
7. Marcano N, Larter S, Snowdon L, Bennett B (2013) An overview of the origin, pathways and controls of H_2S production during thermal recovery operations of heavy and extra-heavy oil. Paper 90187 presented at the GeoConvention 2013, Integration: Geoscience Engineering Partnership, 6–12 May 2013, Calgary, AB
8. Kapadia P, Wang J, Kallos M, Gates I (2012) New thermal-reactive reservoir engineering model predicts hydrogen sulfide generation in steam assisted gravity drainage. J Petrol Sci Eng 94–95:100–111
9. Hyne JB et al (1982) Aquathermolysis of heavy oils. Revis Tec Intevep 2:87–94

10. Cross MM, Manning DAC, Bottrell SH, Worden RH (2004) Thermochemical sulfate reduction (TSR): experimental determination of reaction kinetics and implications of the observed reaction rates for petroleum reservoirs. Org Geochem 35(4):393–404
11. Lewandowski DA (2000) Design of thermal oxidation systems for volatile organic compounds. CRC LLC Lewis Publishers, Boca Raton
12. Groysman A (2010) Corrosion for everybody. Springer, Dordrecht
13. Lai GY (1990) High-temperature corrosion of engineering alloys. ASM International
14. Wright IG (1987) High-temperature corrosion. In: Corrosion, vol 13, ASM Handbook, ASM International, pp 97–103
15. Antos GJ, Aitani AM (eds) (2004) Catalytic naphtha reforming, 2nd edn. Marcel Dekker, New York
16. Wiehe IA (2008) Process chemistry of macromolecules. CRC Press Taylor & Francis Group, Boca Raton
17. Wang W, Watkinson AP (2011) Iron sulfide and coke fouling from sour oils: review and initial experiments. In: Proceedings of international conference on heat exchanger fouling and cleaning. Crete Island, Greece, pp 23–30
18. McConomy HF (1963) High temperature sulfidic corrosion in hydrogen free environment. API Proc 43(3):78–96
19. Couper AS, Gorman JW (1971) Computer correlations to estimate high-temperature hydrogen sulfide (H_2S) corrosion in refinery streams. Mater Prot and Perform 10(1):31–37
20. API RP 581 (2008) Risk-based inspection technology. API International, Washington
21. Corrosion (1987) vol 13, ASM International, p 1271
22. White RA, Ehmke EF (1991) Materials selection for refineries and associated facilities NACE, p 65
23. Study on oil refining and oil markets (2008) Prepared for European Commission by Purvin & Gertz
24. Shah P et al (2015) Advances in processing high naphthenic acid crudes. PTQ Q2 2015, 1–7. www.digitalrefining.com/article/1001119
25. Johnson D, MacAteer G, Zuk H (2002) Naphthenic acid corrosion field evaluation and mitigation studies. ERTC (European Refining Technology Conference), 7th Annual Meeting, Paris
26. Jayaraman A, Singh H, Lefebvre Y (1986) Naphthenic acid corrosion in petroleum refineries. A review. Revue de l`Institute Francais Du Pétrole 41(2):265–274
27. Speight JG (2014) High acid crudes. Elsevier, Oxford
28. Dalmia A (2013) Analysis of naphthenic acids in filtered oil sands process water (OSPW) using LC/TOF with no sample preparation. PerkinElmer, Shelton. http://www.perkinelmer.com/CMSResources/Images/44-154957APP_Analysis_of_Nalphthenic_Acids.pdf
29. Meredith W, Kelland SJ, Jones DM (2000) Influence of biodegradation on crude oil activity and carboxylic acid composition. Org Geochem 31(11):1059–1073
30. Vandecasteele JP (ed) (2008) Petroleum Microbiology, Edition Technip, Paris
31. Grewer DM, Young RF, Whittal RM, Fedorak PM (2010) Naphthenic acids and other acid-extractables in water samples from Alberta: what is being measured? Sci Total Environ 408:5997–6010
32. ASTM D664–11a (2011) Standard test method for acid number of petroleum products by potentiometirc titration. ASTM International, West Conshohocken, PA, USA
33. Da Silva MT et al (2012) Method for determination of the total acid number and naphthenic acid number of petroleum, petroleum cuts and petroleum emulsions of water-in-oil type by mid-infrared spectroscopy. US patent 8,222,605
34. Flego C et al (2014) Evolution of naphthenic acids during the corrosion process. Energy Fuels 28:1701–1708
35. Messer B, Tarleton B, Beaton M, Phillips T (2004) New theory for naphthenic acid corrosivity of Athabasca oilsands crudes. Paper 04634 presented at the CORROSION 2004, NACE International, Houston, TX, USA

36. Groysman A, Penner J (2008) Process for inhibiting naphthenic acid corrosion, WO/2009/053971
37. Lu T (2012) Challenges in opportunity crude processing. NALCO, An Ecolab Company
38. Qing W (2010) Processing high TAN crude: part1, PTQ Q4, http://www.digitalrefining.com/article_1000524.pdf
39. NACE Publication 34105 (2015) Effect nonextractable chlorides on refinery corrosion and fouling. NACE International, Houston, TX, USA
40. Clemente JS, Fedorak PM (2005) A review of the occurrence, analyses, toxicity, and biodegradation of naphthenic acids. Chemosphere (Elsevier) 60(5):585–600
41. Han X et al (2008) Influence of molecular structure on the biodegradability of naphthenic acids. Environ Sci Technol 42:1290–1295
42. Biryukova OV, Fedorak PM, Quideau SA (2007) Biodegradation of naphthenic acids by rhizosphere microorganisms. Chemosphere (Elsevier) 67:2058–2064
43. Herman DC, Fedorak PM, Mackinnon MD, Costerton JW (1994) Biodegradation of naphthenic acids by microbial populations indigenous to oil sands tailings. Can J Microbiol 40:467–477
44. Groysman A et al (2005) Corrosiveness of acidic crude oil and its fractions In: Proceedings EUROCORR 2005, Lisbon
45. Groysman A et al (2005) Corrosiveness of acidic crude oil and its fractions. Mater Perform 44(4):34–39
46. Groysman A et al (2005) Study of corrosiveness of acidic crude oil and its fractions. Paper 05568 presented at the NACE CORROSION 2005. NACE International, Houston, TX, USA
47. Dean FWH, Powell SW (2006) Hydrogen flux and high temperature acid corrosion. Paper 06436 presented at the CORROSION 2006. NACE International, Houston, TX, USA
48. Rangarajan P, Holmes F, Randolph B (2006) Effect of naphthenic acid structure on carbon steel corrosion. Conoco Philips Company, AIChE, Spring Meeting
49. Gutzeit J (2006) Crude unit corrosion guide. PCC Process Corrosion Consultants
50. Piehl RL (1988) Naphthenic acid corrosion in crude distillation units. Mater Perform 27 (1):37–43
51. Kane RD, Chambers B (2011) High temperature crude oil corrosivity: where sulfur and naphthenic acid chemistry and metallurgy meet. Paper 4A presented at the Corrosion Solutions Conference 2011 Proceedings, pp 137–144
52. Kane RD, Cayard MS (2002) A Comprehensive study on naphthenic acid corrosion. Paper 02555 presented at the NACE CORROSION conference. NACE International, Houston, TX, USA
53. Kane R, Trillo E (2006) The state-of-the-art of naphthenic acid and sulfidic corrosion evaluation and prediction. Presentation at the 2006 Spring Meeting & 2nd Global Congress on Process Safety, 9th Topical Conference on Refinery Processing, AIChE
54. Bota GM et al (2010) Naphthenic acid corrosion of mild steel in the presence of sulfide scales formed in crude oil fractions at high temperature. Paper 10353 presented at the NACE CORROSION conference, NACE International, Houston, TX, USA
55. Gutzeit J (1977) Naphthenic acid corrosion in oil refineries. Mater Perform 16(10):24–35
56. Panchal CB et al (1999) Mechanisms of iron sulfide formation in refining processes. In: Bott TR (ed) Understanding heat exchanger fouling and its mitigation. Begell House. New York, pp 291–298
57. Wang W, Watkinson AP (2011) Iron sulphide and coke fouling from our oils: review and initial experiments. In: Proceedings on international conference on heat exchanger fouling and cleaning—2011, June 05–10, Crete Island, pp 23–30
58. Parker RJ, McFarlane RA (2000) Mitigation of fouling in bitumen furnaces by pigging. Energy Fuels 14:11–13
59. Verachtert TA (1980) Trace acid removal in the pretreatment of petroleum distillate. US Patent 4,199,440 A
60. Varadaraj R, Sartori G (2000) Process for neutralization of petroleum acids (LAW810). US Patent 6,030,523 A

61. Gorbaty ML et al (2000) Process for neutralization of petroleum acids using overbased detergents. US Patent 6,054,042 A
62. Sartori G, Savage DW, Ballinger BH, Dalrymple DC (2000) Process for decreased the acidity of crudes using crosslinked polymeric amines (LAW871). US Patent 6,121,411 A
63. Sartori G, Savage DW, Ballinger BH (1997) Process for neutralization of petroleum acids. US Patent 5,683,626 A
64. Sartori G, Savage DW, Gorbaty ML, Ballinger BH (1997) Process for neutralization of petroleum acids using alkali metal trialkylsilanolates. US Patent 5,643,439 A
65. Kittrell N (2006) Removing Acid from Crude Oil, Crude Oil Quality Group, New Orleans Meeting, http://web.archive.org/web/20080528051559/ http://www.coqg.org/20060223TAN%20Merichem.pdf
66. Kelland MA (2014) Production chemicals for the oil and gas industry, 2nd edn. CRC Press Taylor & Francis Group, Boca Raton
67. http://www.statoil.com/en/technologyinnovation/refiningandprocessing/oilrefining/refiningnaphtheniccrudes/pages/default.aspx
68. Drzewicz P et al (2012) Impact of peroxydisulfate in the presence of zero valent iron on the oxidation of cyclohexanoic acid and naphthenic acids from oil sands process-affected water. Environ Sci Technol 46(16):8984–8991
69. Zhang A, Ma Q, Goddard WA, Tang Y (2004) Improved processes to remove naphthenic acids. Annual Technical Progress Report. California Institute of Technology
70. De Conto JF et al (2014) Naphthenic acids recovery from petroleum using ionic silica based hybrid material as stationary phase in solid phase extraction (SPE) process. Adsorption 20:917–923
71. Sartori G et al (2004) Method for inhibiting corrosion using phosphorous acid. US Patent 6,706,669
72. Babaian-Kibala E (1993) Naphthenic acid corrosion inhibitor. US Patent 5,252,254
73. Babaian-Kibala E (1996) Phosphorus thioacid ester inhibitor for naphthenic acid corrosion. US Patent 5,552,085
74. Zetlmeisl MJ et al (1990) Corrosion inhibitors for use in hot hydrocarbons. US Patent 4,941,994
75. Zetlmeisl MJ (1995) A laboratory and field investigation of naphthenic acid corrosion inhibition. Paper 334 presented at the CORROSION/95, NACE International, Houston, TX, USA
76. Edmondson JG (1996) High temperature corrosion inhibitor. US Patent 5,500,107
77. Babaian-Kibala E (1994) Phosphate ester inhibitors solve naphthenic acid corrosion problems. Oil Gas J 92(9):31–35
78. Winslow MC et al (2005) Solutions for processing opportunity crudes. ERTC (European Refining Technology Conference, 10th Annual Meeting, Vienna
79. Jackson T, Winslow MC, Wilson M (2004) Prolonged experience processing high acid crude—cross oil & refining company. ERTC 9th Annual Meeting, Prague
80. Jackson T, Winslow MC, Wilson M (2005) Prolonged experience processing high acid crude cross oil & refining company. EUROCORR 2005, Lisbon
81. Vanhove A (2006) Advances in corrosion monitoring while processing high acidic opportunity crudes. Paper presented at the 7th Israel Conference on Corrosion and Electrochemistry, Bar-Ilan University
82. Babaian-Kibala E et al (1997) Use of sulfiding agents for enhancing the efficacy of phosphorus in controlling high temperature corrosion attack. US Patent 5,630,964
83. Yeganeh MS et al (2003) Method for inhibiting corrosion using certain phosphorous and sulfur-free compounds. US Patent 6,593,278
84. Sartori G et al (2003) Method for inhibiting corrosion using certain aromatic acidic species. US Patent 6,559,104
85. Sartori G et al (2003) Method for inhibiting corrosion using 4-sulfophthalic acid. US Patent 6,583,091

86. Zetlmeisl MJ (1999) Control of naphthenic acid corrosion with thiophosphorus compounds. US Patent 5,863,415
87. Edmondson JG (1997) High temperature corrosion inhibitor. US Patent 5,611,911
88. Edmondson JG (1994) High temperature corrosion inhibitor. US Patent 5,314,643
89. Deruyck F, Brabant JV (2001) Elongated steel object treated with a corrosion inhibiting composition. US Patent 6,329,0734
90. Revie RW (ed) (2006) Uhlig's Corrosion Handbook, 2nd edition. Wiley, pp 1096–1097
91. Groysman A et al (2007) Low temperature naphthenic acid corrosion study. Paper 07569 presented at the CORROSION 2007. NACE International, Houston, TX, USA
92. O'Kane JM et al (2010) Detection and monitoring of naphthenic acid corrosion in a visbreaker unit using hydrogen flux measurements. Paper 10351 presented at the NACE Corrosion Conference. NACE International, Houston, TX, USA
93. Dean FWH et al (2010) Correlation of hydrogen flux and corrosion rate measurements carried out in a severe episode of corrosion-erosion attributable to naphthenic acid. Paper 10178 presented at the NACE Corrosion Conference. NACE International, Houston, TX, USA
94. Groysman A et al (2007) Naphthenic acid corrosion study. Paper presented at the EUROCORR conference, Freiburg, Germany
95. Groysman A et al (2007) Low temperature naphthenic acid corrosion study. Paper presented at the NACE CORROSION 2007 conference. NACE International, Houston, TX, USA
96. Shreider AV, Shparber IS, Archakov YI (1976) Vliyanie vodoroda na khimicheskoye I neftyanoye oborudovaniye (Influence of hydrogen on chemical and oil equipment). Mashinostroyeniye, Moscow
97. Pourbaix M (1974) Atlas of the electrochemical equilibria. NACE, Houston, TX, USA
98. Smialovski M (1967) Influence of hydrogen on properties of iron and its alloys. Zaschita metallov (Protection of Metals) 3(3):267–277
99. Grigoriev VP, Gorbachev VA (1970) Influence of stimulators of hydrogenation on dependence of cathodic polarization of metal on additives nature. Zaschita metallov (Protection of metals) 6(3):306–310
100. Ratajczykowa I, Palczewska W (1968) Hydrogen sulphide as catalytic poison in the kinetics of hydrogen atom recombination on iron. Bull De l`Acad Polanaise des Sci, Serie de sciences chimiques 16(2):77–82
101. Korb LJ, Olson DL (eds) (1987) Metals handbook, vol 13: Corrosion. ASM International, pp 1277–1280
102. Hudson RM (1964) Hydrogen absorption by and dissolution rate of low-carbon steel in sulfuric, hydrochloric, phosphoric and nitric acids. Corrosion 20(8):245t–251t
103. Beloglazov S (2013) Electrochemical hydrogen and metals: absorption, diffusion and embrittlement prevention in corrosion and electroplating (chemistry research and applications). Nova Science Publishers
104. Messer B et al (2008) Role of nascent hydrogen in refinery corrosion. Paper 08549 presented at the CORROSION 2008 conference. NACE International, Houston, TX, USA
105. Rachinsky V, Smialovski M (1969) Influence of different factors on hydrogen embrittlement of steel and iron. Zaschita metallov (Protection of Metals) 5(5):482–490
106. Groysman A (2010) Corrosion for everybody. Springer, Dordrecht
107. Tuttle RN (1970) Selection of materials for use in a sour gas environment. Mater Protect 9 (4):11–14
108. Warren D, Beckman GW (1957) Sulfide corrosion cracking of high strength bolting materials. Corrosion 13(10):33–48
109. Whiteman MB, Troiano AR (1965) Hydrogen embrittlement of austenitic stainless steel. Corrosion 21(2):53–56
110. Owczarek E, Zakroczymski T (2000) Hydrogen transport in a duplex stainless steel. Acta Mater 48(12):3059–3070
111. Parvathavarthini N, Saroja S (2001) Studies on hydrogen permeability of 2.25 % Cr-1 % Mo ferritic steel: correlation with microstructure. J Nucl Mater 288(2–3):187–196

112. NACE International Publication 8X194 (2006) Materials and fabrication practices for new pressure vessels used in wet H_2S refinery service. NACE International, Houston, TX, USA
113. NACE Standard MR0103-2007 (2007) Materials resistant to sulfide stress cracking in corrosive petroleum refining environments. NACE International, Houston, TX, USA
114. ANSI/NACE MR0175/ISO 15156 (2009) Petroleum and natural gas industries—materials for use in H_2S-containing environments in oil and gas production. NACE International, Houston, TX, USA
115. Smialowski M (1962) Hydrogen in steel. Pergamon Press Ltd, Warszawa
116. Catastrophic Rupture of Heat Exchanger (Seven Fatalities) Tesoro Anacortes Refinery (2010) Report 2010-08-I-WA. US Chemical and Investigation Board, Washington
117. Pishko R, McKimpson M, Shewmon PG (1979) The effect of steelmaking on the hydrogen attack of carbon steel. Metall Mater Trans A 10(7):887–894
118. Inglis NP, Andrews W (1933) Effect of H on various steels at 150–500 °C and 200–250 atm. J Iron Steel Inst 128:383
119. Schuyten J (1947) Hydrogen attack on metals at high temperatures and pressures. Corr Mater Perform 4(5):13
120. Nelson GA (1949) Hydrogenation plant steels. In: Proceedings API 29M(III):163-174. American Petroleum Institute, Washington
121. Api RP 941 (2008) Steels for hydrogen service at elevated temperatures and pressures in petroleum refineries and petrochemical plants, 7th edn. American Petroleum Institute, Washington
122. Nelson GA (1951) Metals for high-pressure hydrogenation plants. Trans Am Soc Mech Eng 73(2):205
123. Nelson GA (1968) Corrosion data survey. NACE, Houston, TX, USA
124. Shih HM, Johnson HH (1982) A model calculation of the Nelson curves for hydrogen attack. Acta Metall 30:537–545
125. Parthasarathy TA (1985) Mechanisms of hydrogen attack of carbon and 2.25Cr-1Mo steels. Acta Metall 33(9):1673–1681
126. Van der Burg MWD, Van der Giessen E, Tvergaard V (1998) A continuum damage analysis of hydrogen attack in a 2.25Cr-1Mo pressure vessel. Mater Sci Eng A241:1–13
127. Manna G, Castello P, Harskamp F (2007) Testing of welded 2.25Cr-1Mo steel in hot, high-pressure hydrogen under creep conditions. Eng Fract Mech 74:956–968
128. Shewmon PG (1976) Hydrogen attack of carbon steel. Metall Trans A, Vol 7A February
129. API TR (Technical Report) 941 (2008) The Technical Basis Document for API RP 941, Edition 8, American Petroleum Institute, Washington
130. Groysman A (2009) Corrosion monitoring. Corr Rev 27(4–5):205–343
131. Birring AS, Bartlett ML, Kawano K (1989) Ultrasonic detection of hydrogen attack in steels. Corrosion 45(3):259–263
132. Birring AS, Riethmuller M, Kawano K (2005) Ultrasonic techniques for detection of high temperature hydrogen attack. Mater Eval February 110–115
133. Revie RW (ed) (2006) Uhlig's corrosion handbook. Wiley-Interscience, Wiley
134. ANSI/NACE Standard TM 0284–2003 (2003) Standard test method evaluation of pipeline and pressure vessel steels for resistance to hydrogen-induced cracking. NACE International, Houston, TX, USA
135. Jack BL (1999) Alternatives to HIC steels in sour refinery environments. Paper 99425 presented at the NACE CORROSION/99 conference. NACE International, Houston, TX, USA
136. Danis JI (1999) Wet H_2S Cracking—A Refiner's Experience With HIC Resistant Steels. Paper 99431 presented at the NACE CORROSION/99 conference. NACE International, Houston, TX, USA
137. ASTM A370–15 (2015) Standard test methods and definitions for mechanical testing of steel products. ASTM International, West Conshohocken, PA, USA
138. ISO 10423:2009 (2009) Petroleum and natural gas industries—Drilling and production equipment—Wellhead and christmas tree equipment

139. NACE Standard SP 0472-2008 (2008) Standard practice, methods and controls to prevent in-service environmental cracking of carbon steel weldments in corrosive petroleum refining environments. NACE International, Houston, TX, USA
140. White RA, Ehmke EF (1991) Materials selection for refineries and associated facilities. NACE
141. Groysman A, Erdman N (2000) A study of corrosion of mild steel in mixtures of petroleum distillates and electrolytes. Corrosion 56(12):1266–1271
142. Groysman A (2005) Corrosion of aboveground fuel storage tanks. Mater Perform 44(9):44–48
143. Bottoms RR (1930) Process for separating acidic gases. US Patent 1,783,901
144. Gutzeit J, Merrick RD, Scharfstein LR (1987) Corrosion in petroleum refining and petrochemical operations. In: Corrosion, vol 13, ASM Handbook, ASM International, pp 1262–1287
145. DuPart MS, Bacon TR, Edwards DJ (1993) Understanding corrosion in alkanolamine gas treating plants. Hydrocarbon Processing. Part 1, 72(4):75–79. Part 2, 72(5):89–94
146. Liu HJ, Dean JW, Bosen SF (1995) Neutralization technology to reduce corrosion from heat stable amine salts. Paper 572 presented at the NACE conference Corrosion/95, NACE International, Houston, TX, USA
147. Gregory RA, Cohen MF (1988) Removal of salts from aqueous alkanolamine using an electrodialysis cell with ion exchange membrane. European Patent 286 143
148. Burns D, Gregory RA (1995) The UCARSEP™ process for on-line removal of non-regenerable salts form amine units. Paper presented at the Laurance Reid gas conditioning conference, The University of Oklahoma, Norman, OK, USA

Recommended Literature

149. Pohl HA (1962) Solubility of iron sulfides. J Chem Eng Data 7(2):295–306
150. Rebak RB (2011) Sulfidic corrosion in refineries—a review. Corr Rev 29:123–133, Walter de Gruyter, Berlin
151. Farrel D, Roberts L (2010) A study of high temperature sulfidation under actual process conditions. Paper 10358 presented at the NACE CORROSION 2010 conference, NACE International, Houston, TX, USA
152. Niccolls EH et al (2008) High Temperature Sulfidation Corrosion in Refining. Paper presented at the 17th International Corrosion Congress, October 6–10, 2008, Las Vegas, NACE International, Houston, TX, USA
153. Hucińska J (2006) Influence of sulfur on high temperature degradation of steel structures in the refinery industry. Adv Mater Sci 6(1):16–25
154. Tebbal S, Kane RD (1996) Review of critical factors affecting crude corrosivity. Paper 607 presented at the CORROSION/96, NACE International, Houston, TX, USA
155. Craig HL (1996) Temperature and velocity effects in naphthenic acid corrosion. Paper 603 presented at the CORROSION/96. NACE International, Houston, TX, USA
156. Turnbull A, Slavcheva E, Shone B (1998) Factors controlling naphthenic acid corrosion. Corrosion 54(11):922–930
157. Slavcheva E, Shone B, Thurnbull A (1999) Review of naphthenic acid corrosion in oil refining. Br Corr J 34(2):125–131
158. Wu XQ et al (2004) Study on high-temperature naphthenic acid corrosion and erosion-corrosion of aluminized carbon steel. J Mater Sci 39:975–985
159. Wu XQ et al (2004) Erosion-corrosion of various oil-refining materials in naphthenic acid. Wear 256:133–144
160. Babaian-Kibala E et al (1993) Naphthenic acid corrosion in a refinery setting. Paper 631 presented at the CORROSION/93, NACE International, Houston, TX, USA

161. www.bjservices.com. BJ Chemical Services "RNB 40218 Naphthenic Acid Corrosion Inhibitor Prevents High Temperature Corrosion"
162. Haynes D (2006) Naphthenic acid bearing refinery feedstocks and corrosion abatement, AIChE—Chicago Symposium
163. Laredo GC et al (2004) Naphthenic acids, total acid number and sulfur content profile characterization in Isthmus and Maya crude oil. Fuel 83:1689–1695
164. Jayaraman A, Singh H, Lefebvre Y (1986) Rev Inst Fr Pet 41:265
165. Zeinalov EB, Abbasov VM, Alieva LI (2009) Petroleum acids and corrosion. Pet Chem 49 (3):185–192
166. Laredo GC et al (2004) Identification of naphthenic acids and other corrosivity-related characteristics in crude oil and vacuum gas oils from a mexican refinery. Energy Fuels 18:1687–1694
167. Rangarajan P, Holmes F, Randolph B (2006) Effect of naphthenic acid structure on corrosion. Presentation at the 2006 Spring Meeting & 2nd Global Congress on Process Safety, 9nd Topical Conference on Refinery Processing, AIChE
168. Yépez O (2005) Influence of different sulfur compounds on corrosion due to naphthenic acid. Fuel 84(1):97–104
169. Yépez O (2007) On the chemical reaction between carboxylic acids and iron, including the special case of naphthenic acid. Fuel 86(7–8):1162–1168
170. Kevrešan S et al (2003) Effect of naphthenic acids on formation of adventitious roots in sunflower cuttings. HELIA 26(39):75–82
171. Claesen C, Lordo SA, Scattergood G (2007) Chemical inhibition of high temperature sulfidic corrosion in lab evaluations and petroleum refinery applications. Paper presented at the EUROCORR 2007, Freiburg, Germany

Chapter 5
Corrosion Failures and Solutions at Units

The only lesson to be learned from history is that people do not take any lessons from history.
George Bernard Shaw (1856–1950), an Irish playwright.

Abstract Corrosion and corrosion control in vacuum overhead system, typical and unique cases are given and analyzed. Corrosion case studies, analysis and remedies are described at the following units: hydrodesulfurizer, visbreaker, petrochemical plant, and auxiliary facilities (flare disposal system, bitumen facility for odor removal, dissolved gas flotation, spent soda treatment, water systems, gas scrubber at the ethylene plant, separator, and urea injection). Examples of corrosion problems and solutions occurring with pipes and tanks are also given.

It is very important to collect corrosion cases, and create "banks" of problems and corrosion atlases in the oil refining and petrochemical industry. They help to identify corrosion problems and risk at early stages, namely, to learn, understand, avoid or reduce likelihood, extent and severity of corrosion. Such generalizations were carried out [1–6]. Despite numerous corrosion cases described in literature, information about new corrosion failures are dispersed in reports and papers, some of them are not reported at all and are present as tacit knowledge. In my practical work with oil refinery and petrochemical equipment, about 10–15 % of corrosion events which I encountered I did not find them described in literature. This situation indicates that not only new generation of engineers but also experienced specialists not always know and are familiar with all problems which appear inevitably. Even if I know 1000 corrosion cases, there is always the 1001 case which I have never heard.

Another problem is that there are typical and untypical corrosion failures with oil refining and petrochemical equipment. There are associated facilities and media at auxiliary processes which accompany producing fuels and organic substances at oil refineries and petrochemical plants. These facilities are power station with steam and boiler feed water generation, cooling water system, fire-fighting system, and units related to environmental problems: water, air and soil purification, and use of petroleum product wastes.

A. Groysman, *Corrosion Problems and Solutions in Oil Refining and Petrochemical Industry*, Topics in Safety, Risk, Reliability and Quality 32,
DOI 10.1007/978-3-319-45256-2_5

Strict requirements to pollutions in atmosphere gave rise to erection of units and equipment absorbing acid gases (HCl, H_2S, NO_x) which in its turn resulted in corrosion problems which have not met before. For instance, the alloy Hastelloy B-2 as believed was the best solution for material of an ejector (a Venturi scrubber) for absorbing acid gases in the CCR unit failed after 9 months of service as a result of corrosion-cavitation [7]. Another example is the purification of gaseous emissions from noxious NO_x to the atmosphere. Nowadays emissions of NO_x in oil refineries, petrochemical, and electric power plants are reduced by injecting of liquid solution of urea which decomposes by water vapor to ammonia, and the latter reacts with NO gas producing N_2 gas which is the main component of the atmosphere. Carbon steel is not resistant to urea and only stainless steel must be used at this unit. In addition to correct selection of corrosion resistant material, erosion and cavitation problems can appear in this unit.

Sometimes uncertainty is present in defining the causes of corrosion accidents, when several corrosion phenomena occur simultaneously or different reasons give rise to similar manifestation and failure. For instance, MIC often occurs concurrently with general, under deposit, pitting and crevice corrosion; erosion, cavitation and impingement proceeds with corrosion, etc. Pitting may be caused by dissolved oxygen, chlorides, naphthenic acids, acid dew point corrosion, under deposit corrosion, or cavitation. SCC of carbon steel may occur as a result of caustic, H_2S, H_2 or polythionic acid attack.

Old and new corrosion problems and accidents will be described and analyzed in this chapter. Sometimes they require new solutions concerning changes of technological parameters, construction, and use of suitable materials. What is very important, new approach to corrosion management and proper relationship to corrosion problems and risk must be applied (see Chap. 9).

5.1 Vacuum Distillation System

Overhead stream at the vacuum distillation system contains water vapor, H_2S, HCl and sometimes light organic acids. Vacuum bottom can contain high concentrations of naphthenic acids and sulfur compounds. As a result, following two major corrosion problems can occur at vacuum distillation unit: low-temperature overhead acid corrosion of condensers, valves and pumps; high-temperature corrosion of piping if naphthenic acids and sulfur compounds are present in vacuum bottom (at ~340–350 °C).

Causes of corrosion in the overhead at the vacuum distillation unit are similar to that at the atmospheric column: H_2S and HCl dissolved in condensed water. About 90 % $MgCl_2$ and 10 % $CaCl_2$ are usually hydrolized in desalted crude and produced HCl travels up to preflash and crude columns. The remaining 10 % $MgCl_2$ and 90 % $CaCl_2$ end up in atmospheric residue which is sent to the vacuum heater and reacting with steam, HCl gas is formed. Sometimes waste products containing organic chlorides and entrained seawater containing chloride salts and shipped by tankers can be additional source of chlorides in the vacuum overhead. In addition to similar corrosion components (HCl and H_2S) like in atmospheric overhead system, there are specific corrosion problems. Vacuum system is vulnerable to air entering and further oxidation of sulfur compounds with formation of inorganic (H_2SO_3, H_2SO_4) and organic (acetic) acids which are aggressive to carbon steel. Acetic acid can be formed also as a result of decomposition of naphthenic acids and high molecular weight aliphatic acids.

Some N-organic compounds in atmospheric residue can be converted to ammonia in the vacuum heater and in the case of HCl presence ammonium chloride can be formed in the vacuum overhead. Sometimes amine neutralizers are injected into the vacuum overhead for pH control. As a result, hydrochloride amine salts also can be formed. Both salts dissolve in overhead water condensate and can be present in separator drum.

Thus, corrosion problems in vacuum overhead are caused by HCl, H_2S, CH_3COOH, O_2, H_2SO_3, and H_2SO_4. It is believed that corrosion in vacuum system is less severe than in atmospheric unit. This is deceptive situation. Uncontrolled pH in water condensate (e.g., pH ~3–4) can result in severe corrosion of carbon steel equipment with formation of holes (see Sect. 5.1.2.2). Therefore, corrosion control and monitoring in the vacuum overhead are vital.

Besides corrosion, erosion problems inside of steam ejectors can occur in vacuum overhead. Usually erosion is caused by excessive steam velocities and formation of water drops in steam (droplet impingement). Typical exhaust temperature of steam should range from 120 to 150 °C. Lower temperatures indicate that the steam is wet.

5.1.1 Corrosion Control in Vacuum Overhead System

The choice of materials of construction depends on sulfur and naphthenic acid content of atmospheric residue. Usually carbon steel is used at $T < 230$ °C if sulfur concentration is less than 1 wt% and TAN < 0.5 mg KOH/g. If sulfur concentration is more than 1 wt% and $T > 230$ °C, low-alloy 5Cr steel is recommended. When TAN > 0.5 mg KOH/g and $T > 230$ °C, SS 316L or SS317L are recommended.

Vacuum column internals are often 12Cr SS and the column itself is clad also with 12Cr SS. Aluminum can be used because it is resistant to both sulfur compounds and naphthenic acids (as well as to other organic acids). Admiralty brass CDA 443 and Ti Gr.2 can be used as materials of tubes in condensers.

Corrosion control in both atmospheric and vacuum overhead systems starts with effective tank settling, desalting, and caustic injection to desalted crude. As opposed to atmospheric overhead, no wash water is injected into vacuum overhead system. The main corrosion control method in vacuum overhead is injecting a neutralizer in order to keep pH = 5.5–6.5 of water condensate in separator. Neutralizer can be morpholine, cyclohexylamine, or diethylaminoethanol injected into steam (see Appendix E). In many cases it is enough. However, amine corrosion inhibitor sometimes is injected together with neutralizer. Amine neutralizer should not be injected into dry overhead line to avoid hydrochloride salt deposit formation and then under deposit corrosion. An excess of neutralizer can result in higher pH values (>7), formation of deposits and further corrosion. It is important to prevent air leaks which can cause corrosion. Helium testing can be used to find leaks.

5.1.2 Case Studies

Five corrosion cases are described and analyzed in this section.

5.1.2.1 Corrosion of the 1st Vacuum Condenser

Conditions. Hydrocarbons, acid gases (mostly H_2S) and steam after vacuum distillation column flowed in the shell of vacuum condenser at 40 °C. Cooling water treated with chemicals (corrosion inhibitors, anti-scaling agents, biocides) flowed inside the tubes at 30–35 °C and 4–5 bar. Neutralizer was not injected most period of time (about 4 years) in the overhead of vacuum distillation system.

Materials of construction. Tubes—copper-nickel 90/10 (CDA 706). Tube-sheet—carbon steel ASTM A515 Gr. 70. Tube-sheet—clad copper-nickel 90/10, 10 mm in thickness (from water side). Baffles—carbon steel ASTM A285 Gr. C.

Service period before the failure: 5 years.

Visual examination and findings. Localized corrosion including pits, splitting, porosity and sponginess of the surface was found on carbon steel baffles and tube-sheet from the processing side (Fig. 5.1). Carbon steel also corroded in the area of contact with copper-nickel tubes. Pits as well were found on outer surface of copper-nickel tubes.

Failure phenomenon: *Acid corrosion, pitting corrosion, galvanic corrosion.*

Cause of failure and its explanation. Morphology of failed surface of baffles and tube-sheet manifested that light organic and inorganic acids (including H_2S) which

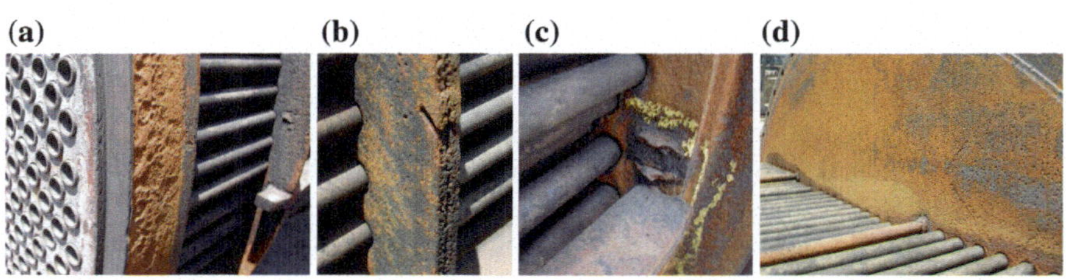

Fig. 5.1 **a** Corroded carbon steel tube-sheet and baffle (process side). **b** Corroded baffle. **c** Completely corroded and destructed welding. **d** Pitted surface of the tube-sheet (process side)

were not neutralized, attacked carbon steel surface in the form of splitting (delaminating), porosity, sponginess, and pits. Nascent hydrogen atoms were liberated in cathodic process, diffused into carbon steel and caused such forms of attack (see Fig. 5.1). H_2S reacted with outer surface of copper-nickel tubes and resulted in pits.

Solutions and recommendations.

(a) It is impossible to repair carbon steel attacked by hydrogen which was formed in corrosion reaction with acids.
(b) To produce tube-sheets and baffles from killed steel or A516 HIC (special carbon steels resistant to hydrogen attack) (see Sect. 4.3.1 and Appendix F).
(c) To inject inhibitor of acidic corrosion (in addition to neutralizer) in the overhead of vacuum distillation system.
(d) To replace copper-nickel tubes with carbon steel, because cooling water is treated with corrosion inhibitors of iron, and there is not necessary to use corrosion resistant alloys in cooling water.

5.1.2.2 Failure of the 1st Step Vacuum Condenser Boot

Conditions. The 1st step vacuum condenser is intended for cooling mixture of hydrocarbons by cooling water. Hydrocarbons containing water vapor flew out the vacuum distillation column. The temperature of condensate which exits from the condenser was 20 °C. Neutralizer was not injected in the vacuum overhead system during the first 4 years, and the bottom of the condenser (30 % area) corroded significantly. Amine neutralizer (4 ppm) was injected into the entrance of the vacuum condenser during the last 4 years before the failure. Streams from three vacuum condensers (the 1st, the 2nd and the 3rd steps) are mixed in the tank. Corrosion coupons for corrosion monitoring installed in the entry to the tank for the condensate did not show the real situation in the 1st condenser.

Materials of construction. Tubes and tube-sheets—Ti Gr.2. Shell—carbon steel.

Service period before the failure: 8 years.

Visual examination and findings. The hole (10 cm in diameter) was detected in the bottom (boot) in the 1st step vacuum condenser (Fig. 5.2).

The upper part of the condenser (entry of hydrocarbons and water vapor) was significantly rusted. Titanium tubes were in good appearance. Deposits were formed on the outer surface of titanium tubes.

Determination of chemical composition of aqueous condensate in the condensate vessel was carried out once a week. The pH values ranged in 3.6–4.1 (during the last year before the failure; instead recommended 5.5–6.5); sometimes pH = 8–9.8. Concentrations of chloride ions ranged in 42–56 ppm (instead recommended 10 ppm maximum).

Carbonate scale was formed inside titanium tubes only in the inlet to the condenser. Opposite side of the condenser (cooling water exit) was in good appearance. Aluminum sacrificial anodes functioned well.

Failure phenomenon: *Acid corrosion.*

Fig. 5.2 **a** General view of the 1st step vacuum condenser. **b** The hole (10 cm) in the bottom (boot). **c** Rusted *upper part* of the condenser. **d** Deposits on outer surface of titanium tubes in the condenser. **e** Scale inside the titanium tubes and on the tube-sheet (cooling water side). **f** Opposite side (exit of cooling water). **g** Aluminum sacrificial anodes in cooling water (inlet channel)

Cause of failure and its explanation. Severe general corrosion of inner surface of the 1st vacuum condenser was caused by aqueous acid solution containing chlorides. Concentration of injected amine neutralizer was not enough. In any case, sometimes overdosage happened which caused high pH = 8.0–9.8 of aqueous condensate, and this resulted in formation of salt deposits.

Solutions and recommendations.

(a) To carry out cladding from Ti Gr.2 of inner surface of the 1st vacuum condenser.
(b) To coat alternatively by epoxy-ceramic coating (600 μm in thickness) or epoxy vinyl ester coating (3 mm in thickness).
(c) Simultaneous injection of neutralizer and corrosion inhibitor.
(d) Measurements of pH and corrosion rates should be performed continuously.

5.1.2.3 Corrosion of the 2nd Step Vacuum Condenser

Conditions. Hydrocarbons, acid gases (mostly H_2S) and steam after the vacuum column flowed at 40 °C in the shell of the 2nd step vacuum condenser. Cooling water treated with chemicals (corrosion inhibitors, anti-scaling agents, biocides)

Fig. 5.3 **a** Corroded inner surface of the shell of the 2nd step vacuum condenser. **b, c** Localized corrosion near the exit of water condensate from the 2nd step vacuum condenser

flowed inside the tubes at 30–35 °C and 4–5 bar. Neutralizer was not injected much time (about 4 years) in the overhead of the vacuum distillation column.

Materials of construction. Shell—carbon steel ASTM A515 Gr. 70. Tubes—copper-nickel 90/10 (CDA 706).

Service period before the failure: 5 years.

Visual examination and findings. Localized corrosion (shallow pits) were found in large area near the exit of water condensate from the 2nd step vacuum condenser (Fig. 5.3).

Failure phenomenon: *Under deposit corrosion.*

Cause of failure and its explanation. Acid deposits forming on the inner surface of the shell near the exit of the condensate caused severe localized corrosion underneath.

Solutions and recommendations.

(a) To repair pits by welding.
(b) To inject proper neutralizer and corrosion inhibitor (low salt) in the overhead system of vacuum distillation column.
(c) To install ER-probes in order to monitor corrosion situation.

5.1.2.4 Corrosion of Tube-Sheets and Baffles in Heat Exchanger

Conditions. Heat exchanger is intended for cooling hydrocarbon gases containing water vapor and H_2S flowing out the vacuum distillation column. Temperature of hydrocarbon gases was ~50 °C. Cooling water flowed inside the tubes at 30–35 °C and 4 bar.

Material of construction. Tubes—Admiralty brass CDA 443 (70Cu–30Zn). Baffles and tube-sheets—carbon steel. Admiralty brass clad was applied on the tube-sheet in contact with cooling water. There was no clad from the processing side (hydrocarbon gases containing water vapor and H_2S).

Service period before the failure: 4 years.

Fig. 5.4 **a** Overhead condenser (vacuum unit). **b** Corroded carbon steel baffle and Admiralty brass tubes. **c** Galvanic corrosion of carbon steel tube-sheet (locations of contact with Admiralty brass tubes). Black copper sulfides were formed on brass tubes as a result of corrosion by wetted H_2S

Visual examination and findings. Localized corrosion of baffles and tube-sheets from the process side in the places of contact with Admiralty brass tubes (Fig. 5.4). Outer surface of Admiralty brass tubes were covered by black copper sulfide scale.

Failure phenomenon: *Low-temperature sulfidic corrosion, galvanic corrosion.*

Cause of failure and its explanation. Hydrocarbon gases contained H_2S and water vapor. The latter was condensed, H_2S was dissolved in water phase and attacked outer surface of Admiralty brass tubes. Galvanic corrosion of carbon steel baffles and tube-sheets occurred in the places of contact with Admiralty brass tubes.

Solutions and recommendations.

(a) To replace the material of tubes and tube-sheets with Ti Gr.2. This material proved itself as a resistant to the medium in the overhead at the vacuum unit (see Sect. 8.1.1). Tube-sheets can be made from carbon steel cladded with Ti Gr.2 from both sides (processing stream and cooling water). Baffles should be made from SS 316 as electrode potentials of SS 316 and Ti Gr.2 are very close and there is no danger of galvanic corrosion.

(b) In the case of the use of Admiralty brass tubes, the same material had to be applied as clad or solid tube-sheets from the processing side (similar to cooling water side).

5.1.2.5 Corrosion of Check Valves

Conditions. Two valves functioned in the overhead vacuum system. Hydrocarbon gases, H_2S and water vapor flowed inside of two valves at 50 °C. Neutralizer was not always injected, and corrosion inhibitor was not applied at all.

Material of construction: Carbon steel ASTM A216 WCB.

Service period before the failure: 1 year.

Visual examination and findings. Two holes (12 and 2 mm in diameter) were found in the body of two valves (Fig. 5.5). Black iron sulfide corrosion products were found inside on the surface of moving disc, bolt, pin and other accessories of two valves. The pH values of water after condensation changed from 3.8 to 7.8, but most time were about 4.

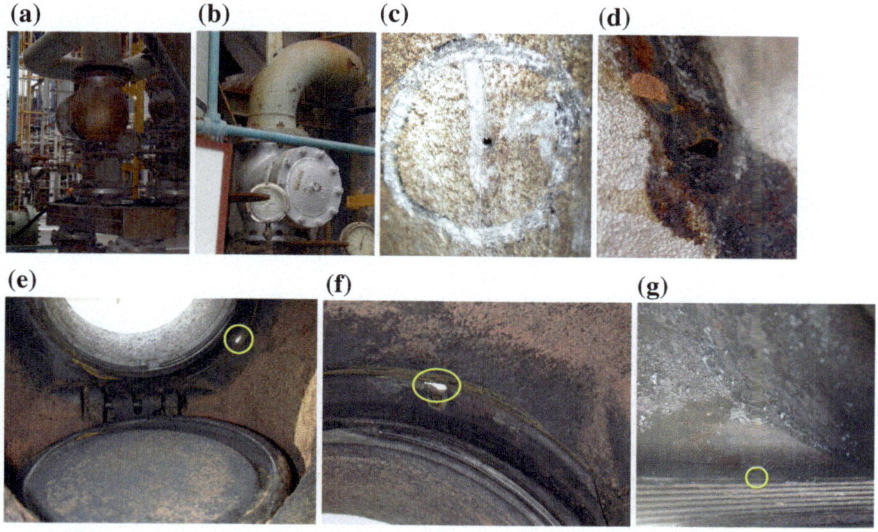

Fig. 5.5 **a**, **b** Locations of two failed valves at the vacuum unit. **c**, **d** Corrosion holes on the valve body (outside view). **e**–**g** Inside view of corroded internals with corrosion holes

Failure phenomenon: *Low-temperature sulfidic corrosion and wear*.

Cause of failure and its explanation. Dissolved H_2S in water condensate formed sour water with pH $\sim$ 4. Inner surface of carbon steel valves and different parts of valves were attacked by H_2S. When dimensions of moving disk diminished as a result of corrosion, contact of this disk with the body of the valve was poor, namely, there was no tightness. Free movement of the disk caused wear of inner surface of the body of the valve and the wall thickness of the body eroded from 12.5 mm to zero (until the formation of holes).

Solutions and recommendations.

(a) To perform correct injection of the neutralizer and corrosion inhibitor with required concentrations in the overhead of distillation vacuum system.
(b) To replace carbon steel of the valve and internals with low-alloy steel (5 % Cr) which is resistant to H_2S.

5.2 Hydrodesulfurizer (HDS)

The main corrosive substance at HDS units is H_2S (see Chap. 1). In addition to its corrosiveness, H_2S can be oxidized to form SO_2, then SO_3 and in the presence of water vapor sulfurous and sulfuric acids can be formed (see Appendix D). Hydrogen attack and ammonium bisulfide also represent corrosion risk at HDS units. Similar to corrosion problems at other units, wrong design, poor operating practices, and solution contamination are the main reasons of corrosion at HDS units.

5.2.1 Case Studies

Six cases are described and analyzed in this section.

5.2.1.1 Corrosion of Trays in Stabilizer (HDS Gasoline)

Conditions. The temperature in the overhead of the stabilizer was 60–80 °C. Gasoline entering to the stabilizer contained 400 ppm H_2S and 200 ppm H_2S in the exit. Thus, in addition to the function of the stabilizer, it functioned as a stripper. The stabilizer was in service 20 years. The feed was naphtha during the first 12 years of service and no corrosion problems were noted. The feed was replaced with the FCC gasoline during the last 8 years. There were a lot of shutdowns during this period, and water wash (with boiler feed water) was used to remove deposits in the tower.

Material of construction. Carbon steel.

Service period before the failure: 8 years.

Visual examination and findings. Severe corrosion of trays with rust formation occurred in the overhead of the stabilizer (Fig. 5.6). The thickness of trays was reduced from original 3.5 to 0–0.5 mm. Sulfates were found in corrosion products.

Failure phenomenon: *General corrosion.*

Cause of failure and its explanation. Dissolved oxygen in hydrocarbon stream caused oxidation of H_2S and formation of sulfurous/sulfuric acids in the presence of wash water. Two acids caused corrosion of carbon steel trays in the tower at 60–80 °C. It is important to emphasize that water is also mostly corrosive to carbon steel at these temperatures.

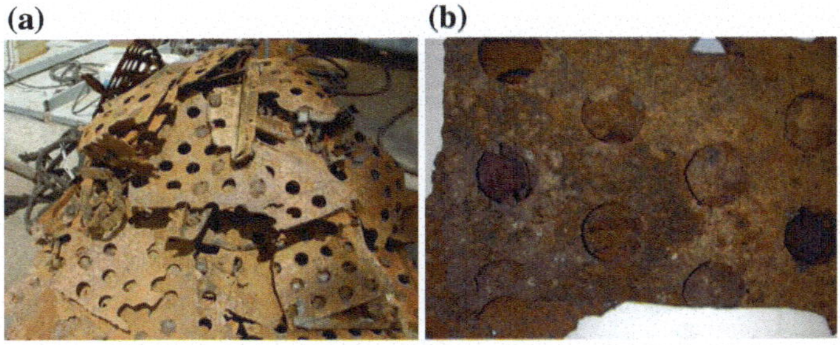

Fig. 5.6 **a** General view of corroded trays. **b** Magnification of the picture **a**

Solutions and recommendations.

(a) To replace the material of trays from carbon steel with austenitic SS Alloy 20Mo-6 or Alloy 20Cb-3.
(b) It is not recommended to use wash water in this case because sulfurous/sulfuric acids can be formed.

5.2.1.2 Corrosion of Air Cooler (SRU)

Conditions. The air cooler is intended for cooling quench water which exits from the quench tower. Quench water containing H_2S and N_2 entered into the air cooler at 80 °C and flowed out at 50 °C.

Material of construction: carbon steel A516 Gr.70.

Service period before the failure: 4 months.

Visual examination and findings. The hole was formed twice between two nozzles in the center of the header. Carbon steel patch (12 mm in thickness) was put the first time, but the hole (4 cm diameter) was formed once more after 4 months exactly in the center of the patch (Fig. 5.7). The corrosion rate is 36 mm/year. Analysis of PI data showed that during two last months before the failure pH of quench water ranged in 2.2–3.0. Black deposits inside the air cooler and blockage of pipes were found. The deposits consisted of corrosion products iron sulfides and sulfates.

Failure phenomenon: *Acid corrosion*, *erosion* (by turbulent flow).

Cause of failure and its explanation. The presence of iron sulfides in deposits show that H_2S dissolved in water caused severe corrosion of carbon steel. However, only the presence of H_2S in the water could not lead to such high corrosion rate (36 mm/year). Decrease of pH values to 2.2–3 indicates that the gas SO_2 which was not reduced to H_2S in the reactor resulted in the formation of two acids, H_2SO_3 and H_2SO_4. These two acids at 80 °C caused severe corrosion of carbon steel. The corrosion rate of 36 mm/year could be only in the presence of these acids at high temperature (80 °C), or when erosion (turbulent flow of acidic solution) occurred. Corrosion products with the stream entered the air cooler through nozzles and

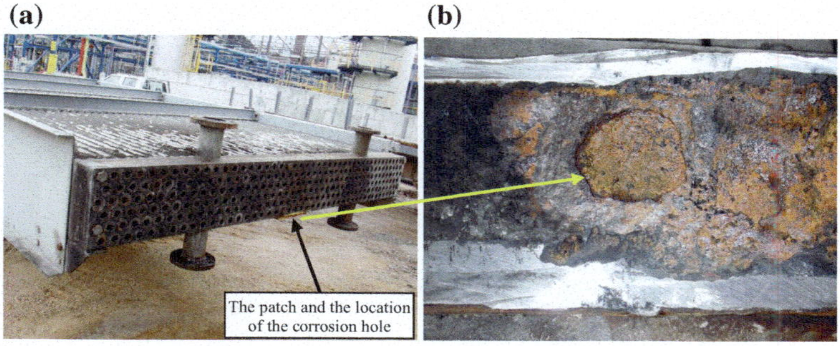

Fig. 5.7 **a** General view of the air cooler. **b** The hole and the patch

Table 5.1 Corrosion rate of carbon steel in H_2SO_4 aqueous solution (95 °C; 4 h; laboratory conditions)

H_2SO_4 (wt%)	pH	Corrosion rate (mm/year)
0.05	2.0	36
0.1	1.7	54
0.2	1.4	117
0.5	1.0	258

clogged pipes. Pipes at the ends of the header were clogged in front of the inlet of quench water (two nozzles). Turbulent flow of acidic aqueous solution occurred in the center of the header which led to accelerated corrosion. The proof of such corrosion mechanism is corrosion rates of carbon steel by H_2SO_4 determined in laboratory conditions: 36–258 mm/year (Table 5.1).

Solutions and recommendations.

(a) Careful water wash of the inner surface of the air cooler and pipes from deposits.
(b) Installation of drainage valve in the center of the header box in order to prevent accumulation of acidic solution.

5.2.1.3 Corrosion of Heat Exchanger (SRU)

Conditions. Heat exchanger is intended for chilling sour water by cooling water at the SRU. Sour water flowed inside the tubes at ~40 °C. Cooling water flowed in the shell at 30–35 °C and 4 bar.

Material of construction. Tubes—SS 304L. Tube-sheet—carbon steel A105 cladded with SS316L (10 mm thickness). Shell—carbon steel A516 Gr.70 cladded with SS316L (3 mm thickness).

Service period before the failure: 3 years.

Visual examination and findings. Corrosion in following locations of heat exchanger was found (Fig. 5.8):

(a) Localized corrosion of tube-sheet close to clad. The morphology of deteriorated area is typical to acid attack (see Fig. 5.8c, d).
(b) Severe corrosion of accessories of the channel from the sour water side (see Fig. 5.8e–g).
(c) Severe corrosion in locations of cooling water inlet into the shell (see Fig. 5.8h).

Corrosion products (iron sulfides) were found on tube-sheets surface from the sour water side.

Failure phenomena:

(a) *Galvanic corrosion* (part of the tube-sheet in contact with SS 316L clad).
(b) *General corrosion* (inner surface of the channel).

Cause of failure and its explanation. The pH value of sour water should be about 8, but according to PI data, pH dropped to 0.5–3 during several months. It is clear

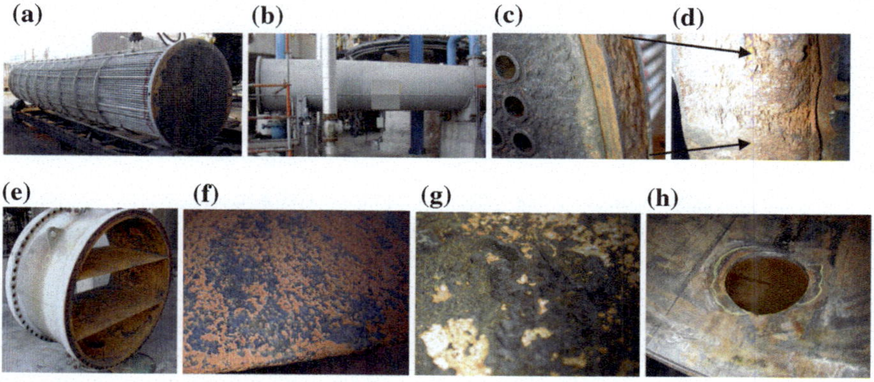

Fig. 5.8 **a** General view of heat exchanger bundle (HDS unit). **b** The shell. **c** Corroded tube-sheet contacted clad SS 316L. **d** Magnification of **c** picture. **e–g** Corroded inner surface of the channel (sour water). **h** Corroded inner surface of the shell inlet (cooling water)

that pH of sour water containing only weak acid H_2S cannot cause such low pH values. When there is no correct ratio H_2S/SO_2 in the HDS unit exit, acid gas SO_2 can penetrate through the reactor TGT (Tail Gas Treatment) (see Chap. 1) and dissolve in sour water giving rise to a sharp decrease of pH (to 0.5–3) because of formation of H_2SO_3 and H_2SO_4 acids. Usually such situation happens when HDS is shutdown or put into operation, or acid gases enter the system. Under these unstable conditions, the air control system could not react to abrupt changes of H_2S concentrations. Acidic sour water was corrosive to carbon steel and reacted with inner surface of the channel. This acidic sour water penetrated to the area of contact of carbon steel and SS 316L clad in tube-sheet and resulted in galvanic corrosion of the former. Corrosion in the area of inlet of cooling water into the shell occurred because of defects in the clad.

Solutions and recommendations.

(a) To replace sour water immediately in the case of pH drop in acidic region ($pH < 7$).
(b) To maintain processing parameters: to ensure correct ratio H_2S/SO_2 at the HDS unit exit.
(c) This heat exchanger functions after the air cooler and reduces sour water temperature on 5 °C. Therefore, it is recommended to put it in bypass and introduce it into the service only with the increasing of load or decreasing in performing of the air cooler.
(d) To carry out the tube-sheet from one piece of SS 316L and not to make clad on carbon steel.
(e) In order to protect the channel from corrosion by sour water, to install sacrificial aluminum/zinc anodes and coat inner surface with epoxy coating resistant to acidic solutions.

5.2.1.4 Corrosion of Heat Transfer Oil Heaters (SRU)

Conditions. Mixture of acid gases H_2S and SO_2 (amount ratio H_2S:SO_2 = 2:1), water vapor and N_2 flowed inside the tubes. Temperature of gases changed from 150 to 195 °C and even to 260 °C. Heat transfer oil flowed outside the tubes at 300 °C.

Material of construction. Tubes—carbon steel ASTM SA 214; wall thickness—2.77 mm. Shell—carbon steel ASTM A516 Gr.70.

Service period before the failure: 2 years.

Visual examination and findings. Severe corrosion of three heat transfer oil heaters from acid gases side in tubes was found (Fig. 5.8). Pitting corrosion was detected in welds. The wall thickness of the tubes and tube-sheets decreased significantly. Large amounts of yellow-black deposits consisted of iron, sulfur, and oxygen were found inside the tubes. After dissolution of these deposits in distilled water, pH = 3.2. Sulfates were detected in large quantities in this solution. Corrosion of outer surfaces of heat transfer oil heaters under thermal insulation was found too (Fig. 5.9).

Failure phenomenon: *Acid dew point corrosion* (inner surface of tubes and tube-sheets); *corrosion under thermal insulation* (outer surface of heat transfer oil heater).

Cause of failure and its explanation. Sulfur oxides (SO_2, SO_3) in the presence of water vapor condensed at 150–180 °C and two acids H_2SO_3 and H_2SO_4 were formed. These two acids caused severe corrosion of tube-sheets and inner surfaces of tubes made from carbon steel.

Solutions and recommendations.

(a) To replace carbon steel tubes with HASTELLOY C 276 or duplex steel FERRALIUM 255
(b) To coat outer surface of heat transfer oil heaters in order to prevent corrosion under thermal insulation.

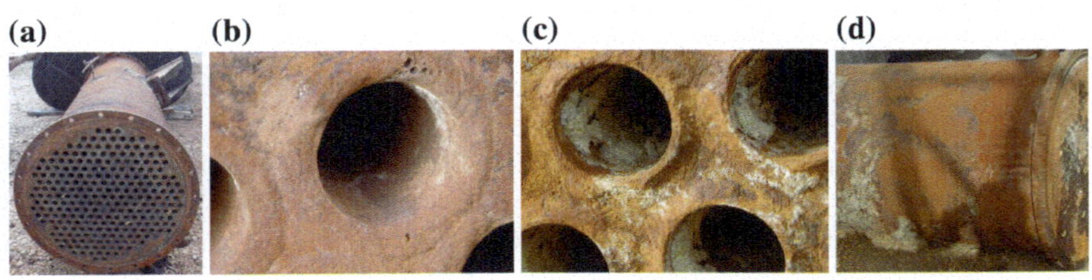

Fig. 5.9 **a** General view of failed heat transfer oil heater. **b** General and pitting corrosion of the tube-sheet. **c** Severe corrosion of the tube-sheet and tubes (deposits are accumulated inside the tubes; the ends of the tubes are out from corroded tube-sheet). **d** Corrosion of outer surface of heat transfer oil heater under thermal insulation

5.2.1.5 Corrosion of Heat Exchanger (SRU)

Conditions. Remains of gases N_2, H_2S, SO_2, vapors of water and sulfur following the Claus process flew inside the tubes. The temperatures of gases were 130 °C at the inlet and 290 °C at the exit respectively. Heat transfer oil flowed outside the tubes at 310–320 °C and 5 bar.

Material of construction. Tubes—carbon steel ASTM SA 214; thickness of tubes —2.77 mm. Shell—carbon steel ASTM A 516 Gr.70.

Service period before the failure: 2 years.

Visual examination and findings. Severe corrosion of the tube-sheet surface (especially welds) and tubes' ends was found at the gas inlet side (Fig. 5.10). The area of the heat exchanger corroded very slightly in the location of gases exit.

PI data showed that temperature often decreased to less than 100 °C because of frequent outages (Fig. 5.11). Nitrogen would have to be circulated during such outages but this circulation was not always realized.

Failure phenomenon: *Acid dew point corrosion.*

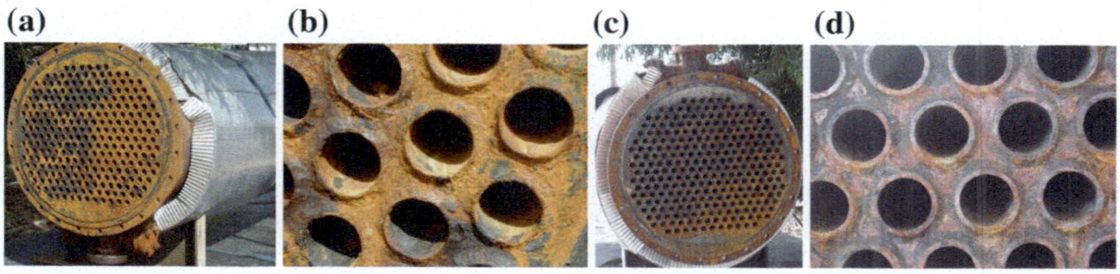

Fig. 5.10 Failed heat exchanger. **a**, **b** Inlet of gases (severe corroded area). **c**, **d** Outlet of gases (slightly corroded area)

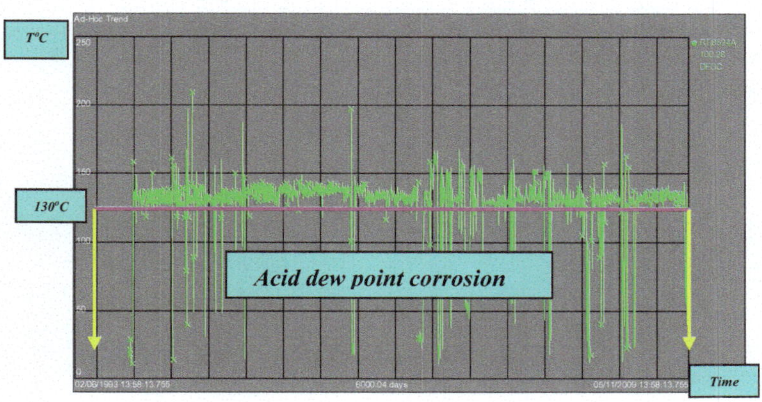

Fig. 5.11 Temperature fluctuations (*green*) of incoming gases to the heat exchanger versus time (PI data)

Cause of failure and its explanation. The findings point out the occurrence of acid dew point corrosion. When temperature drops under 130 °C (acid dew point for H_2SO_3), SO_2 reacts with H_2O and liquid inorganic acid H_2SO_3 is formed. In fact, the dew point temperature when aqueous acidic solution is formed depends on concentrations of SO_2 and H_2O in vapor phase. The gas SO_2 can be oxidized to SO_3, and this process usually occurs in the presence of catalyst which diminishes the activation energy of the oxidation. If SO_3 is formed, it can react with H_2O, and strong acid H_2SO_4 is produced, even at T = 130–180 °C. When the heat exchanger is stopped, the gas temperature drops significantly below the dew point (130–180 °C) for two liquid inorganic acids H_2SO_3 and H_2SO_4, and they are formed. These acids attacked carbon steel tube-sheet and tubes. Few hours were enough for acid dew point corrosion to occur.

Solution and recommendation.

To maintain high temperature (above the dew point, T > 180 °C) in order to prevent formation of liquid H_2SO_3 and H_2SO_4 acids.

5.2.1.6 Corrosion of the Tube for Gas Transportation (HDS Gas Oil)

Conditions. The tube is intended for transporting of gaseous mixture at ambient temperature for burning at HDS gas oil unit. Gaseous mixture contained H_2 (39.7 vol%), CH_4 (16.3 vol%), C_2H_6 (18.7 vol%), C_3H_8 (14.7 vol%), C_4H_{10} (7.4 vol%), and C_5H_{12} (1.3 vol%).

Material of construction: SS 304. Diameter of the tube—12.5 mm.

Service period before the failure: 1 year.

Visual examination and findings. Holes formed from the inner surface, and corrosion products were found in the tube (Fig. 5.12).

ED-XRF examination. Corrosion products accumulated inside the tube contained (wt%): Fe (68.6), Cr (11.4), Ni (8.7), S (7.6), and Cl (3).

Chemical analytical examination of gaseous mixture. In addition to hydrocarbon gases and H_2, chlorides and water vapor were detected in the mixture of gases transporting inside the tube.

Failure phenomenon: *Pitting corrosion.*

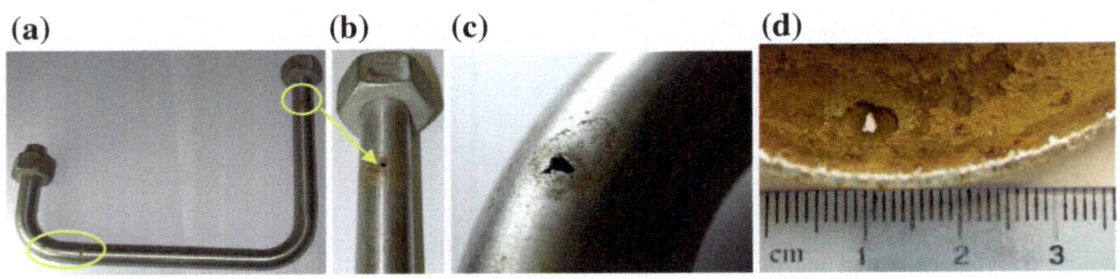

Fig. 5.12 **a** The tube for gas transportation with three corrosion holes. **b**, **c** Magnification of the picture **a**. **d** The view on the corrosion hole inside the tube

Cause of failure and its explanation. Chlorides and water vapor in gaseous hydrogen-hydrocarbon mixture caused pitting corrosion of inner surface of stainless steel tube. Hydrogen chloride was formed as a result of hydrolysis of organochlorine compounds (see Appendix C).

Solutions and recommendations.

(a) To prevent penetration of chlorides and water vapor into the mixture of gases.
(b) In the case of impossibility of prevention of penetration of aggressive corrosive impurities into gaseous mixture, it is recommended to replace the material of the pipe (SS 304) with AL-6XN, or duplex 2205, or SS 317LM.

5.3 Visbreaker

In the visbreaking process, the feedstock (residue from vacuum distillation unit) is heated to 430–510 °C and mildly cracked in a furnace [8]. S-organic compounds containing in residue are destructed to H_2S at these temperatures. High molecular weight organic acids can be destructed to lighter organic acids. Then after the furnace, the cracking products containing H_2S and organic acids (naphthenic acids among them) are sent to a fractionation column. LPG, gasoline, gas oil and bitumen are separated in this column. Following corrosion problems exist in a visbreaker unit: NAC; high- and low-temperature corrosion by sulfur compounds. NAC of fractionation column walls made from carbon steel can occur with the rate 9.6 mm/year [9, 10]. LPG and gasoline exiting from the column usually contain H_2S and water vapor. They are cooled to 60–100 °C and severe low-temperature corrosion by wetted H_2S can occur during 3–4 years. Two cases are analyzed in this section.

5.3.1 *Case Studies: Corrosion of Heat Exchangers*

5.3.1.1 Cooling Gases

Conditions. Heat exchanger is intended for cooling gases containing LPG, H_2S and water vapor flowing from the stabilizer. Sometimes the stabilizer serves as a splitter. Temperature of gases exited from the stabilizer was 60–70 °C; from the splitter was 85–100 °C. Sixty tubes leaked during the third year of service and they were clogged up (Fig. 5.13a). Cooling water flowed inside the tubes at 30–35 °C and 4–5 bar.

Materials of construction. Tubes—Admiralty brass CDA 443. Baffles—carbon steel.

Service period before the failure: 2 years before the first failure and then tube bundle was replaced (without failure analysis) and served 3 years before the second similar failure.

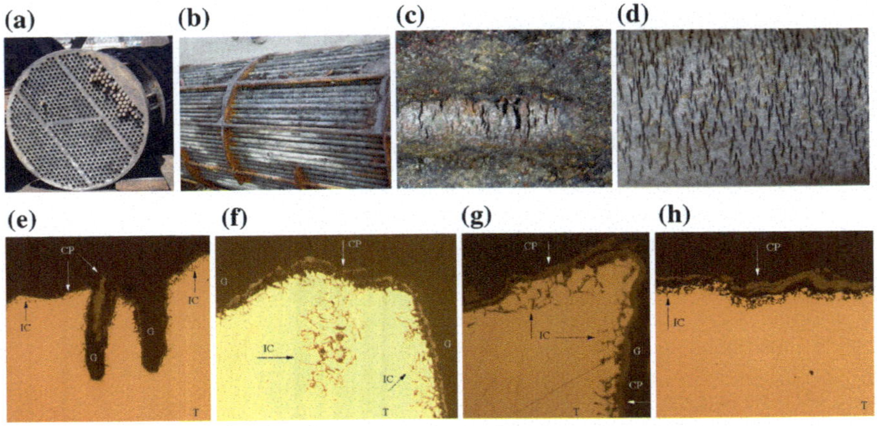

Fig. 5.13 **a** The tube bundle with 60 clogged tubes. **b** Corroded Admiralty brass tubes (with deposits) and corroded carbon steel baffles. **c**, **d** Cracks, holes and grooves on tube surface. **e–g** Metallographic cross-section of grooves with different magnification (**e**: ×100, **f**: ×200, **g**: ×500): IC—Intergranular corrosion (corrosive attack on grain boundaries), CP—corrosion products, G—grooves, T—tube. **h** Metallographic cross-section near the grooves (×100)

Visual examination and findings. Scale and deposits were found outside the tubes (Fig. 5.13b, c). Baffles were severely corroded. Holes, cracks and grooves were detected on outer surface of tubes close to impingement plate. Scale consisted of black copper sulfide. Deposits consisted of copper sulfide and iron sulfide. Metallographic examination. Corrosive attack of grain boundaries was found (see Fig. 5.13e–h).

SEM examination. Following elements were detected inside grooves: Cu, Zn, O, S, and Cl.

Failure phenomenon: *General and intergranular corrosion.*

Cause of failure and its explanation. Hydrogen sulfide and water vapor containing in hydrocarbon gases and flowing outside the Admiralty brass tubes caused severe corrosion. Copper alloys are not resistant to hydrogen sulfide. Localized corrosion at the grain boundaries caused formation of grooves. Carbon steel baffles also corroded by hydrogen sulfide and water vapor which condensed to liquid phase at T < 100 °C.

Solution and recommendations.

To replace the material of tubes and baffles with Ti Gr.2. Tube-sheets from both sides contacting process stream (LPG) and cooling water also must be replaced with Ti Gr.2.

5.3.1.2 Light Naphtha

Conditions. Heat exchanger is intended for cooling of light naphtha which exited at 85 °C from the splitter. Temperature of naphtha was ~100 °C and more. Cooling water flew inside tubes at 30–35 °C and 4–5 bar.

Materials of construction. Tubes—Admiralty brass CDA 443. Shell and channel—carbon steel. Inner surface of carbon steel channel was coated by epoxy paint.

Service period before the failure: 4 years.

Visual examination and findings. Deposits on outer surface of tubes and rust at inlet of naphtha into the heat exchanger were found (Fig. 5.14). Deposits consisted of copper sulfides and sulfates. Total destruction and detaching of epoxy coating and rust were detected in the channel (cooling water side).

Failure phenomenon: *Low-temperature sulfidic corrosion and destruction of epoxy coating.*

Cause of failure and its explanation. Chemical content of deposits (copper sulfides and sulfates) showed that naphtha contained H_2S and H_2O and they gave rise to harsh corrosion of outer surface of Admiralty brass tubes. Copper alloys are not resistant to H_2S.

High temperature (~100 °C and more) caused destruction of epoxy coating and severe corrosion of the surface in the channel contacted cooling water.

Solution and recommendations.

To replace Admiralty brass tubes with Ti Gr.2. Tube-sheets and channels also must be made from Ti Gr.2 or cladded with this alloy.

5.4 Petrochemical Plant

Petrochemical plant consists of units resembling those at oil refinery. And even many media and environment are similar. Raw original organic material comes from refinery. Reformate producing at the CCR and FCC units is fed to petrochemical plant to produce aromatics. These specific organic solvents can dissolve water and oxygen. For instance, the solubility of oxygen in benzene is 70 ppm and in water is 8.5 ppm at 25 °C [2]. Moreover, if the solubility of oxygen in water is

(a) (b) (c) (d)

Fig. 5.14 **a** Corrosion deposits on outer surface of Admiralty brass tubes (naphtha inlet). **b–d** Destruction of epoxy coating and rust (cooling water side). **d** Cooling water inlet

decreased with increasing temperature, that in benzene is increased. Chlorides also can be main corrosive contaminant in reformate coming from the oil refinery. Sulfolane widely used for extraction aromatics can degrade into acidic corrosive products (see Chap. 1). Boiler feed water and cooling water are also vital media at petrochemical units. Typical and specific corrosion problems and their solutions at these units are analyzed in this section.

5.4.1 Case Studies

5.4.1.1 Corrosion of Fin-Fan Air Cooler Tubes at the Overhead of Benzene Column

Conditions. Liquid benzene flowed inside the tubes of fin-fan air cooler at 70–90 °C and atmospheric pressure at BTX unit. The benzene column received feedstock from two sources: (1) Benzene from the stabilizer. This benzene flowed through the Clay treater for removal of olefins; (2) The aromatic mixture benzene, toluene and xylene (BTX) from the extraction unit. After the extraction, aromatic hydrocarbon mixture passed through the column with Clay treater (active soil) for removal olefins and then entered into the benzene tower.

A leak of hydrocarbons was detected from the air cooler in the overhead of the benzene tower.

Material of construction. Carbon steel ASTM A179. The wall thickness of tubes —2.336 mm.

Service period before the failure: 4 years.

Visual examination and findings. Severe corrosion with black-red-brown corrosion products was found inside the air cooler tubes (Fig. 5.15). The wall thickness of tubes decreased from original 2.336 mm to nil. Corrosion rate was 0.58 mm/year, that is ~6 times greater than allowable value 0.11 mm/year.

Chemical analytical examination showed that corrosion products consist of iron sulfides, iron hydroxides and iron oxides. Sulfates were detected in corrosion products. Chlorides were absent. Two samples of benzene feed to the tower showed

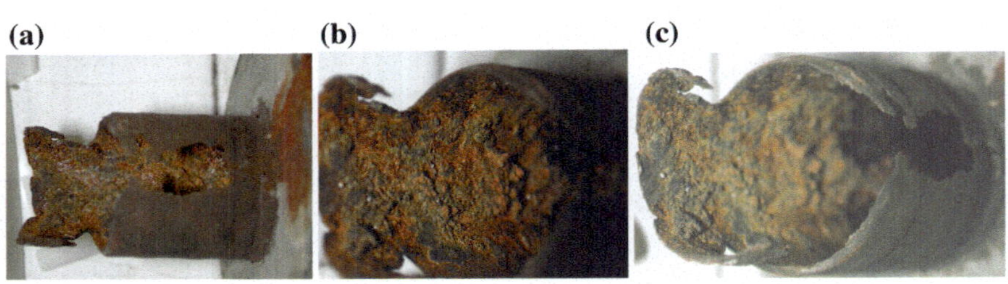

Fig. 5.15 a–c Failed corroded tube in the air cooler at the overhead of benzene tower, 4 years of service (BTX unit, petrochemical plant)

that the benzene bottom from the stabilizer was clear, and the benzene from the Clay treater was very turbid.

Failure phenomenon: *Acid corrosion.*

Cause of failure and its explanation. Corrosion occurred on the inner surface of the air cooler tubes due to the presence of water, dissolved oxygen and acids in benzene at temperature 70–90 °C in the exit from the benzene tower. The presence of sulfates in corrosion products indicates that sulfuric acid also contributed to corrosion.

The feed after the Clay treater brought the corrosion problems, as acids were present in original Clay. According to the supplier of the Clay, its activation occurred by means of sulfuric acid (H_2SO_4) and then was washed with hot water to remove remains of sulfuric acid. When new original Clay intended for use at the unit was washed with distilled water at 20 °C in the laboratory, it was determined pH = 2.9 and above 3000 ppm sulfates in washing water. Washing at 80 °C resulted in pH = 2.6. Thus, water dissolved in benzene swept (washed out) the remains of sulfuric acid from the Clay and then caused severe corrosion of inner surface of carbon steel tubes in the air cooler.

Solution and recommendation.

To require from the supplier of the Clay to wash well from the remains of sulfuric acid.

5.4.1.2 Corrosion of Fin-Fan Air Cooler Tubes at the Overhead of Aromatic Column

Conditions. An air cooler at the overhead of aromatic tower is intended for cooling the mixture of benzene, toluene, hexane, heptanes and pentane by air. The mixture of three last hydrocarbons (feedstock) flew from the CCR unit. The mixture of aromatic hydrocarbons exited from the aromatic tower and flowed inside the tubes at 60–80 °C and vacuum 0.2–0.3 bar. A leak of hydrocarbons was identified from the air cooler.

Material of construction. Carbon steel ASTM A179. The wall thickness of tubes —2.336 mm. Aluminum fins were installed on carbon steel tubes.

Service period before the failure: 11 years.

Visual examination and findings. Severe corrosion was found inside the tubes and aluminum fins (Fig. 5.16). Rust was formed inside the tubes. The wall thickness of carbon steel tubes decreased from original 2.336 mm to zero and tubes were ruptured.

ED-XRF examination. Rust formed inside tubes contained (wt%) Fe (57.5), Cl (21.1), Al (9.9), Si (7.8), and S (2.7).

Failure phenomenon: *General (uniform) corrosion.*

Cause of failure and its explanation. Presence of chlorides and water in the mixture of hydrocarbons caused severe corrosion of inner surface of carbon steel tubes in air cooler. Chlorides when hydrolyzing gave rise to formation of hydrochloric acid which attacked inner surface of carbon steel tubes. Then, when

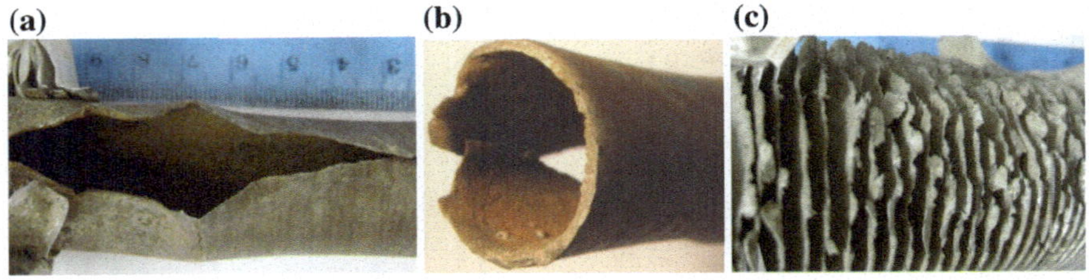

Fig. 5.16 **a**, **b** Ruptured carbon steel tube in the air cooler at the overhead of aromatic tower. **c** Corroded aluminum fins

holes were formed in tubes, corrosive acid solution escaped and corroded aluminum fins.

Solutions and recommendations.

(a) To install chloride absorber in the CCR unit.
(b) To replace carbon steel tubes with Ti Gr.2 which is resistant to chlorides and sulfur compounds.

5.4.1.3 Corrosion of Reflux Pipe of Pre-fractionator (BTX Unit)

Conditions. Hydrocarbons C_5 flowed in reflux pipe at 60–70 °C and 0.5 bar. The leak of hydrocarbons was detected through the hole. The section of the pipe and pump have been replaced due to corrosion-erosion 4 years before the last failure.

Material of construction. Carbon steel. Diameter—150 mm.

Service period before the failure: 2 months.

Visual examination and findings. The hole and rust tubercles were found on the inner surface of the pipe (Fig. 5.17). Deep shallow pits were detected after removing the rust tubercles. When rust tubercles were partly dissolved in distilled water, the pH was 3.6, and sulfate and chloride ions were detected in aqueous solution.

ED-XRF examination. Rust contained (wt%) Fe (97), S (2), O (0.7), and Cl (0.3).

Failure phenomenon: *Under deposit corrosion.*

Cause of failure and its explanation. Dissolved water, chlorides (contaminant incoming with reformate from the CCR unit) and corrosive compounds formed as a result of sulfolane destruction containing in hydrocarbon stream at the BTX unit caused corrosion of inner surface of carbon steel pipe. Uneven layers of corrosion products (rust containing chlorides and sulfates) were formed inside the pipe. These acidic products (sulfate sand chlorides) resulted in the decrease of pH to 3.5 and localized under deposit corrosion.

Solutions and recommendations.

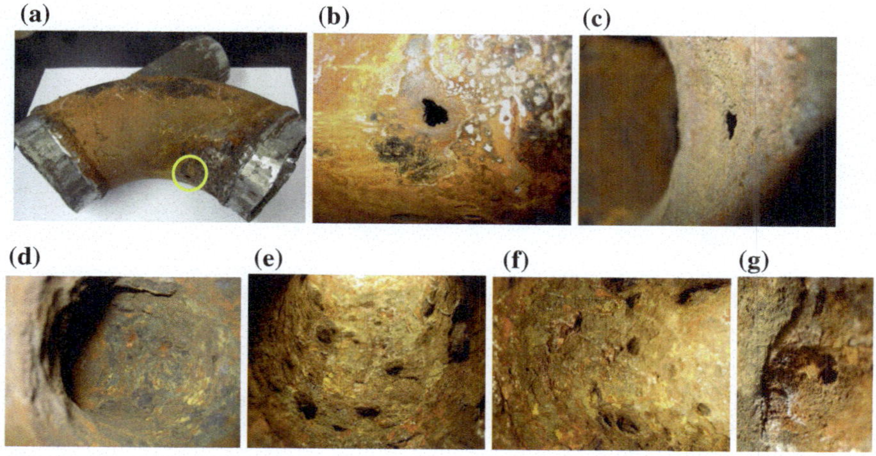

Fig. 5.17 **a** General view of the pipe section with the hole. **b** A view of the hole from outside. **c** A view of the hole from inside. **d–f** Inside corroded surface with rust tubercles, **g** a pit under the tubercle

(a) To absorb chlorides in reformate incoming from the CCR unit.
(b) To inject corrosion inhibitor in sulfolane stream at the BTX unit in order to prevent corrosiveness of destructive compounds.

5.4.1.4 Failure of Sulfolane Feed Pipe to the Recovery Tower

Conditions. Liquid mixture of sulfolane (75 wt%) and BTX (25 wt%) flowed in the pipe at 160 °C and 2 bar. The sulfolane contained 3 wt% H_2O. Corrosion inhibitor MEA was not injected into the sulfolane at the extractor bottom (as it had to be). The pressure in the elbow was 2 bar before the controller, and the vacuum was after it. Most corrosion problems started after the implementation of CCR unit which supplied reformate containing traces of chlorides. The catcher (scavenger) of chlorides at the CCR unit has not been functioned for the first 2 years. The carbon steel pump transporting liquid mixture of sulfolane and BTX failed after a year of the service because of erosion, and the material of the pump was replaced with SS 304. Leak was detected in the pipe elbow going from the stripper to the recovery tower. The elbow was at the distance 20 m from the previously failed pump.

Material of construction. Elbow—carbon steel. Diameter—200 mm.

Service period before the failure: 3 years.

Visual examination and findings. Three holes were found in the elbow right away after the welding cone at the distance 2 m from the inlet to the recovery tower after the pump before the level controller (Fig. 5.18). The thickness of the elbow wall significantly decreased from 6.2 mm (original) to 3.6 mm. Hard thick deposits

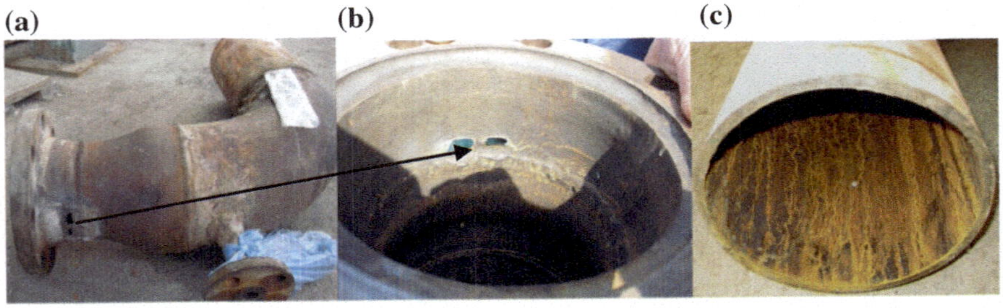

Fig. 5.18 **a** Failed elbow with corrosion holes. **b** The holes at the cone end before the flange. **c** Inlet of liquid mixture of sulfolane and BTX into the elbow

were found on the bottom of the recovery tower. Chlorides were always detected in corrosion products during the period of service before the failure (3 years).

ED-XRF analysis. Deposits in the bottom of recovery tower contained (wt%): Fe (50), S(30) and Cl (20).

Failure phenomenon: *Erosion-corrosion.*

Cause of failure and its explanation. The flow velocity was changing drastically in the area of the cone right away after the welding because the diameter 200 mm of the elbow was decreasing to 150 mm of the flange. The obstacle in the form of welding protrusions also contributed in the changes of flow regime immediately after the weld. Thus, turbulent flow and severe velocity change of liquid mixture of sulfolane and BTX containing corrosive components (chlorides and acid destructive compounds of sulfolane) caused erosion-corrosion. Morphology of failed inner surface around the holes confirms erosion attack.

Solutions and recommendations.

(a) To purify liquid sulfolane from corrosive destructive components.
(b) To remove obstacles (welding protrusions) from the inner side of welds.
(c) To produce elbow and flange of similar diameter (to exclude cone).

5.4.1.5 Corrosion of the Reducer of the Pipeline to the Recovery Tower (BTX Unit)

Conditions. Liquid mixture of sulfolane, benzene and toluene flowed inside the pipe at 150 °C into the recovery tower. The pipe was 152 mm in diameter, and original wall thickness was 8 mm. The pressure controller before the failed section (the reducer) showed the pressure 2 bar. The vacuum inside the tower was 0.5 bar.

Material of construction: Carbon steel.

Service period before the failure: 15 years.

Visual examination and findings. The hole and pits were found in the feed pipeline (the reducer) close to the recovery tower. Welds inside were mostly corroded. The wall thickness of the reducer was diminished from original 8 to 1.5 mm (Fig. 5.19).

(a) (b) (c)

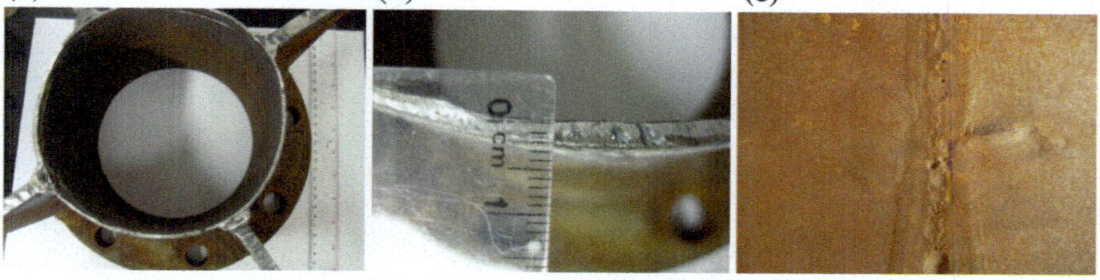

Fig. 5.19 **a** Failed reducer. **b** Corroded reducer wall (from original 8 to 1.5 mm). **c** Corroded weld inside with pits

Failure phenomenon: *Erosion-corrosion.*

Cause of failure and its explanation. Decrease in pressure from the controller to the tower (from 2 to 0.5 bar) caused flashing of liquid and formation of two-phase system liquid + vapor in the reducer. The two-phase mixture during flowing gave rise to mechanical damage of the inner surface of the reducer. In addition, when sulfolane has been used for a long time it is destructed into substances which are corrosive to carbon steel, especially in heterogeneous areas of welds.

Solutions and recommendations.

(a) To increase the diameter of the reducer from 152 to 304 mm in order to diminish the flow velocity.
(b) To decrease corrosiveness of sulfolane by injection of inhibitors.

5.4.1.6 Corrosion of Steam Generator Tubes and Shell

Conditions. Water from three sources entered into the deaerator and went out at 120–170 °C and 5 bar. Oxygen scavenger was injected into the water after the deaerator. This water entered into the steam generator at 90 °C and 5 bar. There was no dissolved oxygen meter in the water line during the first 4 years of service. Therefore, water quality was not known during this period. Oxygen meter installed after 4 years registered high concentrations of dissolved oxygen in deaerated water, up to 200 ppb, when allowable maximum concentration would have to be not more than 30 ppb (Fig. 5.20d).

Service period before the failure: 9 years for the tubes; 5 years for the shell.

Material of construction. U-tubes—carbon steel ASTM A179. Shell—carbon steel ASTM A106 Grade B.

Visual examination and findings.

U-tubes. Deep pits (~1 mm) were found on outer surface of the tubes (see Fig. 5.20b, d). The pits were formed only in the place of water inlet and at the end near the tube-sheet. Uneven red corrosion products were found on the surface of the tubes at the location of water inlet. Physicochemical analysis showed that red

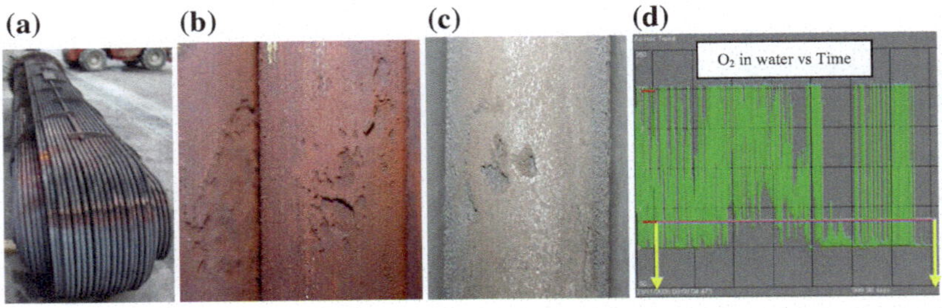

Fig. 5.20 **a** Failed U-tube bundle of the steam generator. **b**, **c** Pits (1 mm deepness) on outer surface of U-tubes. **b** Rust before cleaning. **c** Outer surface after cleaning. **d** Changes of concentration of dissolved oxygen in water with time (PI data). Allowable maximum concentration is 30 ppb

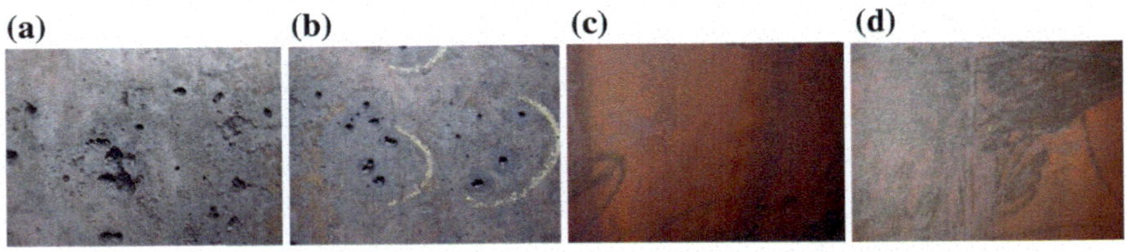

Fig. 5.21 **a**, **b** Pits on the *bottom* of the steam generator shell. **c**, **d** Rust on the inner surface of the steam generator shell wall

corrosion products consist of hematite (Fe_2O_3) and hydroxides (FeOOH). The water level signs were observed on the tube bundle.

Shell (inner surface). Deep, distinct, almost semispherical pits were found at the bottom of inner surface of the shell at the location of water inlet (Fig. 5.21). Red-brown rust was formed on inner surface of the steam generator walls.

Failure phenomenon: *Oxygen corrosion (pitting corrosion* by *dissolved oxygen*).

Cause of failure and its explanation. Red-brown rust (iron oxides and hydroxides) manifested that dissolved oxygen in concentrations significantly greater (6–7 times) than maximum allowable value (30 ppb) was present in deaerated water. Dissolved oxygen resulted in pitting corrosion both of outer surface of tubes and at the bottom in the steam generator shell (in the location of water stagnation). In addition, two-phase system liquid water-steam was present in the area of the water level which caused boiling of water on the tube surface and thus accelerated corrosion.

Solutions and recommendations.

(a) To keep the concentration of dissolved oxygen in deaerated water less than 30 ppb. Possible causes of excessive concentrations of dissolved oxygen are a malfunctioning deaerator, improper feed of oxygen scavengers, or air in-leakage (through valves, flanges, pumps, seals, and expansion joints) [11].

(b) To carry out continuous monitoring of dissolved oxygen concentration in deaerated water.
(c) The only protective oxide magnetite (Fe_3O_4, grey-black) should be present on the inner surface of the steam generator shell. However, excessive dissolved oxygen concentration in deaerated water can cause pitting corrosion. Cracks, breakdown and holidays in the magnetite coating layer induced by thermal and mechanical stresses during start-up, shutdown, outages, and rapid load swings also can result in pitting corrosion by dissolved oxygen.
(d) To maintain the level of water above the upper tubes in the bundle in order to cover them with water and thus to prevent formation of two-phase system liquid water-steam at the border.
(e) To plug damaged tubes (remove them from using).
(f) To remove the red rust by means of pressure water jet.

5.4.1.7 Failure of Tubes in Steam Generator

Conditions. The steam generator is intended for the cooling of the feed (mixture of liquid aromatic hydrocarbons xylene isomers) flowing inside the U-tubes. The boiler feed water (150 °C, 5 bars) flowing outside the tubes, is heated by xylenes stream and turned into the steam. The diameter of the outlet pipe in the steam generator was 76 mm.

Material of construction. U-tubes—carbon steel ASTM A179. The wall thickness of seamless U-tubes was 2.5 mm.

Service period before the failure: 6 months.

Visual examination and findings. The holes of ~2 mm diameter were detected in the top (upper part) of the tube bundle of the steam generator in the location opposite to steam outlet (Fig. 5.22).

The holes were formed under the "step" (stair) which in its turn was formed as a result of local decrease of the tube wall thickness. The wall thickness is relatively uniform at the top of the tube, while there was a lessening on the outer wall thickness of the tube in the location where probably water level was.

SEM examination. Wave-like surface morphology is detected near the holes, which is a typical surface appearance affected by the cavitation.

Failure phenomenon: *Erosion–cavitation.*

Cause of failure and its explanation. The holes were formed due to the erosion–cavitation which occurred on the external surface of the tubes in the upper part (top) of the steam generator tube bundle. The water level would have to be above the surface of the tubes. In other words, water would have to cover the tube surface in the steam generator. The water level dropped, and the top of the tubes was at the water–steam (two phases) border. The water boiled at this border. During the boiling, the shock waves—'explosion' of liquid water occurred, namely the water bubbles turned into the steam. Metallurgical examination and surface morphology of the top of failed tubes indicated cavitation. Small diameter (76 mm) of the steam

Fig. 5.22 **a** General view of failed tube bundle. **b** Failed tube with two holes under the "step" (stair). **c** Macroscopic cross-section perpendicular through the tube near the holes. **d** Failed tube with two holes (~2 mm). **e** Failed tube with a hole (~2 mm)

exit from the steam generator caused rapid flow turbulence resulted in erosion. That is, the phenomenon is erosion-cavitation.

Solutions and recommendations.

(a) To increase the steam outlet diameter from 76 to 152–228 mm (two-three times more).
(b) To keep the water level above the top surface of the tube bundle.

5.4.1.8 Failure of the Pipe in the Furnace (BTX Unit)

Conditions. Liquid mixture of xylenes C_9–C_{15} flowed in pipes in the furnace (first separation). Temperature inside pipes was 350 °C, pressure was 30 bar. Periodic cleaning by means of 'pigs' has been carried out during service period because of clogging of inner surface of pipes by coke. The failed pipe was located the second closest to the flame in the furnace.

Material of construction. Carbon steel ASTM A106 Gr. B. Diameter of pipe—150 mm.

Service period before the failure: 13 years.

Visual examination and findings. Bulging, fissures and splitting (80 mm in length) on the outer surface of the pipe were detected during cleaning by means of 'pigs' from the coke (Fig. 5.23). Thick (1–2 cm) black deposits of coke were found inside the pipe.

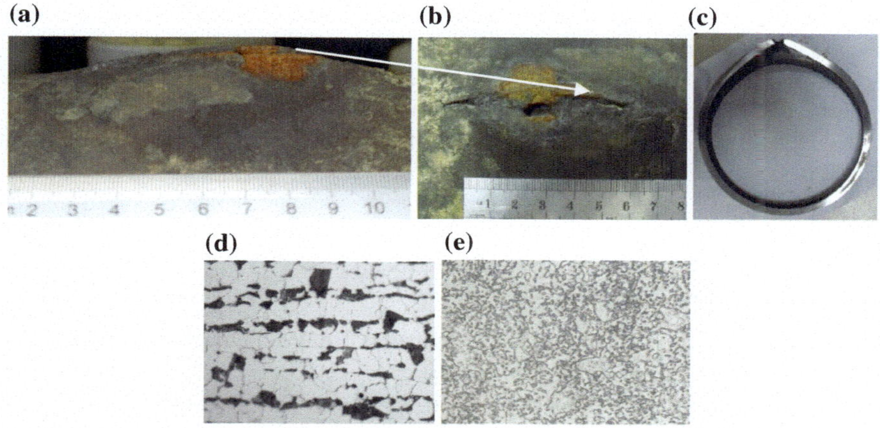

Fig. 5.23 **a** The pipe with the bulging. **b** The bulging, fissures and splitting (crack) on the external surface. **c** Macroscopic cross-section through the failure zone of the pipe. **d** Micro-structure of the pipe, 'cold' side (×600). **e** Micro-structure of the pipe failure area, 'hot' side—spheroidal cementite particles in a matrix of ferrite (×1000)

Microscopic examination. The 'cold' side of the pipe (180° from the place of the failure) has ferrite + pearlite structure which is characteristic of this type of the carbon steel (see Fig. 5.23d). The 'hot' side (the failure area) has the structure of spheroidal cementite in the matrice of ferrite (see Fig. 5.23e). This structure is typical to the type of steel which has undergone overheating to temperature of ~700 °C.

Hardness. Microhardness was examined on metallographic samples according to the ASTM B294-10 [12] (Table 5.2).

Cause of failure and its explanation. The failure occurred as a result of overheating for a long period of 'hot' side of the pipe (close to the flame), when the temperature in this area reached ~700 °C (two times more than the required service temperature 350 °C). The reasons leading to long-term overheating are partial (or total) tube pluggage and excessive fire-side heat input. Thick layer of coke were

Table 5.2 Microhardness of the pipe material in different locations

Location	Hardness[a], $HV_{0.5}$
'Hot' side	
The failure area	134–143
20 cm from the failure	164–166
'Cold' side	
180° from the failure area	154–155
20 cm from the failure	154–158

[a] $HV_{0.5}$—The Vickers microhardness test using a very small diamond indenter with applied load 0.5 kg [13, 14]

accumulated on the inner wall of the pipe and caused its thermal insulation that eventually led to the overheating. Accumulation of coke occurred due to a slow flow of liquid mixture of xylenes in the pipe.

Spheroidal cementite (carbide) microstructure on the 'hot' side of the pipe indicates its overheating during long time. It is required quite a long time to form carbides in steel structure. Hardness on the 'hot' side of the pipe is lower than that on the 'cold' side, that also characterizes the overheating which occurred during long period.

Failure phenomenon: *Long-term overheating.*

Solutions and recommendations.

(a) To maintain the temperature inside pipes not more than 350 °C. When T > 350 °C, xylenes are capable to form coke.
(b) To improve the flow regime (increase flow rate) of the mixture of xylenes in pipes, in order to prevent the accumulation of coke.
(c) To monitor the temperature of pipes. In the case of temperature increase, to inspect the pipes for indication of deposits, and if necessary to carry out internal cleaning of pipes from the coke.

5.4.1.9 Corrosion of the Pipe with Water Condensate

Conditions. Water condensate flowed inside the pipe, and pressure dropped from 24 bar (200 °C) to 4 bar (170 °C).

Material of construction: Carbon steel, 100 mm in diameter.

Service period before the failure: 5 years.

Visual examination and findings. The hole was found in the elbow (Fig. 5.24).

Failure phenomenon: *Erosion.*

Cause of failure and its explanation. Morphology of inner surface around the hole indicated that erosion occurred because of change of direction of water condensate flow in elbow. When pressure dropped ("released") significantly from 24 to 4 bar, a strong change of flow velocity occurred and hammered inner surface of the elbow caused its erosion. This phenomenon is known as *water hammer* (or, more generally, *fluid hammer*) which represents a pressure surge or wave caused when a fluid (usually a liquid but sometimes also a gas) in motion is forced to change

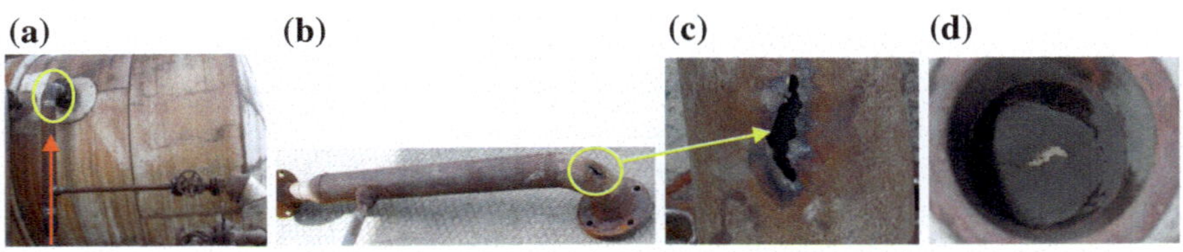

Fig. 5.24 **a** Water condensate entry into the tank. The new bend (elbow) is marked in *yellow*. **b**, **c** The hole in the bend (outside view). **d** The hole in the bend (inside view)

direction suddenly. This pressure wave can cause major problems, from noise and vibration to pipe collapse.

Solutions and recommendations.

(a) To reduce the pressure of the water condensate by fitting a regulator.
(b) To increase the diameter of the pipe with elbow.
(c) To replace carbon steel with SS 316L or to coat inner surface by wear resistant alloys with hardness more than 50 HRC (see Appendix F, Table F9).

5.5 Pipes

Pipelines are intended for transportation of various media: liquids, gases, solids, and their mixtures. Liquids in oil refining and petrochemical industry are crude oils, petroleum products, fuels, organic solvents, different types of water (boiler feed water, demineralised or deionized water), aqueous solutions of electrolytes (sour water, cooling water, spent soda, alkali solutions, potable water, extinguishing water), and amine solutions. Gases are hydrogen (sometimes containing hydrogen sulfide, hydrogen chloride and water vapor) and steam at high temperatures and pressures, nitrogen with hydrogen sulfide, and air. Transported media inside pipes can be homogeneous (one phase) or heterogeneous (several phases). The presence of such impurities as water, H_2S and salts in crude oils, petroleum products and fuels can result in corrosion of inner surfaces of pipes. Naphthenic acids containing in crude oils and petroleum distillates can cause corrosion only at T > 190 °C. Two- and three-phase systems (sand, solid catalysts, solid corrosion products and gases in liquids or drops of liquids in gases) can result in erosion and cavitation. Erosion is the main problem during transportation and treatment of oil sands.

Outer surface of pipes can be exposed to the atmosphere, rain water, soil, mud and underground water, hot gases (in furnaces and stacks), and spillages of petroleum products and fuels. Acid dew point corrosion, corrosion under thermal insulation, under deposit corrosion, fuel ash corrosion, and oxidation (corrosion at high temperature by oxygen in the surrounding air) can occur on outer surfaces of pipes.

Such diversity of media and conditions inside and outside pipes can result in different corrosion problems which should be considered at the stage of design, anti-corrosion measures must be applied and then corrosion situation from both sides of pipes should be controlled during their exploitation. Materials of pipes should be carefully selected according to the environment (inside and outside) and conditions. These materials can be carbon steels, low-alloy steels (with different content of chromium and molybdenum), stainless steels, nickel alloys, copper alloys, and titanium alloys.

Flammable, explosive and toxic substances are transported in pipes. They are exploited in most cases under high temperatures and pressures in oil refining and petrochemical industry. Therefore, pipes are considered as structures of high and particular corrosion risk. They are branched of different diameters and lengths, with many connections, valves and fittings, combined with pumps and compressors. There are a lot of inner and outer factors which can influence corrosion of pipes. Thus, it is hardly to predict when and where failure can occur. Nearly all corrosion phenomena can occur with pipes, e.g., hydrogen attack, sulfidation and NAC, to name a few (see Chap. 4). MIC can occur from both inner and outer surfaces of pipes. Therefore, it is important to analyse and predict all possible corrosion problems at the design stage, to plan anti-corrosion measures, corrosion monitoring and inspection methods. In some cases, type of media and conditions inside pipes significantly influence corrosion. In other cases, a type of media inside pipes does not play essential role.

Typical and untypical corrosion problems and cases happened with pipes at units of oil refineries and petrochemical plants will be considered in this section.

5.5.1 *Case Studies*

5.5.1.1 Failure of T-Shaped Fitting (Visbreaker Unit)

Conditions. The T-shaped pipe fitting as a part of the torch piping, operating at ambient temperature and atmospheric pressure, was located close to the support (Fig. 5.25a). The next support was situated at the distance of 15 m from the fitting.

Material of construction: Carbon steel ASTM A234 Grade WPB.

Service life before the failure: 1 year.

Visual examination and findings. The T-shaped pipe fitting was cracked (see Fig. 5.25a).

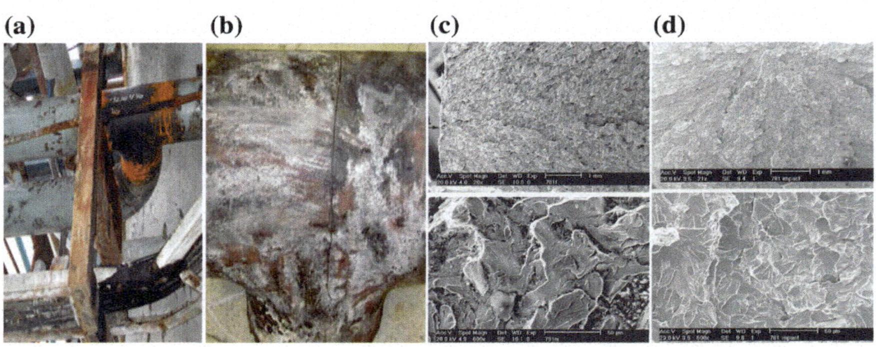

Fig. 5.25 **a** The T-shaped pipe fitting on the support (location of failure in torch piping). **b** The cracked T-shaped pipe fitting. **c** The fracture surface after the opening of the crack (inner surface of the fitting). **d** The fracture surface of the impact specimen (taken from the same pipe) after the Charpy test

Macroscopic examination. The crack initiated from the inner surface of the pipe fitting and propagated through the wall.

Microscopic examination. The crack is perpendicular to inner surface and free from corrosion products.

Examination of mechanical properties. Mechanical properties (yield strength, tensile strength, elongation, and hardness) of failed T-shaped pipe fitting were examined according to standard ASTM A370 [15] and results are shown in Table 5.3.

Examination of fracture surface. A brittle cleavage fracture through the whole cross-section was found (see Fig. 5.25c). According to the chevron fracture, propagated from the inner surface outward. Typical pearlite + ferrite fine structure cannot explain such fracture.

Impact test was performed with Charpy sub size specimen 7.5 × 10.0 mm (Type A) according to ASTM E23 [16]. The impact test results are shown in Table 5.4.

Impact energy values are very low for carbon steel and correspond to brittle range: 8–10 J instead minimum 27 J (ductile range). A brittle cleavage failure was found with the test specimen taken from the same pipe far from the failed fitting (see Fig. 5.25d), identical to the failure of the pipe fitting (compare with the Fig. 5.25c).

Failure phenomenon: *Cracking*.

Cause of failure and its explanation. The cracking of the fitting was caused by the intrinsic steel brittleness, namely, steel was in brittle state at ambient temperature (~ 15–30 °C). This is supported by fractography (the study of fracture surfaces of materials), metallographic examination, and mechanical tests. Mechanical properties of the fitting material do not comply with the ASTM A234 Grade WPB requirements. Maximum tensile strength is too high (642 MPa instead 415–585 MPa), the hardness also is too high (220–230 HB instead 197 HB maximum), the elongation is too low (17 % instead 30 % minimum), and impact energy also is too low

Table 5.3 Mechanical properties of carbon steel T-shaped pipe fitting

Property	Yield strength σ_{YP} (MPa)	Tensile strength σ_{UTS} (MPa)	Elongation ε (%)	Hardness (HB[a])
Result	564	642	17	220–230
ASTM A234 Grade WPB (requirement)	240 min	415–585	30 min	197 max

[a]Hardness Brinel

Table 5.4 Results of impact test (Charpy test) at 20 °C

Impact energy[a]	
Joules	Joules/mm^2
8–10	0.11–0.13

[a]Average data of five specimens (pipe fitting)

(8–10 J instead 27 J). A small stretching, stress or blow of the T-shaped pipe fitting caused the sudden development of the crack of the fitting.

Solution and recommendation.

(a) The fitting specification (mechanical properties) should be adapted to service conditions.
(b) Sometimes engineers think that correct installation of supports to T-shaped pipe fittings can prevent cracking, as unreasonably large distances between supports can make high loads and cause brittle fracture of carbon steel. In this case, study mechanical properties of the fitting material did not meet the terms of standard. The material was in brittle state and distances between supports did not play any role.

Post script. This case resembled the tragedy of Titanic ship in 1912 which sunk because of low *toughness* of carbon steel used for the hull. A metallurgical analysis of steel taken from the hull of the Titanic's wreckage revealed that it had a high ductile-brittle transition temperature, making it unsuitable for service at low temperatures. At the time of the collision of the Titanic with iceberg, the temperature of the ocean water was −2 °C (dissolved salts in water decrease its freezing temperature). We will explain what a ductile-brittle transition temperature for materials is. First we will describe important properties of materials, such as *ductility*, *brittleness*, and *toughness*. *Ductility* is the property of metals to be deformed plastically under tensile stress without fracture. Really, most jewelry are made from gold and silver not only because of beauty but because they have the greatest ductility. It is possible from 1 g of gold to pull the thread 1 km in length. Ductility is explained by specific *metallic bonds* which exist only in metals and alloys. In metallic bonds, outer shell valence electrons are delocalized and shared between many metallic atoms. These electrons allow metal atoms to slide past one another without being subjected to strong repulsive forces that would cause other materials (for instance, ceramics) to shatter. *Brittleness* is the property of materials to broke without significant deformation (strain) when subjected to stress or blow. Brittle materials absorb relatively little energy prior to fracture. Ceramics and glasses represent brittle materials. When metals are subjected to stresses, they also can be brittle. Usually this happens at low temperatures. Some carbon steels are brittle even at ambient temperatures of about 20 °C. There are margins when metals are *ductile* and at lower temperatures become *brittle*. The lower the temperature when carbon steel becomes brittle, the better for constructions, such as shipping, building, bridges and pressure vessels functioning in places with low temperatures. *Toughness* is a resistance to fracture of a material when stressed and is defined as the amount of energy per volume that a material can absorb before rupturing. In other words, *toughness* is the ability of materials suffering various stresses, blows, impacts, etc. This property is very important in the case of the meeting of *Titanic* with iceberg or in our case of the T-shaped pipe fitting subjected to large stresses. How to define the temperature when metal stops to be *ductile* and becomes *brittle*? Seven years before the tragedy of Titanic, in 1905, the French scientist Georges

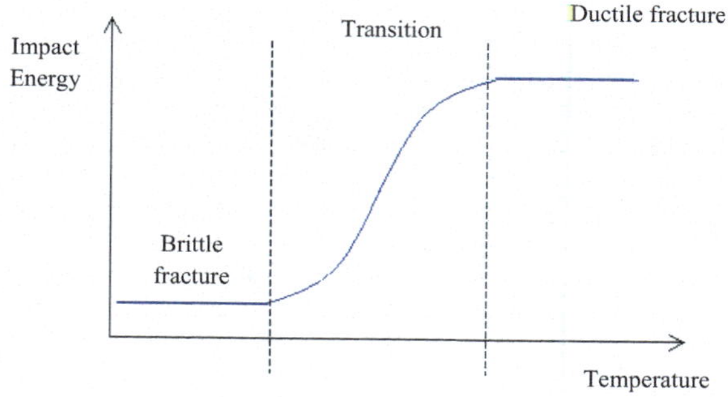

Fig. 5.26 Ductile-brittle transition temperature

Charpy developed the test which now is called in honor 'Charpy impact test'. It determines the amount of energy absorbed by a material during fracture. The ultimate mechanical failure is *fracture*. Fracture can be either ductile or brittle. Little energy is required to fracture brittle materials, such as glass, ceramics, and some types of cast iron. Conversely, tough materials, such as rubber and many steels, absorb considerable amounts of energy in the fracture process. Brittle fracture requires energy to separate atoms and expose new surfaces along the fracture path. Carbon steel lost toughness when the temperature drops below its ductile to brittle transition temperature. A *ductile-brittle transition temperature* is a temperature where the energy needed to fracture the material drastically changes and is the lowest temperature at which a material can be used (Fig. 5.26).

The ductile-brittle transition temperature determined at the impact energy of 20 J is −27 °C for modern carbon steels and 32–56 °C for the carbon steel from the Titanic hull plate. It is apparent that the steel used for the hull was not suited for the service at low temperature −2 °C of the Atlantic ocean. In our case, carbon steel T-shape pipe fitting became brittle at ambient temperatures (~20 °C) because impact energy was too low (8–10 J instead required 27 J). A small stretching, stress, blow or impact could cause the sudden development of the crack. And this really happened when Titanic struck an iceberg.

5.5.1.2 Corrosion of Pipes for Transportation LPG and BTX Under Road Crossing in Ditch

Conditions. The pipes are intended for LPG and BTX transportation at the oil refinery plant. They had no outer coatings and were located in the ditch under the road crossing. The ditch was filled with rain water and mud during winters. Leak of hydrocarbons was detected in the ditch.

Material of construction. Carbon steel ASTM 106 Gr. B. Diameter—150 mm.

Service period before the failure: 30 years.

Visual examination and findings. Severe corrosion of outer surfaces of pipes with holes (6–8 mm in diameter) was detected (Figs. 5.27, 5.28). Thick rust

Fig. 5.27 **a** General view of corroded pipe for LPG transportation. **b** The hole and rust on outer surface of corroded pipe. **c** The location (in the wetted mud) of buried pipes under the road

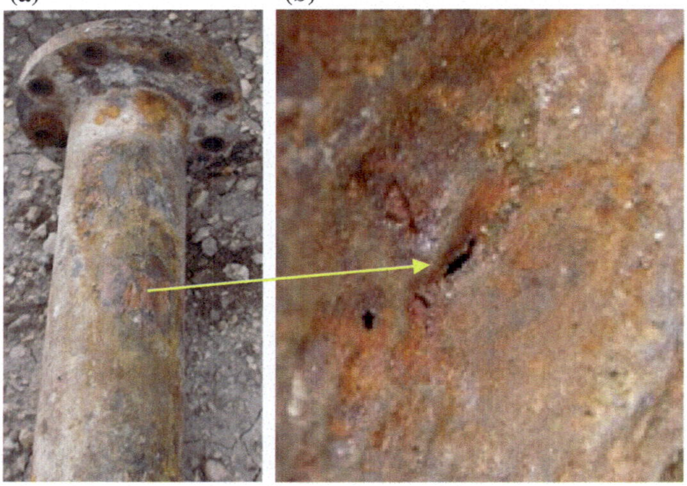

Fig. 5.28 **a** Corroded pipe for BTX transportation. **b** The hole in corroded pipe

consisted of black and brown corrosion products. They contained iron sulfides, iron hydroxides, and chlorides.

Failure phenomenon: *General and pitting corrosion in wetted soil (mud) under the road.*

Cause of failure and its explanation. The outer surface of unprotected pipe was in prolonged contact with rain water and mud saturated with dissolved oxygen. Water and dissolved oxygen were the cause of electrochemical corrosion of outer surface of carbon steel pipes and gave rise to general and pitting corrosion and then to the hole. The detection of iron sulfides in corrosion products show that H_2S in mud was one of the causes of corrosion. Two possible sources of H_2S in mud could exist: SRB in mud or leakage of hydrocarbons containing H_2S. LPG could contain small amounts (several ppm) of chlorides which could appear in mud as a result of leak. We can only wonder that corrosion failure occurred after 30 years of service and not earlier.

Solutions and recommendations.

(a) To relocate the pipes above the ground and to paint by coating resistant to the atmosphere (250 μm in thickness) because it is easier to monitor corrosion situation of aboveground pipes.

(b) Outer surface of pipes located under road crossing in trenches must be controlled by means of coats and cathodic protection (namely, sacrificial anodes) from possible wetted soil corrosion.

5.5.1.3 Corrosion of Pipe in Sleeve

Conditions. The pipeline was intended for transportation of hydrocarbons C_5–C_9 and was located in the sleeve (Fig. 5.29). Pipe had no outer anti-corrosion coating.

Material of construction: Carbon steel ASTM 106 Gr. B. Diameter—150 mm.

Service period before the failure: 2 years.

Visual examination and findings. The hole and severe corrosion of outer surface of the pipe in the sleeve were found (see Fig. 5.29).

Failure phenomenon: *Soil corrosion (general and localized).*

Cause of failure and its explanation. This is a typical case of use of cased carrier pipe for pipes crossing under railroads. When underground carbon steel pipe without coating was in the sleeve, differential aeration cells gave rise to localized corrosion of outer surface (see Fig. 5.29).

Solution and recommendations.

Pipes in sleeve (the case of steel-cased metallic pipeline) must be designed, fabricated, protected, installed, maintained and monitored according to the standard practice [17]. Thus, for pipelines smaller than 200 mm in diameter, the diameter of the casing is a minimum of 50 mm larger than that of the carrier pipe. Outer surface of the carrier pipe should have concrete, epoxy-polymer concrete, polyurethane or polyurea coating. Casing end seals shall be designed to prevent ingress of water and debris. It is recommended to fill the casing annulus with petrolatum wax systems (containing corrosion inhibitors, plasticizers, and thermal extenders). Alternative option is to fill with a corrosion-inhibiting gel filler (containing multi-phase vapor corrosion inhibitors). The corrosion inhibitor should not be harmful to the environment.

(a) **(b)**

Fig. 5.29 **a** Corroded pipe in the sleeve. **b** The hole (10 mm in diameter) formed in the sleeve

5.5.1.4 Corrosion of Spent Soda Pipe

Conditions. The pipeline was intended for transportation of spent soda and was in service without outer coating in the duct full of soil. Leak from this pipe was detected.

Material of construction: Carbon steel ASTM 106 Gr. B. Diameter—150 mm.

Service period before the failure: 12 years.

Visual examination and findings. The holes (5–7 mm in diameter), pits, corrosion 'stairs' and severe general corrosion of outer surface of pipe were found (Fig. 5.30).

Failure phenomenon: *Soil corrosion (general and localized).*

Cause of failure and its explanation. Carbon steel pipe without coating was installed in the duct full of wetted soil. This situation gave rise to typical soil corrosion.

Solutions and recommendations.

(a) Relocation of pipe above ground with suitable coating resistant to refinery atmosphere (250 μm in thickness).
(b) It is mandatory using outer coatings (epoxy, polyurethane or polyurea) for underground pipes and cathodic protection.

5.5.1.5 Failure of Pipe in Furnace (CCR Unit)

Conditions. Naphtha flowed inside the pipe in the furnace at 310–330 °C. Skin temperature: 510–720 °C. Temperature on inner surface of the pipes: 500 °C.

Material of construction: 9 Cr alloy steel (ASTM A 213 Grade T9, Sch 80).

Service period before the failure: 9 years.

Visual examination and findings. Longitudinal cracks and fissures on the pipe surface were found in lower coil in the furnace (Fig. 5.31). In addition to cracks, local bulging of the outer surface of the hot side of the pipe occurred in the failure area. A thick, brittle oxide layer near the failure was found.

(a) (b) (c)

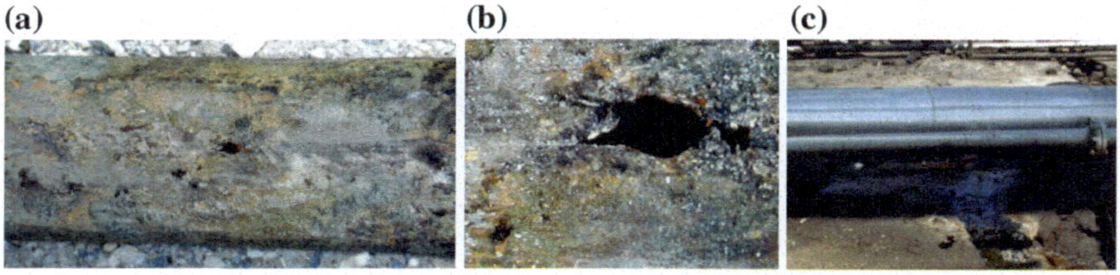

Fig. 5.30 **a** Corroded pipe with holes, pits and 'stair'. **b** The magnification of the hole (7 mm in diameter). **c** General view of spent soda pipes in the duct with flowing water

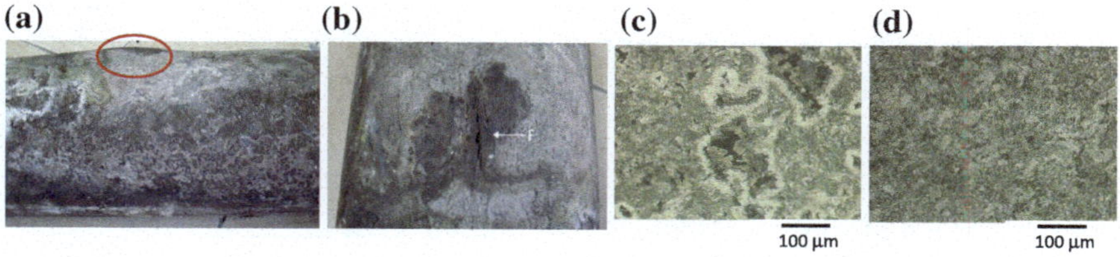

Fig. 5.31 **a** Failed pipe with local bulging (in *red ellipse*). **b** Cracks in the failed area 'F' (upper part of bulging). **c**, **d** Metallographic cross-section of the pipe far from failed area: **c** Hot side. **d** Cold side

Mechanical examination of hardness of pipe surface showed values 400–430 $HV_{0.5}$ (Vickers hardness when using the load 0.5 kg) when maximum allowable value is 190 $HV_{0.5}$ for this alloy.

Metallographic examination. A significant decrease in the thickness of hot area of the pipe in relation to the cold side occurred. The failure nature of the outer and inner hot pipe surface are similar. Oxide layer covered the entire external surface of the crack. The structure of the pipe material from the hot side is martensite while from the cold side is bainite.

EDS analysis. Corrosion products formed on inner and outer surface of the hot pipe consisted of iron, oxygen, chromium, and sulfur (10–36 wt% S).

Failure phenomenon: *High-temperature corrosion (oxidation)—Long-term overheating*.

Cause of failure and its explanation. The maximum skin temperature (720 °C) was close to values (770 °C) at which thermal oxidation of 9 Cr alloy steel becomes excessive [11]. This occurred during long time (~9 years). Original 9 Cr alloy steel usually has tempered bainite structure with hardness 190 $HV_{0.5}$. This structure remained for cold side of the tube whereas hot side structure was changed on martensite. This alloy had relatively large concentration of Cr (9 wt%) and when overheated (because of high temperature) and then quenched the martensite structure was formed with hardness 400–430 $HV_{0.5}$. Local overheating happened because of incorrect flame pattern and overfiring. Tempered alloy steel 9 Cr with hardness >400 $HV_{0.5}$ is vulnerable to *high temperature corrosion and cracking*. The failure of the pipe in the furnace occurred as a result of corrosion (oxidation—long-term overheating) which began from outer surface of the pipe wall on the hot side. Tube-wall thinning on the hot side resulted from cyclic thermal oxidation and spalling. The bulge was formed as a secondary phenomenon.

Solutions and recommendations.

(a) The hottest areas should be inspected for evidence of scale and deposits. Excess deposits should be removed by mechanical or chemical cleaning and prevented from recurring [11].
(b) To check firing procedures and in-service furnace temperatures near overheated areas.

(c) To measure hardness of pipes material before the service and during shut-downs. If hardness prevails 190 $HV_{0.5}$, it is mandatory to replace with suitable pipes having acceptable hardness.

5.5.1.6 Failure of Pipes in Water Recirculation Area of the Boiler (Power Station)

Conditions. The boiler worked 20 years. Temperature: 320–340 °C. Pressure: 130 bar. Burning material was fuel oil. The last test was made 3 years before the failure, and the boiler was in good state. The two pipes of 57.15 mm in diameter and 5.5 mm in wall thickness were exploded in the water circulation area in the boiler.

Material of construction: 0.5 Mo steel ASTM A209-03 (seamless carbon-molybdenum alloy-steel boiler and superheater tubes).

Service period before the failure: 3 years after the last test.

Visual examination and findings. The failed pipe segment was welded on both ends to segments from the original piping system. Wide hole surrounded by deformed tube wall due to bulging was found in the pipe closest to the fire (Fig. 5.32).

Corrosion around the hole and irregular deposits of grey, white and yellow color were detected outside. Uneven thick black scale was found inside the pipe in the area around the hole. Even (smooth) dense black-grey scale was found inside the pipe in the area opposite to fire. The defect leading to the hole originated on the outer tube surface. A significant decrease on the wall thickness of hot area of the tube occurred in relation to the cold side. The hot side-inner surface close to the hole had oxide layer 300–400 μm in thickness. The cold side-outer surface had dense layer of corrosion products. The inner surface showed slight non-uniform oxide layer.

(a) (b) (c) (d)

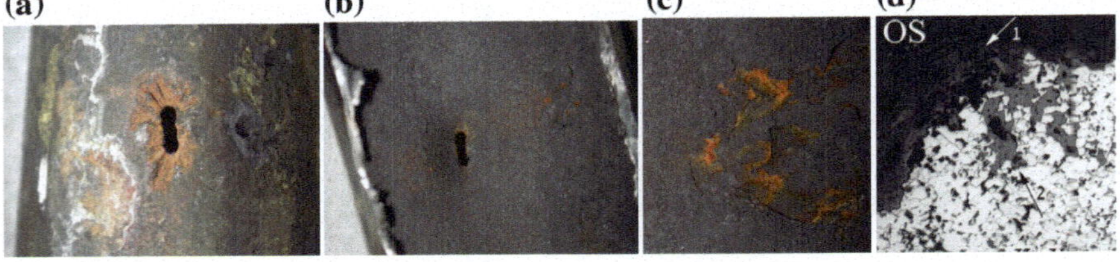

Fig. 5.32 **a**, **b** The view of the hole in the pipe from: **a** outer side; **b** inner side. **c** The inner surface of the tube near the hole close to the fire: one can see uneven thick scale of corrosion products. **d** Metallographic cross-section (×240) close to the hole in failed tube (hot side): *OS*—outer surface, *1*—corrosion products, *2*—intergranular corrosion

Metallographic examination. Loose layer of corrosion products cover the outer surface and significant intergranular corrosion attack was progressing over 0.15 mm beneath the surface.

For comparison, unused tube segment from the storage (reference), similar to the failed one was covered by dense rust layer on the outer surface 150 μm in thickness with no signs of intergranular corrosion attack.

EDS and ED-XRF analysis. Deposits from outer surface contained Fe, O, V, Ni, Ca, Na, S, Si (typical elements when fuel oil was burned).

Failure phenomenon: *Oil (liquid) ash corrosion.*

Cause of failure and its explanation. The material of the tube is not 'blame' for the current failure. A thick scale of magnetite (Fe_3O_4) (300–400 μm in thickness) was formed on the inner surface of only two tube segments close to the fire. They were replaced during the last shutdown 3 years before the failure. Inner surface of tubes was not clean adequately from rust and passivation process for creation magnetite layer was not carried out well. The thickness of the magnetite layer should not exceed 100 μm.

The mechanism of failure. Salts and oxides containing Fe, O, V, Ni, Ca, Na, S, and Si were formed as a result of fuel oil burning. Non-uniform thick (~400 μm) scale of iron oxides on the inner surface of the failed tube indicates that temperature was high. The formation of this thick scale was caused by original rusted surface of the tube before installation which then created insulation layer under operation conditions. As a result, heat transfer in these areas was not good, and overheating began on the outer surface of the segments. When the temperature reached the melting point of the salts formed as a result of the burning of fuel oil, liquid (melted) salts reacted with the iron [2]. Thus, *oil (liquid) ash corrosion* occurred. Corrosion started on outer surface of the pipes. Decrease in wall thickness of the pipes due to corrosion caused a loss of mechanical properties, and an explosion because of high internal pressure of boiler feed water.

Solution and recommendations.

To clean inner surface of pipes from rust and carry out passivation adequately with creation of even (smooth) layer of magnetite 100 μm (maximum) in thickness.

5.5.1.7 Corrosion and Crack of Fire-Fighting Pipe

Conditions. Potable water was filled inside the fire-fighting pipe under stagnation conditions at ambient temperature and pressure 3–5 bar. Extinguishing water was circulated once a week by means of the pump at pressure 10–11 bar for the test of the service ability of the system. The maximum water pressure in the pipe is not greater than 16 bar during possible fire accident. The pipe cracked during the start-up of water circulation.

Material of construction. Carbon steel API 5L Gr. B. Diameter—30.48 cm. Thickness—6 mm.

Service period before the failure: 40 years.

Visual examination and findings. Longitudinal crack of 2 m in length and 2 cm of opening was found (Fig. 5.33). Rust and leaks were found on the outer surface of other segment of the fire-fighting pipe. Uneven thick rust was detected inside the pipe.

Metallographic examination. Corrosive attack occurred from the inner surface of the pipe. The lack of welding (~ 80 % of the wall thickness) was found inside the pipe.

Failure phenomenon: *Water corrosion*; *cracking*; *pitting corrosion*.

Cause of failure and its explanation. The lack of welding inside the pipe wall served as a crevice (crack) which progressed with time through the wall as a result of corrosive attack by stagnant extinguishing water inside the pipe. When the crack progressed to a depth of ~ 90 % of the wall thickness, the pipe lost mechanical properties and the final fracture occurred. Pit were formed in the welds inside the pipe. Some pits penetrated through the pipe wall and caused leaks with formation of rust on outer surface of the pipe (see Fig. 5.34).

Solutions and recommendations.

(a) To use NDT techniques (ultrasonic testing, eddy current, radiography or computer tomography scanning using X-rays or gamma rays, or acoustic emission) for identification defects in fire-fighting welding pipes.
(b) To stop using welding pipes and to use only seamless pipes.
(c) To use galvanized steel for fire-fighting pipes.
(d) To inject corrosion inhibitor (1000 ppm $NaNO_2$) into stagnant extinguishing water. This inhibitor will protect inner surface of carbon steel from corrosion during long period.

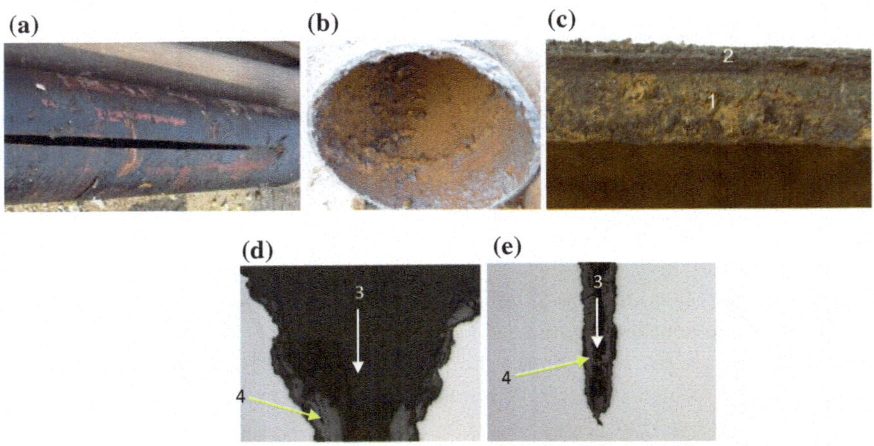

Fig. 5.33 **a** General view of failed fire-fighting pipe with longitudinal crack. **b** Inner rusted surface of the pipe. **c** Different regions of the fracture: *1*—there was no weld (~ 80 % of the wall thickness), *2*—weld after crack. **d, e** Metallographic cross-section through the pipe relatively far from the failure: **d** the crack beginning from the inner surface of the pipe. **e** The crack edge: *3*—crack, *4*—corrosion products

Fig. 5.34 Fire-fighting pipe with rust and leaks. Pits were formed in the welds inside the pipe and penetrated through the pipe wall

5.5.1.8 Corrosion of Pipe Outlet from the Absorber and Valves (CCR Unit)

Conditions. The hydrogen gas (mixed with nitrogen, hydrocarbons and traces of hydrogen chloride) after membranes entered the absorber tower at 44–60 °C (instead recommended 66 °C minimum) and washed by aqueous caustic soda solution (would have to be 11 wt% NaOH) in order to remove hydrogen chloride gas which was present as contaminant in the hydrogen gas. The hydrogen gas left the tower at 35 °C. The pipe (non painted outside) was covered with thermal insulation of mineral wool.

Material of construction. Pipe and valves—carbon steel. Diameter—100 mm.

Service period before the failure: 1 year.

Visual examination and findings. Holes, crack and severe corrosion were found in welds (HAZ) of pipes' outlet and inside valves where hydrogen gas flowed out the tower (Figs. 5.35, 5.36). Thick layers of rust were detected both inside and outside the pipe including valves. Seventy meters of the pipe length were replaced with new one because of severe inner corrosion.

Fig. 5.35 **a** The tower for absorption HCl by aqueous caustic soda solution (CCR unit) (the failed pipe is white luminous covered with aluminum jacket). **b–d** The pipe with circumferential hole in the weld (HAZ): **b** outside view. **c** Inside view. **d** The bend with two holes in the weld. **e** Corroded inner surface of the weld (HAZ) in the bend (with holes)

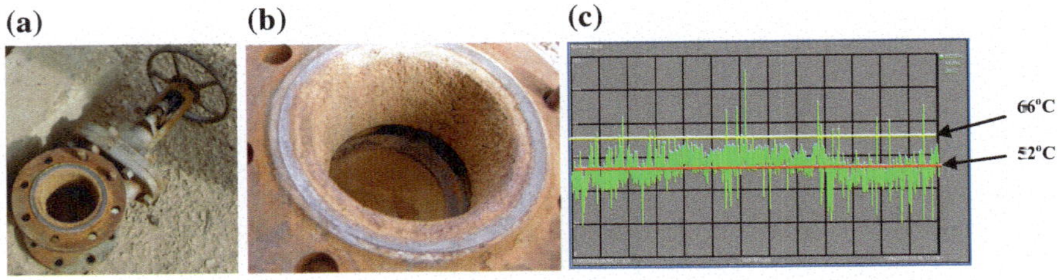

Fig. 5.36 **a, b** Corroded valve inside. **c** Temperature fluctuations of gases entering into the tower (PI data—*green*); *red*—average (44–58 °C); *yellow*—required minimum temperature (66 °C)

The data of chemical analysis of aqueous caustic soda solution showed that concentration of soda decreased many times to zero and ranged in 0–5 wt% NaOH during a year (instead recommended 11 wt% NaOH).

ED-XRF analysis. Rust (inside and outside in the area of weld) contained Fe, O and Cl (up to 30 wt%).

Cause of failure and its explanation. When concentration of caustic soda solution decreased less than 11 wt% NaOH, hydrogen chloride gas was not completely absorbed from the hydrogen gas in the tower. Thus, complete neutralization of acid impurity (HCl) did not occur. Remains of HCl together with water drops left the tower caused severe corrosion of inner surface of carbon steel pipe, especially in HAZ of welds (because of their electrochemical heterogeneity).

The temperature of hydrogen gas with HCl impurities entering the tower should be more than 66 °C. Otherwise, there is the danger of foaming in the tower and clogging of beds in the tower and filters in the pipe because of condensation of hydrocarbons which come with the hydrogen gas. The PI analysis showed that the temperature of gases entering the tower ranged in 44–60 °C instead 66 °C minimum. Therefore, the filters were clogged by condensed hydrocarbons and aqueous soda solution did not absorb HCl well in the tower.

Failure phenomenon: *Acid general corrosion.*

Solutions and recommendations.

(a) To maintain the concentration of caustic soda solution 11 wt% NaOH. If concentration decreases less than 9–10 wt% it is necessary to add fresh caustic soda solution into the tower in order to absorb HCl completely from the hydrogen gas.
(b) To keep the temperature of hydrogen gas entering the tower not less than 66 °C. For this, it is recommended to reduce rotating rate of ventilator (fan) of air cooler which is located before the tower. Alternatively steam accompanying this pipe section could also increase temperature of entering hydrogen gas.
(c) To improve the chemical analysis of HCl detection in hydrogen gas left the tower, because even small remains of HCl (several ppm) together with water can cause severe corrosion of inner surface of carbon steel pipe and valves.
(d) To replace carbon steel with duplex steel SAF 2205.

5.5.1.9 Failure of Kerosene Pipeline

Conditions. Underground pipe for the transportation of kerosene from the oil refinery to the port was in service 65 years. There was no coating on outer surface of the pipe. The failed part of the pipe was located on reinforced concrete supports in underground concrete duct under the road. Cathodic protection was applied during the last 35 years of service, but did not function most time because there was no electrolyte in the duct around the pipe. There was no insulation between the pipe and concrete supports. The ditch space with the pipe in the underground concrete duct had to be empty and blocked from both sides of the wall by blocks and concrete from possible penetration of rain or ground water. Rainwater from the road could collect during winter in the ditch, cover all pipe surface and thus accelerate corrosion over years. Following periodical pressure test inside the pipe, the leak was detected at the top of the pipe in the duct.

Service period before the failure: 65 years.

Material of construction. Carbon steel AISI 1025. Diameter—30.5 cm. Thickness—9.2 mm.

Visual examination and findings. Severe corrosion of outer surface and crack (10 mm in length) on the upper part of the pipe in the concrete ditch were found (Fig. 5.37). Numerous pits were revealed and were mostly concentrated on the upper third of the inner surface of the pipe (kerosene side).

Metallographic examination. Macroscopic examination showed significant diminishing of original wall thickness of upper part comparing with the lower part

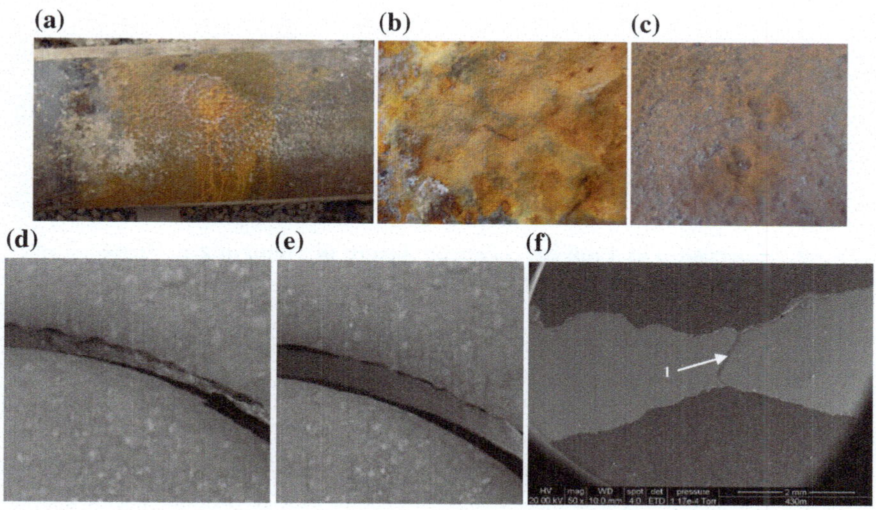

Fig. 5.37 **a** General view of the failure on the kerosene pipe. **b** External corroded surface with the crack. **c** Inner corroded surface (kerosene side). **d, e** Macroscopic cross-sections of the pipe: **d** Close to the crack. **e** Far from the crack. **f** Metallographic cross-section through the pipe, crack area (SEM photo): *1*—crack

of the pipe. Moreover, this reducing of wall thickness occurred from both outside and inside but mostly from outer surface of the pipe. General corrosion occurring from both sides of the pipe during 65 years was the cause of the wall thickness diminishing. The wall thickness of the pipe far from the crack and in the lower part of the pipe was almost the same: 8–8.5 mm. The examination of the wall thickness in the vicinity of the crack appears to be a critical reducing of the pipe wall thickness. Microscopic examination revealed numerous corrosion products on both outer and inner surfaces including the crack inside. The angle of the crack is 45° relative to the surface which is typical fracture (crack) generated under overloading. No indication (signs) of fatigue mechanism responsible for cracking was found.

Failure phenomenon: *Stress corrosion cracking, general and pitting corrosion.*

Cause of failure and its explanation. The concrete ditch space was not hermetically sealed from the penetration of rain water and air. Therefore, inner space of the ditch was full with rain water during winter periods and dry during summer. These periods 'wetting—drying' caused corrosion of outer surface of unprotected pipe in the concrete ditch. Corrosion occurred from both outer and inner surface of the pipe. Outer surface corroded in contact with rain water during the service period of 65 years. The presence of dissolved water and oxygen in kerosene resulted in general and pitting corrosion of inner surface of the pipe (accordingly to electrochemical mechanism) [2].

The crack occurred due to overloading, where the pipe wall thickness decreased locally to a thickness which could not withstand the stresses acting on the pipe. The crack was the final stage of the failure, which occurred only after prolonged corrosive attack of the pipe surface.

Solutions and recommendations.

(a) To coat outer surface of the pipe and put it in the polymeric sleeve in the concrete ditch.
(b) To close hermetically the ditch (with the pipe inside) from both sides from the environment.

5.5.1.10 Corrosion of Pipe for Transportation of Neutralized Brine from Soda Treatment Unit to Separator

Conditions. The pipeline is intended for transportation of neutralized brine from soda treatment unit to separator at ambient temperature and 8 bar.

Material of construction. Carbon steel. Diameter—45 mm, wall thickness—5 mm.

Service period before the failure: 3.5 years.

Visual examination and findings. Dense deposits were accumulated on the inner surface of the pipe and clogged it. The holes were found under the deposits inside the pipe (Fig. 5.38). Pitting corrosion rate is 1.7 mm/year.

ED-XRF analysis. Deposits contained (wt%): Fe (45), S (40), Al (8), and Si (7).

Fig. 5.38 Corroded inner surface with holes in the pipe for transportation of neutralized brine from soda treatment unit to separator

Chemical analysis. Neutralized brine solution contained different salts (mostly Na_2SO_4) and pH changed in 4.5–12.5 during service period. Iron sulfates and iron hydroxides were detected in deposits formed on the inner surface of the pipe.

Cause of failure and its explanation. Sulfate salts in neutralized brine solution resulted in corrosion of inner surface of carbon steel pipe. Corrosion products of iron (iron hydroxides and iron sulfates) were deposited and under deposit corrosion occurred which in its turn gave rise to pits as a result of differential aeration cells functioning under deposits.

Failure phenomenon: *Under deposit corrosion.*

Solutions and recommendations.

(a) To replace carbon steel with HDPE. It will resist to 16 bar at 20 °C and to 10 bar at 40 °C. To carry out accessories (valves, fittings) from polypropylene PN 10 (resistant to 10 bar at 25 °C).

(b) Alternatively carbon steel pipe with inner polymer coating can be used.

5.5.1.11 Failure of the Pipe for the Transportation of Raffinate

Conditions. Underground pipe (welded) is intended for the transportation of raffinate and was installed in the ditch containing 'sweet' (no salts) sand (2 m deepness).

A common raffinate stream in petroleum processing is known as the C_4 stream ('butane-butene mixture' containing butane, isobutane, butene, isobutene, and butadiene) from catalytic cracker. They (raffinate components) are valuable as raw materials for the production of polymers and fuels.

The outer surface of the pipe was coated with polyethylene tape, and proper cathodic protection (impressed current) functioned. The leak of hydrocarbons was detected in the soil in the area of bending (where the pipe did turn) (Fig. 5.39).

Service period before the failure: 20 years.

Material of construction. Carbon steel API Specification 5L. Diameter—150 mm. Original wall thickness—9.2 mm.

Visual examination and findings. The hole was detected in the bend at the 'hour 6' (see Fig. 5.39). Rust and shallow pits underneath were detected on the inner surface of the bend. Rust was also detected on the inner surface of the straight pipe far from the failure. The wall thickness of the failed pipe section (bend) diminished from original 9.2 to 3 mm. The wall thickness of the pipe section far from a failure

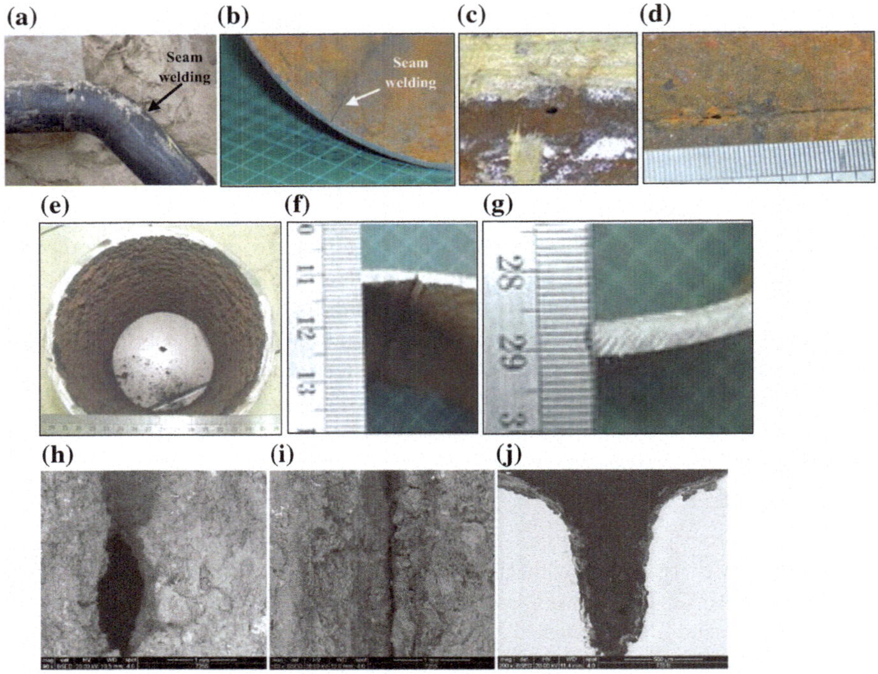

Fig. 5.39 **a** The pipe bend with the hole at the 'hour 6' in the ditch. **b** The seam welding along the pipe. **c** The leak area (outer surface). **d** The leak area (inner surface). **e** The view on the pipe wall thickness (rust inside). **f** Top-pipe segment with the seam weld. **g** Lower pipe section. **h** SEM image of the inner surface of the pipe section (the failure area). **i** SEM image of the inner surface of the pipe section (near the failure zone). **j** SEM photo of metallographic cross-section through the pipe section (the weld zone)

decreased from original 9.2 to 6–7 mm. Thus, the corrosion rates were 0.46 mm/year in the location of the failure, and 0.01 mm/year in the location far from the failure.

EDS analysis. The main elements in the rust in the failure location: Fe, O, Cl (~2 wt%) and small concentrations of Ca and Mg. Chlorides were not detected in the rust located far from the failure.

Failure phenomenon: *Corrosion-erosion*.

Cause of failure and its explanation. Corrosion occurred on the inner side of the pipe transporting raffinate over a period of 20 years. Pure raffinate (mixture of light hydrocarbons) is not corrosive to carbon steel. Rust formed inside the pipe proves that dissolved water, oxygen and chlorides in raffinate caused corrosion of inner surface of the pipe according to electrochemical mechanism. Presence of chlorides in the rust shows that raffinate flowed out the CCR unit contained chlorides. The presence of calcium and magnesium in the rust indicates that the residual water remained after hydrostatic examination of hermeticity of pipeline by means of potable water pressure. A significant decrease in the wall thickness of the pipe bend with a hole in comparison with the wall thickness of the pipe far from the hole indicates that the flow regime was the most important and prevailed factor in occurring of the failure. Erosion due to the flow regime change of the raffinate in the bend gave rise to intensive destruction of the weld. Thus, corrosion-erosion caused severe deterioration of the weld in the pipe bend and the hole formation. Outer surface of the pipeline had no changes.

Solutions and recommendations.

(a) To use coalescers in order to reduce the concentration of water in raffinate. To install catcher (scavenger) of chlorides in raffinate at the CCR unit.
(b) To replace the material of bend from carbon steel with SS 316L.
(c) To install isolated flanges between these bends and carbon steel pipes.
(d) There is no danger of corrosion of straight pipes because corrosion rate is 0.01 mm/year (during 20 years) which is low and reasonable.

5.5.1.12 Corrosion of Welds in Pipes (Spent Soda Treatment Unit and Ferric Sulfate Solution Storage)

Conditions. Spent soda containing chlorides (1000 ppm) flowed inside the pipes at ambient temperature. Storage tank and pipes (around the tank) containing aqueous solution of ferric sulfate $Fe_2(SO_4)_3$ was located at this unit.

Material of construction: SS 316L. Passivation of welds after welding was not carried out.

Service period before the failure: 1 year.

Visual examination and findings. The leaks were detected in welds of pipes containing spent soda and aqueous solution of ferric sulfate (Fig. 5.40). Deposits were found inside the pipes. Severe localized corrosion was found inside the welds. It was defined that dimensions of HAZ were larger in failed welds, than in not failed welds.

Metallographic examination. Macroscopic examination showed severe corrosion beginning at the border between weld and HAZ (Fig. 5.40e). Asymmetry was found in some welds. Microscopic examination revealed corrosive attack at the border of weld seams and a base metal, which started inside (Fig. 5.40g).

Microstructure. Average grain size was defined equal to 2–3, while standard grain size should be 5–8 for austenitic stainless steel (SS 316L).

Chemical analysis of pipe and welding electrode materials. Optical emission spectrometry showed discrepancy between chemical content of welds and requirements according to the standard (Table 5.5).

EDS analysis. Corrosion products inside welds contained (wt%): Fe (51), O (28), Ni (9.5), S (5), Mo (4), Cr (1.5), and Si (1). Chlorine was not found in corrosion products.

Failure phenomenon: *Localized corrosion of welds.*

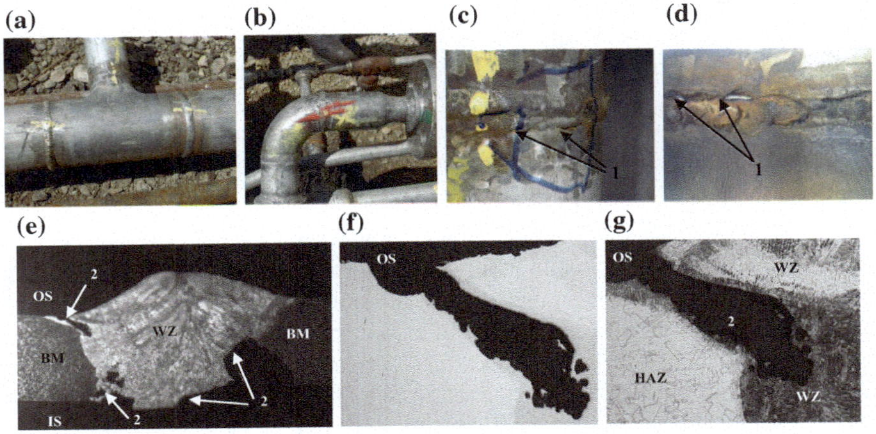

Fig. 5.40 **a**, **b** General view of welded pipes. **c**, **d** Failed welds (with holes 1): **c** outer surface; **d** inner surface. **e** Macroscopic cross section of the weld: *2*—location of corrosion; *WZ*—weld zone, *BM*—base metal, *OS*—external surface, *IS*—inner surface. **f**, **g** Metallography of weld in the location of failure (×60): **f** before etching. **g** after etching: *HAZ*—heat-affected zone

Table 5.5 Chemical composition of welds (SS 316L)

Location of pipe (weld)	Element (wt%)							
	Cr	Ni	Mo	C	Mn	Si	S	P
Spent soda unit	**13.2**	**8.7**	**1.8**	0.035	1.61	0.44	0.007	0.020
Tank $Fe_2(SO_4)_3$	**8.8**	**5.6**	**1.3**	**0.042**	1.46	0.43	0.008	0.025
SS 316L (ASTM 312)	16.0–18.0	10.0–14.0	2.0–3.0	0.035 max	2.0 max	1.0 max	0.030 max	0.045 max
Typical electrode for welding SS 316L	18.0	11.5	2.8	0.025	0.80	0.80		

Note The values that do not match requirements of standards are in bold

Cause of failure and its explanation. The main reason of failure of welds was that electrodes used for welding were of significantly different chemical composition from the recommended composition of SS 316L by standard. There was not enough Cr, Ni and Mo in welding material (see Table 5.5). Base metal (pipe) was SS 316L. This discrepancy in chemical composition created significant electrochemical heterogeneity between weld material and base metal which in its turn resulted in localized corrosion at the border between the weld seams and base metal in contact with electrolyte (spent soda or ferric sulfate). Corrosion started from the inner surface of welds, namely, in the center of weld roots. The second reason of failure was that passivation of internal and external surfaces of welds after welding was not carried out. The third reason was that poor geometry of welds lead to stresses inside and thus accelerated failure. According to the characterization of microstructure, there was an increase in grain size in HAZ (2–3 instead 5–8), and this fact indicates overheating of the welding process which could lead to failure too.

Solutions and recommendations.

(a) To use suitable electrodes for welding SS 316L (see Appendix F).
(b) To carry out welds with proper geometry in order to decrease stresses in welds.
(c) Passivation of internal and external surfaces of welds after welding should be carried out according to ASTM A380 [18].

5.5.1.13 Pipeline Repair Coating

Conditions. Underground pipeline (400 mm in diameter) with polyurethane coating and cathodic protection is intended for crude oil transportation (1.2 m below the ground level).

Material of construction: Carbon steel with outer polyurethane coating (450–550 μm in thickness).

Service period before the failure: 17 years.

Visual examination and findings. Destruction of polyurethane coating (Fig. 5.41).

Failure phenomenon: *Coating destruction*.

Cause of damage and its explanation. Ageing of polyurethane coating in underground conditions.

Solutions. Following repair was carried out:

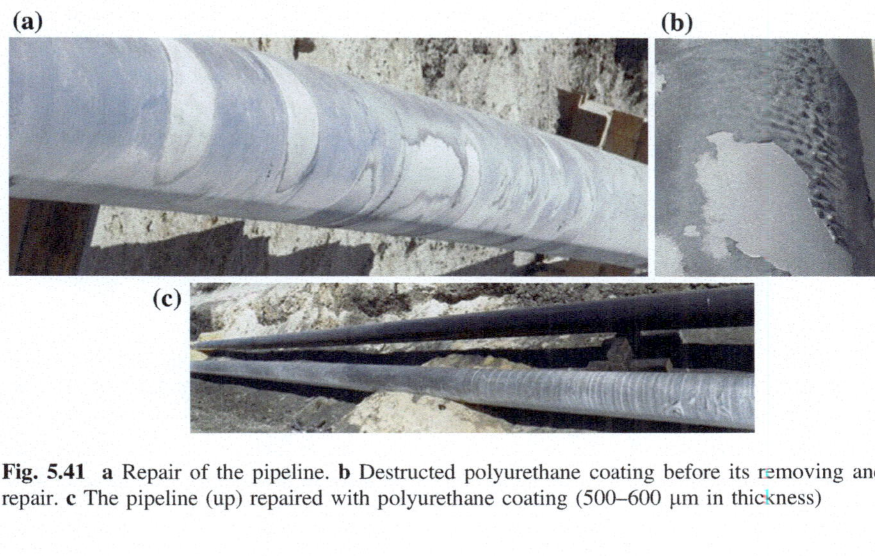

Fig. 5.41 **a** Repair of the pipeline. **b** Destructed polyurethane coating before its removing and repair. **c** The pipeline (up) repaired with polyurethane coating (500–600 μm in thickness)

(a) Aggregate cleaning with basalt particles to Sa2.5 according to EN ISO 8501-1 [19].
(b) Application of polyurethane coating (500–600 μm in thickness) (repair).

Conclusions.

A. Pipes for refinery and petrochemical plants are usually designed to ASME/ANSI B31.3 [20]. Special attention should be paid to welds, supports, bends, and any connections.
B. Any failure of pipes should be registered, studied and analyzed.
C. There is a tendency to relocate underground pipelines to aboveground at refineries and petrochemical plants. There are pros and cons.
Ca. Aboveground pipelines. It is convenient to monitor their condition, repair, replace, and repaint. However, they are vulnerable to different sudden hits from maintenance crews.
Cb. Underground pipelines. It is more complicated to monitor their condition, repair and replace. Anti-corrosion protection is also more complicated because it should include both coatings and cathodic protection.

5.5.1.14 Bright Lights like Crickets (Shining) on Pipe Surface in Furnaces (CCR Unit)

Conditions. Hydrocarbon gases were burned in three furnaces, and the flue gases contacted outer surface of the pipes. The temperature inside the furnaces was ~500 °C, but in the 'glowing/luminous' places could reach up to 700 °C.

Material of construction. 9Cr-1Mo steel. The wall thickness—5.7 mm.

Service period before the phenomenon: 6 years.

Visual examination and findings. Bright lights like crickets were detected on outer surface of pipes during normal service (Fig. 5.42). A dense scale of 0.4–0.5 mm in thickness was formed on the surface of the pipes. This scale was well adhered and it was difficult to remove it. The area of the scale which was in direct contact with the surface of the pipe was grey, and the part of the scale which was in contact with the flue gases was red (see Fig. 5.42b). The scale consisted of four layers distinct in morphology. Two layers close to the pipes` surface contained the mixture of iron and chromium oxides. Cracks were detected in the second oxide layer. The third and fourth layers coming in contact with flue gases consisted only of iron oxides (mainly Fe_2O_3).

EDS analysis: Fe, Cr, O, S, Si, Ca, and Mn were detected in the scale.

Phenomenon: *Sparkling (bright lights).*

Cause of sparkling and explanation The phenomenon of luminous/glowing knots (or incandescence—to 'glow white' from the Latin) is a physical phenomenon related to the excitation of atoms and electromagnetic waves at high temperature. Following mechanism of gleaming could occur on outer surface of pipes (9Cr-1Mo) in furnaces. Steel surface has different colors dependent on temperature (Fig. 5.43).

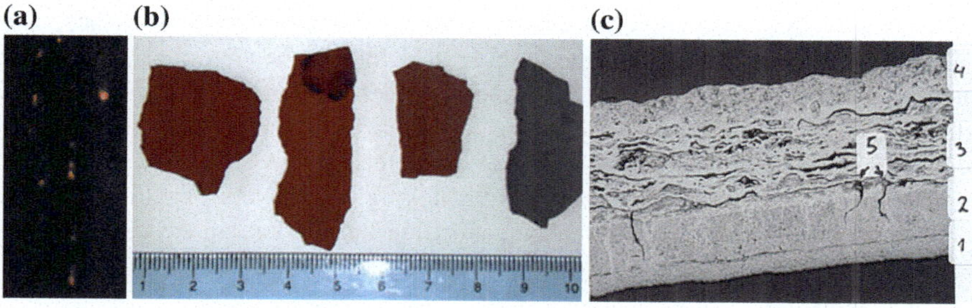

Fig. 5.42 **a** 'Glows'/'hot luminous spots' on the outer surface of the tubes in the furnace (during service). **b** Scale formed on the surface of the 9Cr-1Mo pipes in the furnace: *red*—contact with combustion gases, *grey*—contact with the pipe surface. **c** SEM photo (magnification ×100) of metallographic cross-section of oxides: *1–4*—four oxide layers (0.4–0.5 mm in thickness), *1*—contact with the pipe surface, *5*—cracks in the oxide layer no. 2

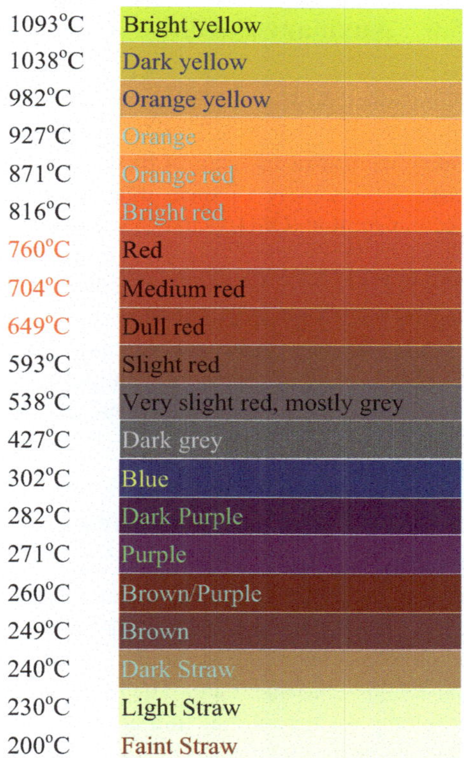

Fig. 5.43 Colors of steel (iron) surface at different temperatures

At temperatures above 427 °C, 'glowing/luminous' are formed. Atoms in the steel are in high energy state due to heat energy, and photons are emitted from the surface of the steel pipes. When the steel pipe is at a high temperature in the furnace, the scale of iron and chromium oxides is formed on the steel surface. The thickness of the scale layer and interference of colors as the 'returned light' are a function of temperature and time. The color of this heated steel changes from dull red through orange and yellow to white, and can be a useful indicator of steel temperature. Therefore, different colors of steel surface have been used for long time for thermal treatment of steels. During flaking, thin layers of oxides (cracks were detected in oxide layers), small areas inside the oxides are exposed to heat energy flow. Glowing is a returned light that we see in these places.

Conclusions.

(a) The 'hot spots/luminous' are not dangerous for use of pipes in furnaces.
(b) Oxidation rate of 9Cr-1Mo steel in furnaces is very low ($\sim$0.06 mm/year) and there is no danger for their use too.

5.6 Auxiliary Facilities

Numerous corrosion problems concerned not directly to running crude, petroleum products, fuels, organic liquids, and sulfur producing and obtaining in units of oil refineries and petrochemical plants but related to environmental, health and safety risks. The activity of such organizations as U.S. Occupational Safety & Health Administration (OSHA) and U.S. Environmental Protection Agency (EPA) resulted in regulations on both safety and emissions control, such as smoke, noise, and odor. Most of these regulations is carried out at facility units where corrosion problems or destruction of polymeric equipment are not an exception. Such facilities are flare disposal system, equipment for odor treatment at bitumen unit, DGF, acid gas scrubber, urea injection for neutralization NO_x, separation unit, and spent soda treatment unit. Six case studies are described and analyzed in this section.

5.6.1 Flare Disposal System

Flares, which are provided to ensure the safety of each refinery, can pose one of the most serious safety hazards to the facility if improperly operated. The result can be explosion. The safety of the flare system is dependent upon the prevention of air infiltration into the flare. The most effective method for the prevention of air infiltration is the introduction of purge gas. Flare stacks are designed to dispose of flammable gases safely by ensuring the combustion of these gases at the exit of the stack (flare tip). It is undesirable to have a flarestack filled with a mixture of combustible gas and air (oxygen) within the flammable limits because of the dangers of internal flash-back or explosion.

Flare disposal system collects and dischargers gas from atmospheric or pressurized process components to the atmosphere to safe locations for final release during normal operations and abnormal conditions (emergency relief). Flare systems typically have a pilot or ignite device that ignites the gas exiting the system because the discharge may be either continuous or intermittent. One of the important component of this system is 'molecular seal'. Molecular seal is the accessory of a flare system that protects the stack from flame front or flash back. Molecular seal causes flow reversal. It is located below the flare tip and serves to prevent air entry into the stack. The molecular seal works by relying on the density difference between purge gas (hydrocarbon mixture) and air to produce a process barrier to air entering the flare (preventing the formation of an explosive mixture in the system). Molecular seal is also named a 'density or labyrinth seal'. The seal is a gas inversion device causing the gas normally flowing in an upward direction to be turned through 180° in the original direction of the flow.

Usually hydrocarbon gaseous mixtures at oil refineries contain hydrogen sulfide and water vapor which can be corrosive to carbon steel equipment at definite conditions. One of such cases is described in this section.

5.6.2 *Case Studies*

5.6.2.1 Corrosion of the Drain Pipe in the Molecular Seal (Flare Disposal System)

Conditions. The flare disposal system is intended for burning 'flare gas' (containing mostly hydrocarbons, 4–5 vol% H_2S, and water vapor) and 'fuel gas' (containing only hydrocarbons). The 'flare gas' passes through a water layer (90 cm height) in which H_2S is absorbed and raises through a riser to the 'molecular seal'. The drain pipe line goes from the bottom of the 'molecular seal' at the angle 45° and goes down vertically. The pipe exit from the 'molecular seal' was clogging with deposits during the service. In order to remove these deposits, steam was injected upwards in portions.

Material of construction. Carbon steel; outer diameter of the pipe—90 mm; wall thickness—7 mm.

Service period before the failure: 5 years.

Visual examination and findings. Severe corrosion of the drain pipeline from the 'molecular seal' and its inner parts was detected (Figs. 5.44, 5.45). The wall thickness of the pipe diminished from original 7 mm to zero, and the pipe was cracked because of loads formed as a result of the reducing of the wall thickness. The part of the pipe at the angle 45° was corroded mainly inside. Gray scale well adhering to inner surface of the pipe was formed. Chemical analysis of this scale showed that it contained iron sulfide, ferric and ferrous sulfate salts.

Failure phenomenon: *Low-temperature sulfidic corrosion (acid corrosion by H_2S).*

Cause of failure and its explanation. Hydrogen sulfide containing in flare gas at concentrations of 4–5 vol% passing through the water layer of 90 cm height was

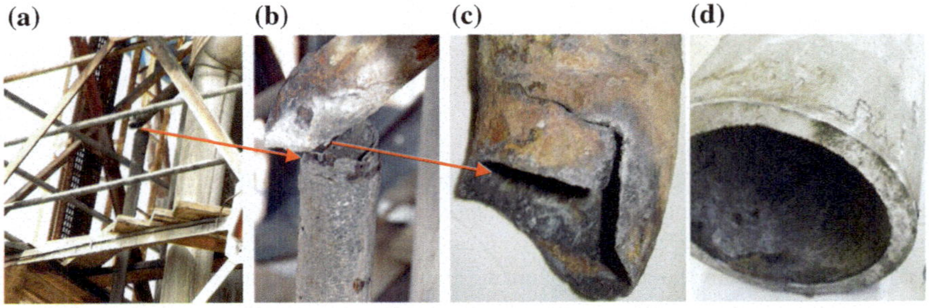

Fig. 5.44 **a** Corroded cracked drain pipe in the 'molecular seal'. **b** Corroded cracked pipe (magnification of the picture **a**). **c** Corroded cracked pipe after the failure. **d** Drain pipe far from the failure

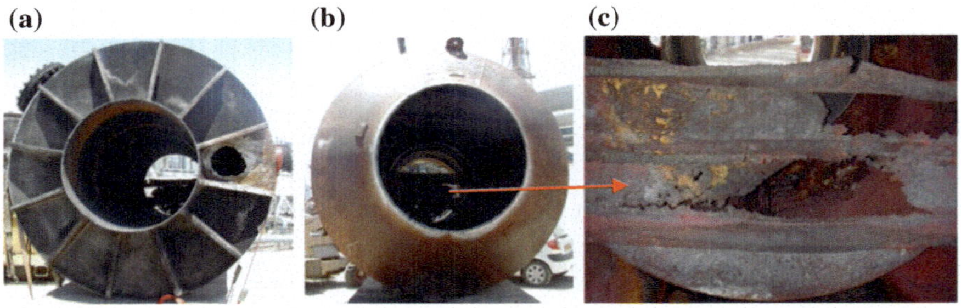

Fig. 5.45 **a** Corroded lower part of the 'molecular seal' (the location of corroded cracked drain pipe). **b** Upper part of the 'molecular seal' (corroded parts inside). **c** Corroded parts inside the 'molecular seal'

absorbed. As a result, aqueous acidic solution was formed and attacked inner surface of drain pipe and lower part of the 'molecular seal' to form iron sulfide deposits. The wall thickness at the angle 45° of the pipe was reduced to such an extent that loads caused crack and the collapse of the pipe. When flare disposal system was shutdown and open to the atmosphere, iron sulfides were oxidized to form ferric and ferrous sulfate acid salts (see Appendix D).

Solutions and recommendations.

To replace critical carbon steel parts of the drain pipe with duplex stainless steel Alloy 7Mo-PLUS (UNS S32950), or austenitic stainless steels Alloy 20Mo-6 (UNS N08026), or Alloy 20Cb-3 (UNS N08020). The corrosion rate of the first alloy in boiling 5 wt% H_2SO_4 is less than 0.12 mm/year [21]. It is not recommended to use SS 304 and SS 316 because their corrosion rates in H_2SO_4 (>0.5 wt%) are high (Table 5.6).

Table 5.6 Corrosion rates of SS 304 and SS 316 in H_2SO_4 [21]

H_2SO_4 (wt%)	pH	T (°C)	Corrosion rate (mm/year)	
			SS 304	SS 316
0.05	2.0	95	<0.025	<0.025
0.25	1.3	60	0.9	<0.025
0.50	1.0	90	20	0.125
1.00	0.7	100	7.5	1.25

5.6.2.2 Failure of Polyethylene Vessels Containing Hydrogen Peroxide and Caustic Soda Solutions (Bitumen Facility of Odor Removal)

Conditions. The polyethylene vessel (capacity 500 L) contained aqueous solution of hydrogen peroxide (10 wt% H_2O_2). This solution was injected into the wet scrubber for deodorization at the bitumen facility (bitumen processing). Similar polyethylene vessel contained caustic soda (10 wt% NaOH) solution. The dosing pump and electro-agitator of total weight 55 kg were installed on the top of the vessel (Fig. 5.46). This system was placed in secondary containment with concrete walls. Temperature changed in 10–45 °C during a year.

Material of construction: MDPE (medium density polyethylene, 935 g/dm^3; black-colored). The wall thickness of the vessel was 1.5 mm. According to the supplier of these vessels, MDPE is resistant to hydrogen peroxide solution at 20 °C and aqueous caustic soda solution at 20–60 °C, and the vessel of 500 L can contain liquids with density up to 1.6 kg/dm^3.

Service period before the failure: 1.5 years.

Visual examination and findings. The vessel containing H_2O_2 was suddenly broken during its service. Due to the falling of the dosing pump and the electro-agitator during destruction of the vessel, the part of concrete wall was broken too. The top and the wall of the polyethylene vessel were fully destructed (see Fig. 5.46b, c). A crack was detected in similar tank containing caustic soda solution (see Fig. 5.46d, e).

Failure phenomenon: *Fatigue and oxidation.*

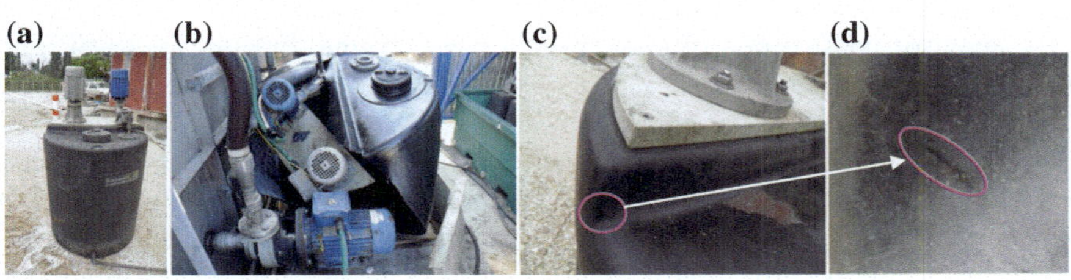

Fig. 5.46 **a** General view of storage vessel made from MDPE. **b** The vessel containing hydrogen peroxide (10 wt% H_2O_2) after the collapse. **c**, **d** The crack in the vessel containing aqueous caustic soda (10 wt% NaOH) solution

Cause of damage and its explanation. Vibration and dynamic forces of the dosing pump and mixer caused fatigue of the polyethylene vessel, crack formation and total destruction. Black-colored polyethylene (if no antioxidants are added) absorbs sunlight and warms more than white-colored polyethylene. We should point out that the material of the vessel (MDPE) is resistant to hydrogen peroxide solution (10 wt% H_2O_2) at 20 °C. However temperature was greater and reached 45°C. Hydrogen peroxide decomposes into oxygen and water and higher temperature stimulates this decomposition. The presence of strong oxidants hydrogen peroxide and oxygen as a product of decomposition of the former in the polyethylene vessel and warming are the factors which could hasten the destruction of the polyethylene container.

Solutions and recommendations.

(a) To remove the dosing pump and mixer from the top of the polyethylene vessel and install them separately from the container.
(b) To use LDPE with metallic knitwear and wall thickness 2 mm for construction material of the vessel.
(c) The vessel must be covered with tarpaulin in order to be protected from sunlight.

5.6.2.3 Corrosion and Coating Failure at the Dissolved Gas Flotation (DGF) Unit

Conditions. Dissolved Gas Flotation (DGF) system is used for removing remains of hydrocarbons and suspended solids from wastewater formed at oil refineries' and petrochemical units. The process floats hydrocarbons, solids and other contaminants to the surface of the liquids and then skimmed off and removed from the liquids. The principle of this separation process is based on the density of the droplets and creation of millions of millions of very small bubbles. The process takes place in large metallic pool. The impeller in the form of snail in the DGF pump is designed with dual sides. Liquid mixture of water, dissolved oxygen and other corrosive components can attack carbon steel surface of inner parts of DGF system. Usually organic protective coatings are used as anti-corrosion action. DGF system works at ambient temperature.

Material of construction. Carbon steel. Epoxy coating 250 μm thickness was applied in the DGF unit.

Service period before the failure: 2 years.

Visual examination and findings. Severe corrosion (up to holes) of inner surface in DGF system was detected, especially at the top location of the pool where water vapor and air (oxygen) contacted coated carbon steel surface (Fig. 5.47). Epoxy coating was fully destroyed in the upper part of the pool. Blisters in coatings and severe pits underneath were detected. Rust tubercles and shallow pits were formed underneath too.

Failure phenomenon: *General and pitting corrosion, coating destruction.*

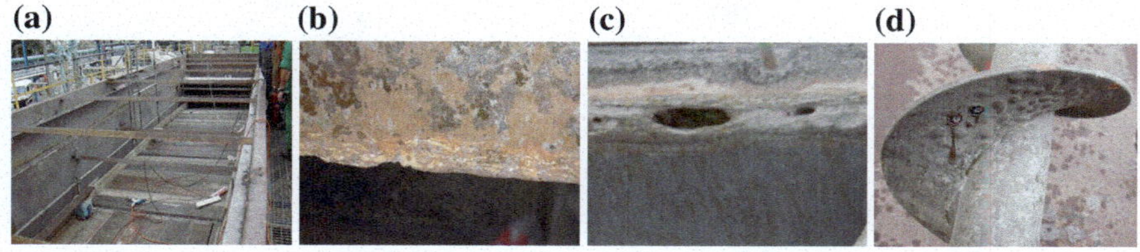

Fig. 5.47 **a** General view of DGF pool. **b** The upper corroded beam in the pool. **c** Corrosion holes in the DGF pool construction. **d** Large blisters in epoxy coating on impeller snail

Cause of failure and its explanation. Thickness 250 μm of epoxy coating was not enough for the protection of carbon steel under immersion conditions in wastewater saturated with air. Blisters were partly formed in the coating but coating was mostly destroyed. Then, severe corrosion of carbon steel surface occurred.

Solution and recommendations.

To apply epoxy, or polyurethane, or polyurea coating system (550 μm thickness).

5.6.2.4 Failure of Gas Scrubber (Ethylene Plant)

Conditions. The gas scrubber is intended for absorption of acid gases forming in regeneration in five reactors, for prevention pollution in the atmosphere. Three acid gases (SO_2, SO_3, and CO_2), and water vapor entered the scrubber at 270 °C. Then temperature in scrubber decreases to 33 °C. The height of the scrubber—5.8 m; the diameter—1 m. Gas inlet pipe is 4.8 m in length and 0.8 m in diameter. The scrubber works twice a year and each time for ~48 h. Water flowing through the cooling tower was pumping into the scrubber for absorbing of acid gases. Thus, absorption occurred in the inlet pipe to the scrubber. Water going out the scrubber had pH = 1–2.

History of failures. Inner surface of lower part of the gas scrubber was painted with epoxy but this coat was destroyed during several months. Significant decrease of wall thickness of lower part of scrubber was detected in the first year of service. Many holes were found in pipes around the scrubber. Severe corrosion of inner surface of the scrubber was detected after 3 years of service. Partial replacement of damaged from corrosion sections and patches were carried out after 5, 7 and 10 years of service. There was an attempt to inject ammonia for neutralization of acid gases, but without result.

Material of construction. Carbon steel.

Service period before the failure: 2 years.

Visual examination and findings. Leaks and severe corrosion with holes were found on inner surface of the scrubber (Fig. 5.48).

Failure phenomenon: *Acid corrosion.*

(a) (b) (c)

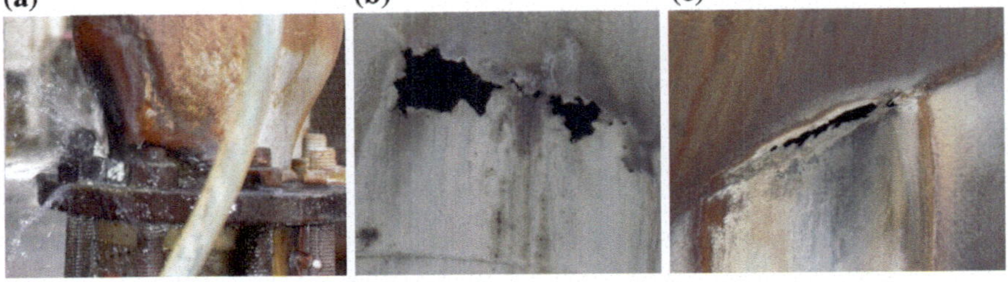

Fig. 5.48 **a** The leak in the pipe connection from the scrubber to the pump. **b** The hole in the wall (the location of acid gases escape). **c** The collapse of the tank shell in corroded area due to extremely low wall thickness

Cause of failure and its explanation. Acid gases (SO_2, SO_3, and CO_2) were absorbed by water, and, as a result, inorganic acids (H_2SO_3, H_2SO_4, and H_2CO_3) were formed. Carbon steel is not resistant to these acids and corrodes fast.

Solutions and recommendations.

(a) Implementation of neutralization of acid gases. Do not use any water, but a dilute aqueous solution of caustic soda (pH = 8–9). It is desirable to cool acid gases (to ~50 °C) before absorption in order to exclude possible caustic embrittlement of carbon steel at high temperature.
(b) Using alloys more resistant to acid attack: HASTELLOY C-276 or C-22, Carpenter 20Cb-3, AL6XN or 254 SMO.
(c) Using acid resistant coatings: epoxy-ceramic (epoxy phenol novolac + SiO_2 + Fe_2O_3); epoxy vinyl ester reinforced with fiberglass; or polysiloxane.
(d) Replacement of inlet pipe of acid gases (270 °C) with one of alloys resistant to acids (see b), and material of the scrubber with glass fiber reinforced polymer (GFRP).

5.6.2.5 Corrosion of Connection (T) of Urea Solution Injection Pipe and Bend into the Furnace

Conditions. The system is intended for neutralization NO_x polluted the atmosphere. Aqueous urea solution was injected into fuel gases containing NO_x at 170 °C and 6 bar. Thermal insulation would have to be on outer surface of injection pipe and bend.

Material of construction. Carbon steel. Diameter of the pipe—60 mm; wall thickness—6 mm.

Service period before the failure: 7 days.

Visual examination and findings. The holes (1–2 cm in diameter) were found right away after the obstacle in the connection pipe (T) and the bend of injection of urea solution into the furnace (Fig. 5.49). Thermal insulation on the tube bend was found broken.

Cause of failure and its explanation. The holes were formed in the place of change of urea flow regime. From the beginning, liquid urea without steam flowed some period and only then steam flowed separately. Changes of flow rate after the obstacle resulted in severe turbulence. The lack of thermal insulation on the tube bend gave rise to condensation of steam (170 °C → 100 °C) and cavitation. Morphology of surface of inner area of carbon steel pipe around the holes shows that holes were the result of erosion (because of severe turbulence flow) and cavitation (because of condensation of steam).

Failure phenomenon: *Erosion and cavitation.*

Solutions and recommendations.

(a) To replace carbon steel "T" with SS 316L (more resistant to erosion and cavitation) and to insulate it with Teflon from other carbon steel parts in order to prevent expected galvanic corrosion as aqueous urea solution is an electrolyte.
(b) To change the geometry of the "T" (to cancel 90° angle) and remove obstacles inside the pipe in order to change the flow regime and velocity, and to prevent turbulence in this part.
(c) To keep thermal insulation undamaged and carry out its periodical inspection.

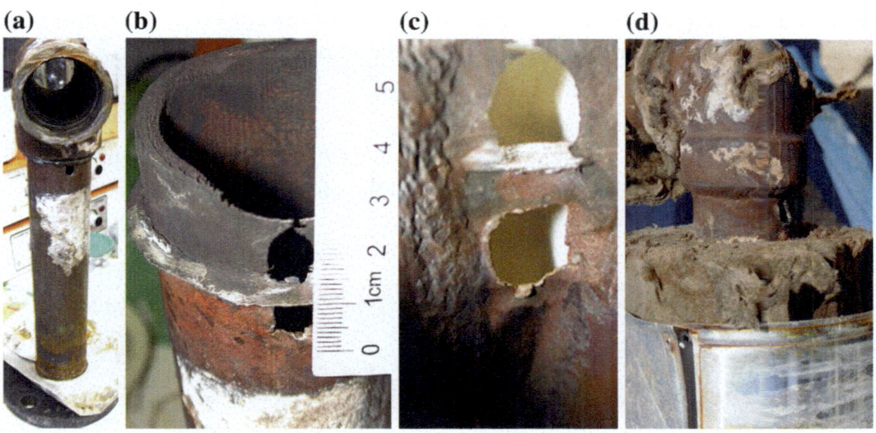

Fig. 5.49 **a** The tube connection (T) with the hole. **b** Holes (outside view). **c** Holes view from the inside (signs of erosion around holes). **d** The hole in the tube bend with lack of thermal insulation

5.6.2.6 Fretting Corrosion of Impeller of Submergible Pump (Waste Water Treatment Plant)

Conditions. Water from three separators was pumped into the equalizing pit. The submergible pump is intended for pumping of excess water from the equalizing pit. The pH values of the water varied from 2 to 14. The submergible pump worked with vibration.

Material of construction. Impeller—SS 316.

Service period before the failure: 3 years.

Visual examination and findings. Shallow pits were detected on the surface of the impeller (Fig. 5.50).

Failure phenomenon: *Fretting corrosion.*

Cause of failure and its explanation. Pit shape and morphology of deteriorated impeller surface shows that *fretting corrosion* occurred. *Fretting* is *wear* that occurs between tight-fitting surfaces subjected to a cyclic relative motion of extremely small amplitude. *Fretting corrosion* is a corrosion phenomenon that occurs at the contact area between two materials under load and subject to minute relative motion by vibration in the presence of corrosive medium. Thus, *fretting corrosion* is the fretting that occurs if the environment participates in metal destruction, or it is combining of *corrosion* and *wear*, or the *abrasive* effect of corrosion debris often occurring in equipment with moving or vibrating elements. During vibrating, the protective film (Cr_2O_3) on the stainless steel is removed by a rubbing action and exposes fresh, active metal to the corrosive action of the corrosive water from the separator. In this case, there was no ring on the impeller.

Solutions and recommendations.

(a) To apply (assemble) ring on the impeller.
(b) To eliminate vibration.
(c) To fill pits by welding and then to apply the chromium coating.

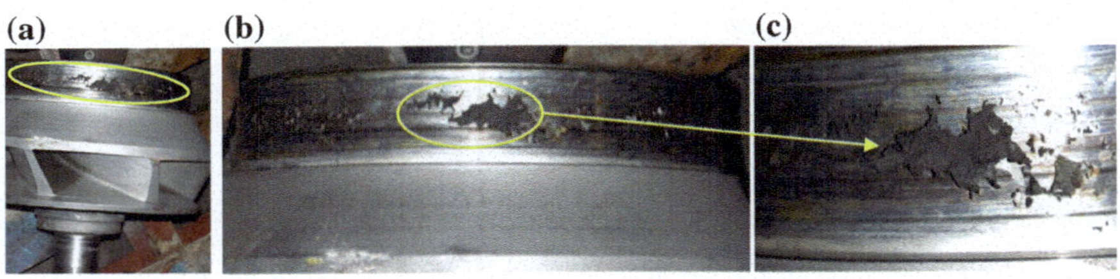

Fig. 5.50 a–c Submergible pump impeller (fretting corrosion)

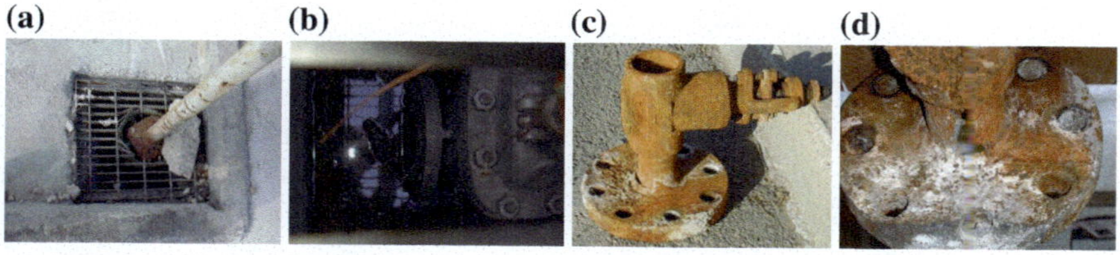

Fig. 5.51 **a** Drainage area: the grid and drainage tap inside the pit. **b** Drain pit with rain water. **c**, **d** Corroded tap for drainage of concentrated sulfuric acid

5.6.2.7 Corrosion of the Tap for Drainage of Concentrated Sulfuric Acid in Secondary Containment (Filling Point of Concentrated Sulfuric Acid at Spent Soda Treatment Unit)

Conditions. The tap was installed in the drain pit and contacted spilled concentrated sulfuric acid (Fig. 5.51). Rain water flowed down through the grid which was installed on the floor above the sump and accumulated inside the pit.

Material of construction: Tap—carbon steel.

Service period before the failure: 2 months.

Visual examination and findings. Holes and severe corrosion of the tap were found (Fig. 5.51).

Failure phenomenon: *Acid corrosion.*

Cause of failure and its explanation. Concentrated sulfuric acid (>93 wt%) does not cause corrosion of carbon steel (see Appendix D). However, carbon steel corrodes intensively in diluted aqueous solution of sulfuric acid. When concentrated sulfuric acid spilled nothing could happen with carbon steel tap. However, because of the presence of rain water in the pit sulfuric acid became diluted and gave rise to severe corrosion of the tap made from carbon steel. Morphology of corroded surface of the carbon steel tap indicated typical acid attack.

Solution and recommendations.

To replace the tap material with one of synthetic fluoropolymers resistant to sulfuric acid of different concentrations: PTFE (Teflon), PFA, TFE, FEP, ETFE, or ECTFE.

5.7 Tanks

Tanks occupy a special niche among carbon steel structures at oil refineries and petrochemical plants. They have diverse constructions and dimensions, and can contain different media: crude oil, petroleum products, fuels, organic solvents, water, slops, acids and alkalis, wastewater, and sour water. Each refinery has about

150–200 tanks of different capacities. Outer surface of tanks can contact atmosphere, rain water, soil, and thermal insulation. Thus, both inner and outer surfaces of tanks are susceptible to numerous corrosion phenomena. Corrosion problems for tanks are well known and corrosion mechanisms are established [2]. Therefore, standards for design, manufacture, anticorrosion control measures, monitoring, maintenance, inspection, repairs and demolition were developed and introduced into practice [22–56]. In spite of these guidelines and codes of practice, corrosion of inner and outer surface of tanks containing different media take place and three case studies are described in this section.

5.7.1 *Case Studies*

5.7.1.1 Corrosion of the Bottom of the Aboveground Storage Tank Containing Asphalt

Conditions. The bottom of the aboveground storage tank containing asphalt was founded on the sand and had no anti-corrosion protection from both sides. The temperature inside the tank was ~200 °C. Thermal insulation (mineral wool) and aluminum jacketing were applied on outer surface of the tank.

Material of construction: Carbon steel. The thickness of the bottom sheets—9 mm.

Service period before the failure: 13 years.

Visual examination and findings. Severe corrosion of outer surface of the bottom contacting sand was found. No corrosion inside the tank bottom was detected. Rust was found inside the upper part of the tank and outside under the thermal insulation (Fig. 5.52).

Failure phenomena: *General and localized corrosion by hot water* (outer surface of the bottom contacting sand); *general corrosion by water vapor and oxygen* (inner surface of upper part of the tank); *corrosion under thermal insulation* (outer surface of the tank).

Causes of failure and explanation.

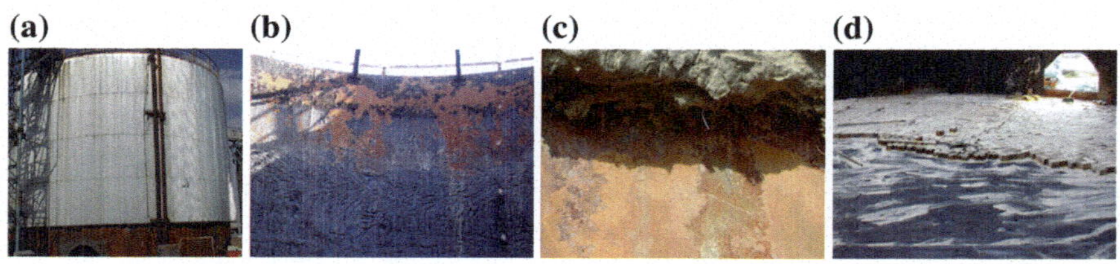

Fig. 5.52 **a** General view of the aboveground storage tank containing asphalt. **b** Inner surface of the tank. **c** Outer surface of the tank (corrosion under thermal insulation). **d** Polyethylene sheets and bricks during the reconstruction (repair) of new bottom

Outer surface of the bottom. Rain water penetrated under the bottom and caused severe corrosion of carbon steel outer surface contacting sand at high temperature (~ 100 °C).

Inner surface of upper part of the tank. Inner surface of strips which contacted asphalt did not corrode because asphalt was adsorbed on the carbon steel surface and protected it from corrosion. However, upper strips under the roof did not contact asphalt (as the tank was not completely filled by asphalt) and were exposed to the air containing water vapor. Typical shelter atmospheric corrosion occurred [2, 57].

Outer surface of the tank. Corrosion under thermal insulation occurred [2, 57].

Solutions and recommendations.

To apply following system of anti-corrosion protection of the bottom: to assemble the polyethylene sheeting (HDPE, 5 mm in thickness), bricks, concrete, and then carbon steel sheets with the thickness of 10 mm.

There is no need for protection against corrosion of outer surface of the tank bottom because carbon steel contacts concrete which has alkali property (pH ≈ 12–12.5 because of the presence of $Ca(OH)_2$ in the concrete). Thus, the steel will be passive (in the case of water penetration).

Note. When double bottom in tanks or new pipelines are installed in the soil, the space between two bottoms and around the pipelines is filled by sand. It is important to use 'sweet' sand which does not contain salts. In no case sea sand must be used. In such cases, sand should be analyzed on the presence of electrolytes in order to prevent possible corrosion of carbon steel tank bottoms or pipelines. For this analysis, sand is mixed with distilled water (1:1), vigorously agitated and aqueous extract is analyzed. Requirements to chemical composition of aqueous extract from a sand are shown in Table 5.7. Sand that meets the requirements of this table is named 'sweet' sand.

5.7.1.2 Corrosion of the Tank (Kerosene Treatment Unit)

Conditions. The tank (volume 50 m^3) is intended to absorb residual water from kerosene or naphtha by solid sodium chloride (NaCl). The wall thickness of the tank—18 mm. Temperature—ambient.

Material of construction: Carbon steel A516 Gr.70.

Service period before the failure: 12 years.

Visual examination and findings. Shallow pits of different forms with deepness to 15 mm on inner surface of the tank were found (Fig. 5.53). All inner surface which contacted salt solution was corroded and pitted.

Table 5.7 Requirements to chemical composition of aqueous extract from a sand ('sweet')

Component/property	Unit	Value
pH		7–8.3
Conductivity	μS/cm	≤200
Cl^-	ppm	≤10
SO_4^{2-}	ppm	≤15

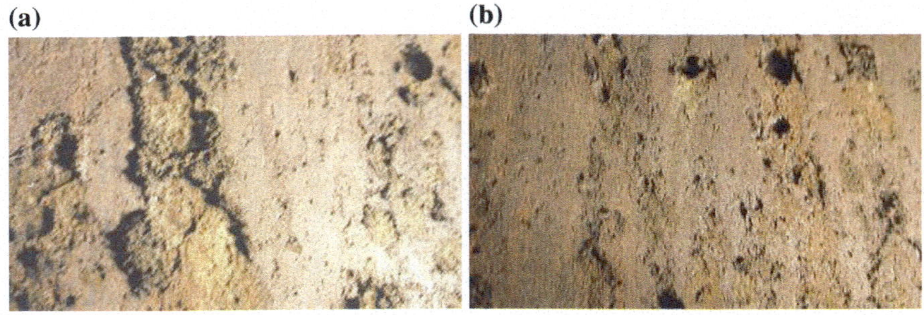

Fig. 5.53 **a**, **b** Corroded and pitted inner surface of the tank (50 m^3) containing solid sodium chloride (NaCl) for absorption residual water from kerosene or naphtha

Failure phenomenon: *General and pitting corrosion (in contact with aqueous NaCl solution).*

Cause of failure and its explanation. Aqueous solution of sodium chloride is aggressive to carbon steel and caused general and pitting corrosion according to electrochemical mechanism.

Solutions and recommendations.

(a) Such tanks are designed for 25 years of service, and failure after 12 years is a half time of planned service life. Anti-corrosion protection of the tank's inner surface had to be planned on the stage of design.
(b) It is impossible to continue using the tank in current situation and it is impossible to repair inner surface of the tank after such harsh pitting corrosion.
(c) To replace failed tank with new one and to paint inner surface of the tank by epoxy coating (550–600 μm in thickness).
(d) To carry out regular inspection (every 5 years) of inner surface of the tank.

5.7.1.3 Corrosion of the Sedimentation Tank (Power Plant)

Conditions. The tank served for purifying of water from carbonates (CO_3^{2-}) and sulfates (SO_4^{2-}) anions by addition burnt lime (CaO) into potable water, namely, for sedimentation. There were no corrosion problems during 14 years of such service. Severe corrosion occurred inside the tank after the changing of its role and starting using of water after biological treatment and addition of flocculent sodium aluminate.

Material of construction: Carbon steel.

Service period before the failure: 1.5 years.

Visual examination and findings. Rust and biofouling were detected inside the tank (Fig. 5.54). Support construction and stair fell because of corrosion of bolts in connecting structures.

Failure phenomenon: *MIC and general corrosion.*

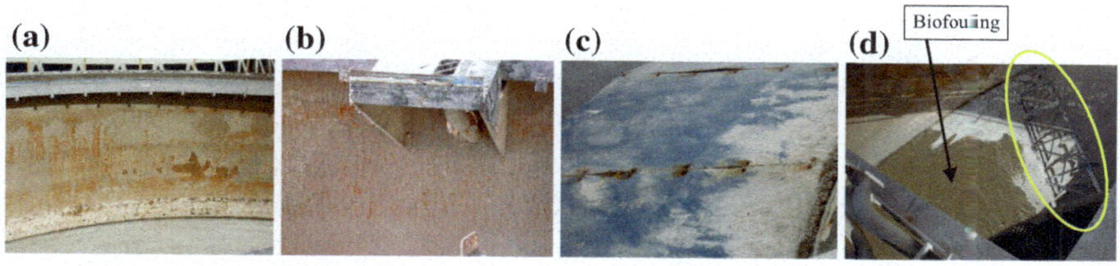

Fig. 5.54 **a**, **b** Inner surface of the tank. **c** Corroded rusted bolts. **d** Biofouling on the bottom of the tank and support construction (circled by *yellow*) which fell due to corrosion of bolts

Cause of failure and its explanation. Diverse microorganisms containing in water after biotreatment caused severe corrosion of inner surface of the tank and accessories made from carbon steel.

Solution and recommendations.

To coat inner surface of the tank and accessories by epoxy, or polyurethane, or polyurea coating.

5.8 Water Systems

Like living organisms cannot exist without water, oil refineries and petrochemical plants cannot function without water too. Water is used in cooling water systems, for preparation steam, and in fire-fighting system. Petroleum products and fuels obtaining after distillation and different catalytic and cracking processes should be chilled. Water is the best media for this. Different water sources bring numerous corrosion problems which need careful attention and treatment. In spite of more narrow range of conditions of water use (10–40 °C and 4–5 bar) than crude oil and petroleum products, all known corrosion phenomena take place with equipment contacting water. On the one hand, theory of metal corrosion in water is known well and broad spectrum of anti-corrosion actions exists and can be applied. On the other hand, there are so many factors, mainly diverse physicochemical compositions of waters and their changes, which influence corrosion of metals in water that make prediction of corrosion difficult. Sometimes fire-fighting pipelines are filled with potable water treated by biocides. Water does not flow in such pipelines and is present under stagnation during long time. In spite of preliminary chlorination of potable water severe MIC (mostly by iron-bacteria) can occur up to through-holes in pipe wall with thickness 6 mm after 2–3 years of service. Such cases are a long way from safety because the absence of water pressure in fire-fighting pipeline during the fire at oil refinery can result in catastrophe.

The corrosion rate of carbon steel in the majority of aqueous media decreases with time because of formation of corrosion products on the surface (similar to atmospheric corrosion). The main problems in water systems are corrosion, scaling, and biofouling. The latter in most cases results in MIC (see Sect. 6.1 and Chap. 7).

Nearly all cooling water systems are treated by chemicals which include corrosion inhibitors, biocides, and anti-scaling agents. Protective coatings and sacrificial anodes are used for corrosion control of inner surfaces of heat exchangers contacting cooling water. Their efficiency is checked every 4–5 years during shutdown. Sometimes designers do not take into consideration all technological parameters when they select coatings. For instance, cooling water temperature usually ranges in 30–35 °C. However, steam at 200 °C is supplied several hours in many heat exchangers to wash inner surface from hydrocarbons before dismantle during shutdown. Most water resistant coatings are not resistant to such high temperature, are destroyed and should be repaired. Sacrificial anodes mounted inside of heat exchangers also do not always serve effectively because of incorrect taking into account electroconductivity of water, especially if water condensate is mixed with hydrocarbons. Corrosion monitoring should be applied in all water systems.

Typical and untypical corrosion problems and their solutions (including corrosion monitoring) in water systems at refineries and petrochemical plants are described in this section.

5.8.1 Case Studies

5.8.1.1 Corrosion of Heat Exchanger Accessories (Isomerization Unit)

Conditions. Cooling water flowed inside the tubes at 30–35 °C and 4–5 bar. LPG from the stabilizer and after the air cooler would have to be flow outside the tubes at 40–60 °C and 12 bar in accordance with processing requirements. Really the temperatures of LPG changed in 80–100 °C. Leak of hydrocarbons into the cooling water was detected in the heat exchanger because of difference in pressure from processing and cooling water sides. The cooling water was treated by chemicals (corrosion inhibitors, anti-scaling agents and biocides) only during the last 1.5 years.

Materials of construction. Tubes—CDA 706 (90Cu-10Ni). Accessories (floating head shell and inlet channel) of heat exchanger—carbon steel.

Service period before the failure: 3 years.

Visual examination and findings. Deposits and severe corrosion underneath were detected on floating head shell and inlet channel which contacted cooling water (Fig. 5.55).

ED-XRF. Following elements were detected in deposits (wt%): Zn (58–66), Ca (22–26), Fe (6–7), Cl (1–3), P (3–6), Si (0.7–1.2), S (0.9–1.0), Al (0.6–0.8), and Cu (0.3–0.7).

Physicochemical analysis. Deposits contained (wt%) organic (10) and inorganic (90) substances, and among the latter chlorides (10).

Failure phenomenon: *Under deposit corrosion* (differential aeration cells).

Cause of failure and its explanation. Real high temperature of LPG (80–100 °C) caused formation of irregular deposits from the cooling water side consisting mostly

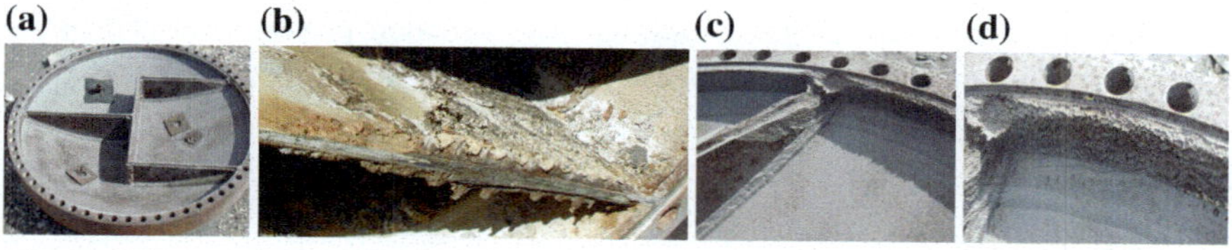

Fig. 5.55 Failed floating head shell and inlet channel (cooling water side). **a** Inner surface of floating head shell after cleaning from deposits. **b** Thick deposits on the surface of the channel. **c**, **d** Corroded areas of floating head shell (under deposit corrosion)

of phosphates and carbonates of zinc, calcium, and iron. Chlorides were also present in these deposits. Corrosion occurred under deposits as a result of action of differential aeration cells.

Solutions and recommendations.

(a) To repair carbon steel surface of the floating head shell and inlet channel by welding.
(b) To ensure requirements of technological process in order to prevent the increase of temperature of LPG stream above 60 °C.
(c) To paint inner surfaces of floating head shell and inlet channel with coatings (epoxy or polyurethane) resistant to water and temperature.

5.8.1.2 Clogging and Corrosion of Heat Exchanger (Power Station)

Conditions. Cooling water flowed outside the tubes. Air flowed inside the tubes. Cooling water was untreated by corrosion inhibitors, anti-scaling agents and biocides. Maximum temperature of air was 90 °C.

Material of construction. Carbon steel.

Service period before the failure: 5 years.

Visual examination and findings. The heat exchanger was shutdown and open. Clogging by deposits was found outside tubes (Fig. 5.56).

Cause of failure and its explanation. The water flow velocity was very slow outside the tubes, heat exchange was poor, and that gave rise to heating which in its turn resulted in deposit formation outside the tubes. Corrosion products consisted of rust.

Fig. 5.56 Failed heat exchanger with deposits

Heat exchanger has been clogged long ago by corrosion products formed in the pipeline with untreated cooling water.

Failure phenomenon: *Clogging (fouling) and general corrosion.*

Solutions and recommendations.

(a) To remove deposits and clean outer surface of tubes.
(b) To use cooling water only treated by corrosion inhibitors, anti-scaling agents and biocides.

5.8.1.3 Corrosion in Coolers of Compressor (Power Station)

Conditions. There are three coolers of the 1st, the 2nd and the 3rd stages of compressor. The purpose of coolers is to cool compressed air. Water treating with corrosion inhibitor sodium silicate flowed inside the tubes. Compressed air flowed in the shell and was heated to 140 °C.

Materials of construction. Coolers—copper tubes and copper fins. Shell—carbon steel. Inner surfaces of shells were coated by epoxy paint.

Service period before the failure: 4 years.

Visual examination and findings. During the opening of coolers in order to define the reasons of the diminishing of the cooling efficiency of the air, severe corrosion (rust) and destruction of coating was found on the inner surface of the shell in coolers of the 2nd and the 3rd stages (Fig. 5.57).

Cause of failure and its explanation. When air is compressed, the water vapor containing in air condenses. The temperature of air rises to 110–140 °C as a result of compression. As a compression stage is higher, the air temperature is also higher. Water vapor is partly condensed. Thus, inner surface of the shell contacted two-phase system hot liquid water and water vapor containing oxygen. Epoxy coating using inside the shell was not resistant to hot water (~ 100 °C) under immersion conditions. Carbon steel corroded by mixture of hot liquid water and water vapor containing oxygen. Galvanic corrosion of the inner surface of carbon steel shell in contact with copper tubes and fins in the presence of condensed hot water also contributed in severe corrosion.

(a) (b) (c)

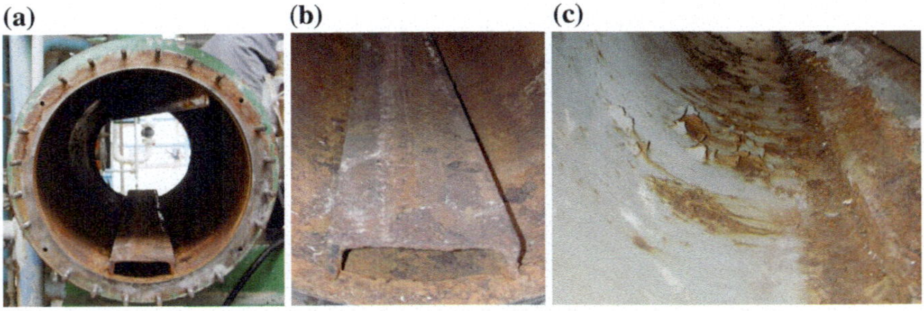

Fig. 5.57 **a** The cooler (the 3rd stage of the compressor). **b** Inner rusted surface (magnification of the picture **a**). **c** Failed epoxy coating (the 2nd stage of the compressor)

Sometimes fins are made from aluminum or its alloy. They contact copper tubes and in the presence of water aluminum will work as sacrificial anode, will corrode quickly and will turn into dusted corrosion products of aluminum (hydroxides and oxides) clogging the cooler (see Fig. 5.65). This can occur during 1–2 years. If the fins are made of copper like tubes, neither they nor tubes will corrode, but dust coming with dirty refinery air in any case will clog the space of the cooler, reducing its cooling efficiency.

Failure phenomenon: *Hot water corrosion, galvanic corrosion, destruction of epoxy coating.*

Solutions and recommendations.

(a) To dry the air from water vapor in the inlet into compressor.
(b) To install filter for catching dust in the entering air.
(c) To paint inner surface of carbon steel shell by coating system resistant to high temperature (~100 °C) under immersion conditions: epoxy novolac (450 μm in thickness).

5.8.1.4 Corrosion of Tubes in Heat Exchanger

Conditions. Cooling water (treated with corrosion inhibitors, anti-scaling agents and biocides) flowed outside the tubes at 50 °C. Cooling water contained ~250–500 ppm Cl^-. Sour water flowed inside the tubes. Leaks were revealed in 130 tubes among total 1300 tubes in heat exchanger.

Material of construction. SS 304.

Service period before the failure: 4 years.

Visual examination and findings. Holes and pits formed from the outer surface of the tubes were found (Fig. 5.58).

Cause of failure and its explanation. The failure began from the outside of the tubes contacting cooling water, and occurred as a result of the development of pitting corrosion under deposits containing chlorides.

Failure phenomenon: *Pitting corrosion.*

Solutions and recommendations:

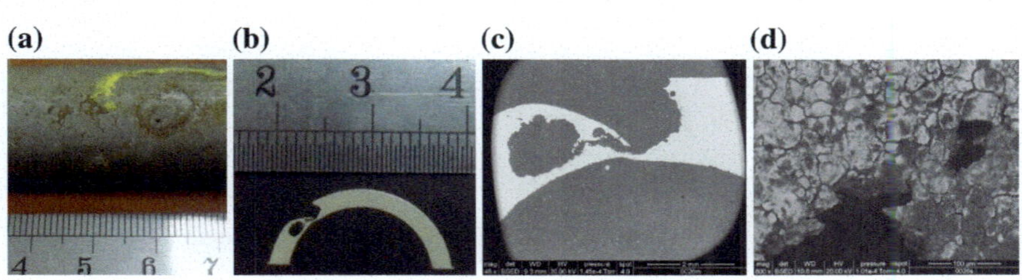

Fig. 5.58 **a** The tube (SS304) with the pit. **b** Macroscopic cross-section through the tube with the pit. **c** SEM photo of metallographic cross-section of the tube with pit. **d** SEM photo of the pit

(a) To prevent formation of deposits on outer surface of the tubes: increase flow velocity of cooling water (not less than 1.2 m/s).
(b) To replace SS 304 with SS 317L which is more resistant to pitting corrosion.

5.8.1.5 Pitting Corrosion of SS 410 Tubes in Heat Exchanger (CCR Unit)

Conditions. Two new heat exchangers were tested for hermeticity (tightness) by means of water pressure test (uncontrolled chloride content) and have been leaving for 2 years for storage without draining. Water was remaining inside and outside tubes. Tubes leaked during the commissioning at the CCR unit.

Materials of construction. Tubes—martensitic SS 410 (12 % Cr). The manhole and tube-sheets—carbon steel.

Service period before the failure: 2 years.

Visual examination and findings. Rust was found on the surfaces of tubes, tube-sheets and inlet nozzle (Fig. 5.59). Pits and holes were detected on the surface of several tubes. The holes were formed from the outside of the tubes.

SEM examination. Pit on outer surface of the tube and corrosion products at the pit bottom were detected (see Fig. 5.59d).

EDS analysis of corrosion products: Fe, Cr, O and Cl ($\sim$1.5 wt%) were found inside the pit.

Metallographic examination. Large pits and development of small pits inside were found.

Failure phenomenon: *Pitting corrosion* (of martensitic SS 410) and *general corrosion* (of carbon steel).

Cause of failure and its explanation. Failure was caused because of stagnation of water containing chlorides. It is important to recall that *stainless steel* got its name in 1913 because of its high resistance to pure atmosphere. Martensitic SS 410 contains passive Cr_2O_3 film and is susceptible to pitting corrosion by chlorides in stagnant water. Critical maximum concentration of chlorides in water is 5 ppm without pitting corrosion of SS 410 [57]. This stainless steel is also susceptible to pitting corrosion in atmosphere containing chlorides.

(a) (b) (c) (d)

Fig. 5.59 **a** General view of heat exchanger with failed tubes. **b** Rusted and pitted tubes. **c** Rusted inlet nozzle. **d** SEM photo (magnification ×80) of the pit covered with corrosion products

Solutions and recommendations.

(a) To use pure water containing less than 5 ppm chlorides for hydrostatic pressure test of equipment made from martensitic SS 410 and to fill with diesel fuel for long time storage.
(b) In the case of use of water containing more than 5 ppm chlorides, immediately to drain heat exchanger after the pressure test, to wash it with pure water (less than 5 ppm chlorides) and to fill with diesel fuel for long time storage.
(c) Passivation of SS 410 tubes according to requirements of standards [18, 58].

5.8.1.6 Crevice Corrosion in Heat Exchanger (Amine Treating Unit)

Conditions. Cooling water (untreated with corrosion inhibitors) flowed inside the tubes of two heat exchangers at 30–36 °C.

Material of construction: Tubes and tube-sheets—SS 316.

Service period before the failure: 4 years.

Visual examination and findings. Localized corrosion was detected under polymer tube-sheet gasket (located between the tube-sheet and channel) from the cooling water side (Fig. 5.60).

Failure phenomenon: *Crevice corrosion.*

Cause of failure and its explanation. The polymer gasket was not tightly installed between the channel and tube-sheet stainless steel. Thus, the gap (crevice) was formed between gasket (non-metal) and tube-sheet stainless steel surface (metal). This crevice was wide enough to permit cooling water entry, but sufficiently narrow to maintain the stagnant zone. In the case of stainless steel, crevice corrosion was initiated by changes in local chemistry within the crevice (increase of Cl^- content or decrease pH). As a result *crevice corrosion* occurred due to lack of tightness between the gasket and the tube-sheet stainless steel surface [59–61].

Solution and recommendations.

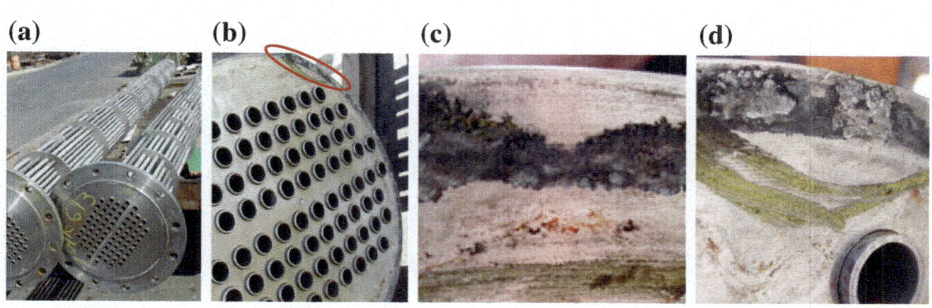

Fig. 5.60 **a** General view on stainless steel bundles of heat exchangers. **b** Tube-sheet with failed area as a result of crevice corrosion (*red*). **c, d** shallow pits (crevice corrosion)

(a) To prevent using wetting polymeric materials and use only nonabsorbent (non-porous) gaskets (e.g., Teflon, PTFE).
(b) To decrease dimensions of the gap (to increase the extent of tightness) in order to prevent the formation of crevice and cooling water infiltration.

5.8.1.7 Corrosion and Deposits in Oil Cooler of Turbo-Generator (Power Station)

Conditions. The oil cooler of turbo-generator is intended for the chilling oil by water. Cooling water (untreated by corrosion inhibitors) flowed inside the tubes during 4 years. This water was replaced with new cooling water treated by zinc-phosphate-phosphonate corrosion inhibitor, biocide and anti-scaling agent during a year before the failure. Inner surface of the channel was coated with epoxy coating. Aluminum sacrificial anodes were installed on the tube-sheet for its and shell protection from corrosion in cooling water.

Material of construction. Tubes and tube-sheet—Admiralty brass CDA 443. Sacrificial anodes—aluminum alloy.

Service period before the failure: 5 years.

Visual examination and findings. White deposits were found on the surface of the tube-sheet of the oil cooler (Fig. 5.61). Chemical analysis showed that the deposits consisted of phosphates and hydroxides of Al, Mg, Ca, and Zn.

Failure phenomenon: *Accelerated corrosion of aluminum sacrificial anode and formation of deposits.*

Cause of failure and its explanation. When cooling water (untreated by corrosion inhibitors) was used, aluminum sacrificial anodes functioned, namely, dissolved without formation of deposits. When the injecting of zinc-phosphate-phosphonate corrosion inhibitor into cooling water was begun, white deposits of aluminum phosphates and hydroxides were formed. Connecting an aluminum anode to Admiralty brass plate is extremely dangerous combination (see Sect. 5.8.1.10). The purpose of the aluminum sacrificial anode is to protect the defects in painted surface of the channel. The difference of electric potentials of aluminum anode and

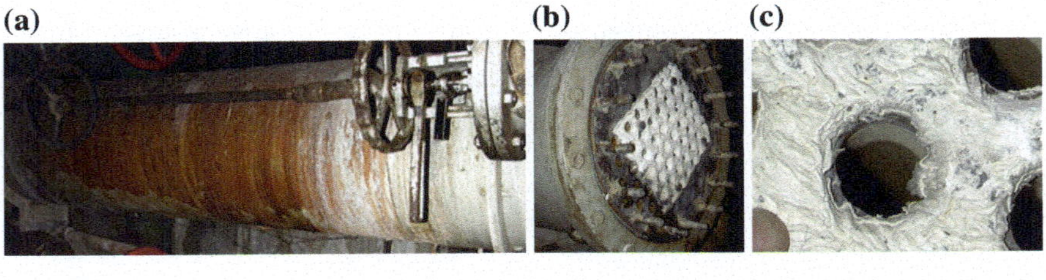

Fig. 5.61 **a** General view of the oil cooler (power station). **b** White deposits accumulated on sacrificial aluminum anode. **c** Magnification of the picture **b**

Admiralty brass is ~1 V which caused intensive dissolution (corrosion!) of aluminum. Deposits consisting of phosphates and hydroxides of Al, Mg, Ca, and Zn were accumulated on the surface of aluminum sacrificial anodes and reduced their activity.

Solutions and recommendations.

(a) Do not use aluminum sacrificial anodes in contact with Admiralty brass.
(b) It is recommended to paint inner surface of channel by epoxy or polyurethane coatings.
(c) To install sacrificial anode from zinc in the channel.

5.8.1.8 Galvanic Corrosion of Heat Exchanger Accessories (HDS HVGO Unit)

Conditions. Heat exchanger is intended for cooling of light naphtha which flowed out at 85 °C (this temperature would have to be) from the splitter at HDS HVGO unit. Real temperature of naphtha was ~100 °C and more. Cooling water flowed inside tubes at 30–35 °C and 4–5 bar.

Materials of construction. Tubes—Ti Gr.2. Tube-sheet—carbon steel coated with epoxy paint.

Service period before the failure: 8 years.

Visual examination and findings. Total destruction of epoxy coating and localized corrosion of tube-sheets in contact with the edges of titanium tubes were detected on cooling water side (Fig. 5.62). Outer surface of titanium tubes which contacted naphtha were in fine appearance (see Fig. 5.62d).

Failure phenomenon: *Galvanic corrosion.*

Cause of failure and its explanation. Galvanic corrosion occurred due to contact between the edges of the titanium tubes and carbon steel tube-sheet coated with epoxy paint. Carbon steel of the tube-sheet was an anode and its area was decreased significantly after painting with epoxy. Titanium tubes were cathode. After painting, ratio of anode area (carbon steel tube-sheet) to cathode area (titanium tubes)

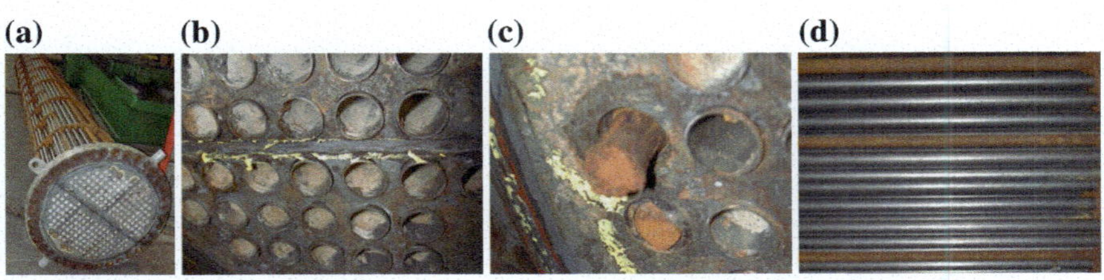

Fig. 5.62 **a** General view of heat exchanger bundle (Ti Gr.2 tubes). **b**, **c** Carbon steel tube-sheet (cooling water side): corrosion of carbon steel tube-sheet around the titanium tubes. **d** Titanium tubes in good appearance (naphtha side)

was reduced drastically and such situation gave rise to accelerated corrosion in defects of epoxy coating (small anode regions underneath coating on carbon steel tube-sheet). The mistake was the painting of the carbon steel tube-sheet (anode).

Solutions and recommendations.

(a) To carry out clad on carbon steel tube-sheets from Ti Gr.2 or prepare tube-sheets from one solid material Ti Gr.2.
(b) If it is impossible to carry out clad from Ti Gr.2, it is better to paint inner surface of titanium tubes ($\sim$30 cm length) and not to paint the carbon steel tube-sheet surface.

5.8.1.9 Galvanic Corrosion of Heat Exchanger Accessories (HDS Unit)

Conditions. Heat exchanger is intended for cooling 'treated water' (containing absorbed gases formed in Tail Gas Treating at HDS unit) by cooling water. 'Treated water' flowed inside the tubes, and the temperature changed from 54 to 43 °C. Cooling water flowed outside the tubes (in the shell) and the temperature changed from 30 to 35 °C.

Material of construction. Tubes—SS 304. Shell, baffles, tie rods, spacer tubes, bolts-nuts—carbon steel A106 Gr.B.

Service period before the failure: 3 years.

Visual examination and findings. Severe corrosion of carbon steel baffles, tie rods, spacer tubes, bolts-nuts and inner surface of the shell were found in the heat exchanger (Fig. 5.63). Stainless steel tubes had good appearance.

Failure phenomenon: *Galvanic corrosion.*

Cause of failure and its explanation. Cooling water is a good electrolyte. There is a difference between electrode potentials of SS 304 (−0.10 V, cathode) and carbon steel (−0.44 V, anode) in cooling water. The area ratio of cathode (SS304 tubes) to anode (carbon steel parts baffles tie rods, spacer tubes, bolts–nuts) is large. Combination of different alloys (carbon steel and stainless steel) in contact in cooling water (strong electrolyte) gave rise to galvanic corrosion of carbon steel elements.

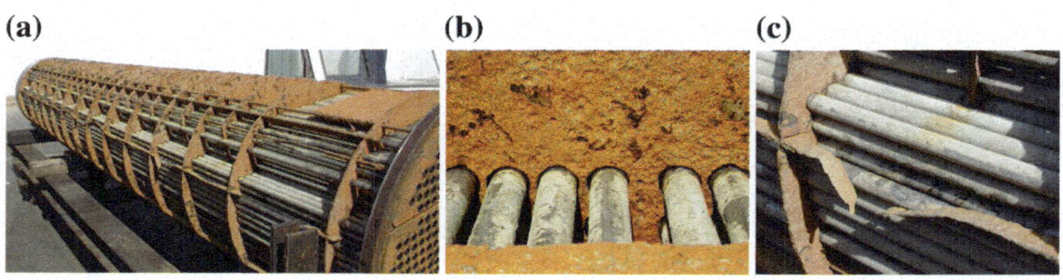

Fig. 5.63 **a** General view of heat exchanger bundle. **b**, **c** Corroded carbon steel parts

Solution and recommendations.

Galvanic corrosion is very old phenomenon and occurs frequently in oil refining and petrochemical industry because of wrong design. All accessories (baffles, tie rods, spacer tubes, bolts-nuts) must be made from material similar to that of the tubes—SS 304.

5.8.1.10 Very Dangerous Galvanic Corrosion

Heat exchangers can be made from different alloys, for instance, carbon steel and Admiralty brass. In order to have the galvanic corrosion occurred, it is not always necessary two different alloys to be in physical contact. Galvanic corrosion can be caused by the presence of metal cations in water. If copper corrosion inhibitors are not injected in cooling water, copper may corrode in water in the presence of dissolved oxygen, and then deposit on carbon steel surface. As a result, severe pitting corrosion takes place under copper deposits (Fig. 5.64).

If heat exchanger consists from copper tubes and aluminum fins installed inside of carbon steel shell, catastrophic failure can occur even when water condensate is formed (Fig. 5.65). This heat exchanger was intended for cooling of air for compressor. Water flowed inside the copper tubes. The air flowed outside the copper tubes among aluminum fins. Water vapor in the air were condensed and caused severe galvanic corrosion of aluminum fins. The conclusion is that in no case to use copper and aluminum in one system or equipment.

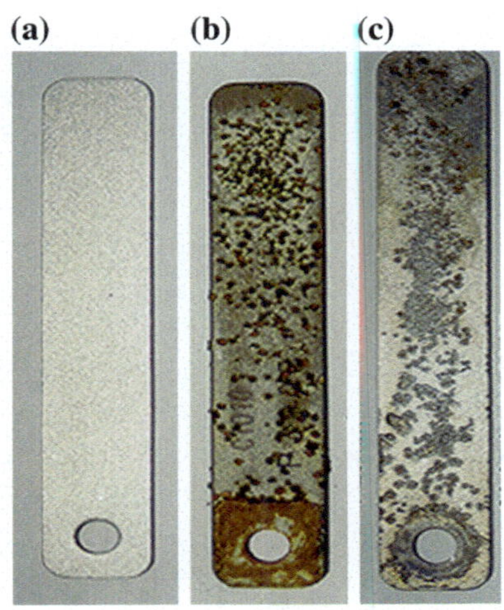

Fig. 5.64 **a** Original carbon steel coupon. **b** Coupon with adsorbed copper particles (after 30 days in cooling water). **c** The same coupon after removing copper particles. Average corrosion rate of carbon steel coupon—0.06 mm/year. Pitting corrosion rate—1.8 mm/year

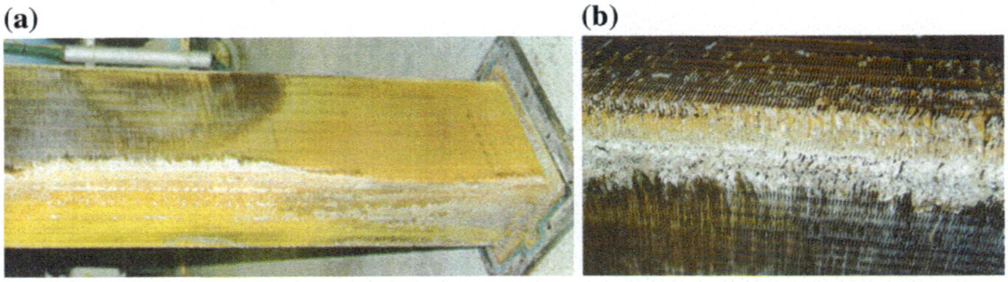

Fig. 5.65 **a** General view of failed heat exchanger (4 years of service in the atmosphere of the oil refinery): tubes—copper; fins—aluminum; shell—carbon steel (not shown). **b** Aluminum fins corroded and turned into white-grey dust

5.8.1.11 Corrosion of Concrete Wall in Cooling Tower

Conditions. Concentrated (98 wt%) sulfuric acid is injected into cooling water for keeping its pH neutral. Injection of acid is carried out in the vicinity ($\sim$20 cm) from the concrete wall of cooling tower.

Material of construction. Concrete wall. Polyethylene pipe for injection of sulfuric acid.

Service period before the failure: 7 years.

Visual examination and findings. Severe corrosion of the concrete wall in the vicinity of acid injection was found in cooling tower (Fig. 5.66).

Fig. 5.66 **a** Polyethylene pipe for injection of concentrated sulfuric acid. **b** Corroded concrete in the vicinity of sulfuric acid injection point

Cause of failure and its explanation. Normal concrete is not resistant to acids. The concrete wall corroded by concentrated (98 wt%) sulfuric acid in the area of its injection.

Failure phenomenon: *Acid corrosion of concrete.*

Solutions and recommendations.

(a) Concrete wall must be repaired. To apply coating system resistant to acids. Since the concrete wall will be wet (even after cessation of cooling water flow), a moisture tolerant coating system (including primer and vinyl ester reinforced with glass flaked coating) should be applied.
(b) To perform sulfuric acid injection pipe from polypropylene and make longer it to 1 m from the concrete wall.

5.8.1.12 Corrosion of Metal Constructions (Fans' Axis) in Cooling Tower

Conditions. Cooling tower. Cooling water at ambient temperature.

Material of construction. Carbon steel.

Service period before the failure: 15 years.

Visual examination and findings. Severe corrosion of fans in cooling tower was detected (Fig. 5.67).

Cause of failure and its explanation. Chemical plant producing fertilizers and acids functioned in the vicinity (several hundred meters) to cooling towers. Such closeness gave rise to corrosive atmosphere in this area. Rains and wetness produced liquid acidic solutions which caused severe corrosion of carbon steel fans in cooling water.

Failure phenomenon: *General corrosion.*

Solutions and recommendations.

(a) To replace corroded fans (axis, base and other elements).
(b) To paint metal constructions by coating systems resistant to aggressive wetted atmosphere [2].

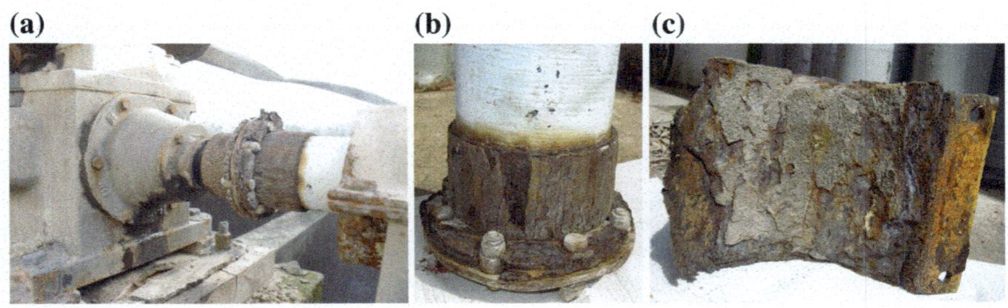

Fig. 5.67 **a**, **b** Corroded axis of the fan (cooling tower). **c** Corroded base of the fan

References

1. API RP 571 (2011) Damage mechanisms affecting fixed equipment in the refining and petrochemical industries, 2nd edn. American Petroleum Institute, West Conshohocken, PA, USA
2. Groysman A (2014) Corrosion in systems for transportation and storage of petroleum products. Springer, Dordrecht
3. Heidersbach R (2011) Metallurgy and corrosion control on oil and gas production. Wiley, Hoboken, New Jersey, USA
4. Garverick L (ed) (1994) Corrosion in the petrochemical industry. ASM International, USA
5. Esaklul KA (ed) (1992) Handbook of case histories in failure analysis, vol 1, 2. ASM International, USA
6. Powell GW (1986) Failure analysis and prevention, Metals Handbook, vol 11, 9 edn. ASM International, USA
7. Groysman A et al (2000) Cavitation corrosion in a continuous catalytic reformer unit. Mater Perform 39(6):62–66
8. Speight JG (2014) High acid crudes. Gulf Professional Publishing, Houston, TX, USA
9. O'Kane JM et al (2010) Detection and monitoring of naphthenic acid corrosion in a visbreaker unit using hydrogen flux measurements. Paper 10351 presented at the CORROSION 2010 conference, NACE International, Houston, TX, USA
10. O'Kane JM et al (2010) Correlation of hydrogen flux and corrosion rate measurements carried out during a severe episode of corrosion-erosion attributable to naphthenic acid. Paper 10178 presented at the CORROSION 2010 conference, NACE International, Houston, TX, USA
11. Port RD, Herro HM (1991) The Nalco Guide to Boiler Failure Analysis. McGraw-Hill Inc., New York
12. ASTM B294-10 (2010) Standard test method for hardness testing of cemented carbides. ASTM International, West Conshohocken, PA, USA
13. ASTM E384-16 (2016) Standard test method for microindentation hardness of materials. ASTM International, West Conshohocken, PA, USA
14. ASTM E92-16 (2016) Standard test methods for Vickers hardness and Knoop hardness of metallic materials. ASTM International, West Conshohocken, PA, USA
15. ASTM A370-15 (2015) Standard test methods and definitions for mechanical testing. ASTM International, West Conshohocken, PA, USA
16. ASTM E23-12c (2015) Standard test methods for notched bar impact testing of metallic materials. ASTM International, West Conshohocken, PA, USA
17. NACE SP0200-2014 (2014) Standard practice steel-cased pipeline practices. NACE International, Houston, TX, USA
18. ASTM A380-13 (2014) Standard practice for cleaning, descaling and passivation of stainless steel parts, equipment, and systems. ASTM International, West Conshohocken, PA, USA
19. EN ISO 8501-1:2007 (2007) Preparation of steel substrates before application of paints and related products—visual assessment of surface cleanliness
20. ASME/ANSI B31.3 (2014) Process piping guide. ASME, New York
21. Dillon CP (1995) Corrosion resistance of stainless steels. Marcel Dekker Inc., New York
22. API Standard 620 (2002) Design and construction of large, welded, low-pressure storage tanks, 10th edn. American Petroleum Institute, Washington, D.C., USA
23. API Standard 650 (2007) Welded tanks for oil storage, 11th edn. American Petroleum Institute, Washington, D.C., USA
24. BS 2654 (1989) Specification for manufacture of vertical steel welded non-refrigerated storage tanks with butt-welded shells for the petroleum industry. British Standard Institution, London
25. DIN 4119-1 (1979) Above-ground cylindrical flat-bottom tank installations of metallic materials; fundamentals, design, tests. Deutsches Institut Fur Normung E.V. (German National Standard)

26. DIN 4119-2 (1980) Above-ground cylindrical flat-bottom tank structures of metallic materials; Calculation. Deutsches Institut Fur Normung E.V
27. UL 142 (1998) Steel aboveground tanks for flammable and combustible liquids. Underwriters Laboratories Inc. (UL), Northbrook
28. API Spec 12B (1995) Specification for bolted tanks for storage of production liquids, 14th edn. American Petroleum Institute, Washington, D.C., USA
29. API Spec 12D (2008) Specification for field welded tanks for storage of production liquids, 11th edn. American Petroleum Institute, Washington, D.C., USA
30. API Spec12F (2008) Specification for shop welded tanks for storage of production liquids, 12th edn. American Petroleum Institute, Washington, D.C., USA
31. API RP 651 (1997) Cathodic protection of aboveground petroleum storage tanks, 2nd edn. American Petroleum Institute, Washington, D.C., USA
32. API RP 1632 (2002) Cathodic protection of underground petroleum storage tanks and piping systems, 3rd edn. American Petroleum Institute, Washington, D.C., USA
33. API RP 1615 (2011) Installation of underground petroleum storage systems, 6th edn. American Petroleum Institute, Washington, D.C., USA
34. API RP 652 (1997) Lining of aboveground petroleum storage tank bottoms, 2nd edn. American Petroleum Institute, Washington, D.C., USA
35. API RP 1631 (1993) Interior lining and periodic inspection of underground storage tanks, 5th edn. American Petroleum Institute, Washington, D.C., USA
36. UL 1746 (2007) External corrosion protection systems for steel underground storage tanks, 3rd edn. Underwriters Laboratories Inc. (UL), Northbrook
37. API Standard 653 (2009) Tank inspection, repair, alteration and reconstruction, 4th edn. American Petroleum Institute, Washington, D.C., USA
38. API RP 575 (2005) Inspection of existing atmospheric and low-pressure storage tanks, 2nd edn. American Petroleum Institute, Washington, D.C., USA
39. API 510 (2006) Pressure vessel inspection code: in-service inspection, rating, repair, and alteration, 9th edn. American Petroleum Institute, Washington, D.C., USA
40. EEMUA Publ 159 (2003) Users' guide to the inspection, maintenance and repair of aboveground vertical cylindrical steel storage tanks, 3rd edn
41. KWA (1999) Recommended practice for inspecting buried lined steel tanks using a video camera, 1st edn. Ken Wilcox Associates, Inc. (KWA), USA
42. NFPA 326 (2010) Standard for the safeguarding of tanks and containers for entry, cleaning, or repair. USA
43. NLPA Standard 631 (1991) Entry, cleaning, interior inspection, repair, and lining of underground storage tanks, 3rd edn. National Leak Prevention Association, USA
44. NFPA 30 (2012) Flammable and combustible liquids code. USA
45. PEI/RP100 UST Installation (2011) Recommended practices for installation of underground liquid storage systems. Petroleum Equipment Institute, Tulsa, USA
46. API RP 1621 (2001) Bulk liquid stock control at retail outlets, 5th edn. American Petroleum Institute, Washington, D.C., USA
47. API RP 1595 (2006) Design, construction, operation, maintenance, and inspection of aviation pre-airfield storage terminals, 1st edn. American Petroleum Institute, Washington, D.C., USA
48. API/IP RP 1540 (2004) Design, construction, operation and maintenance of aviation fueling facilities, 4th edn. American Petroleum Institute, Washington, D.C., USA
49. API 2610 (2005) Design, construction, operation, maintenance, and inspection of terminal and tank facilities, 2nd edn. American Petroleum Institute, Washington, D.C., USA
50. UL 58 (1996) Standard for safety: steel underground tanks for flammable and combustible liquids, 9th edn. Underwriters Laboratories Inc. (UL), Northbrook
51. API Spec 12P (2008) Specification for fiberglass reinforced plastic tanks, 3rd edn. American Petroleum Institute, Washington, D.C., USA
52. UL 1316 (1994) Glass-fiber-reinforced plastic underground storage tanks for petroleum products, alcohols, and alcohol-gasoline mixtures, 2nd edn. Underwriters Laboratories Inc., Northbrook

53. PEI/RP900 (2008) Recommended practices for the inspection and maintenance of UST systems. Petroleum Equipment Institute, Tulsa, USA
54. API RP 1626 (2010) Storing and handling ethanol and gasoline-ethanol blends at distribution terminals and filling stations, 2nd edn. American Petroleum Institute, Washington, D.C., USA
55. API RP 1627 (1986) Storage and handling of gasoline-methanol/cosolvent blends at distribution terminals and service stations, 1st edn. American Petroleum Institute, Washington, D.C., USA
56. EEMUA Publ 154 (2002) Guidance to owners on demolition of vertical cylindrical steel storage tanks and storage spheres, 2nd edn
57. Groysman A (2010) Corrosion for everybody. Springer, Dordrecht
58. ASTM 268 (2004) Standard specification for seamless and welded ferritic and martensitic stainless steel tubing for general service. ASTM International, Washington, D.C., USA
59. Halkyard D (2014) Refinery based case study of a novel gasket designed for use in problematic shell and tube heat exchangers. In: 18th ISC (International Sealing Conference), Stuttgart, Germany, vol B15, pp 543–561
60. Worden K (2014) Flange corrosion: prevention and mitigation through better gasketing. Paper 125f presented at the AIChE Spring Meeting & 10th Global Congress on Process Safety, New Orleans, LA, USA
61. Turnbull A (1998) Impact of graphite gasket/duplex stainless steel couples on crevice chemistry and likelihood of crevice attack in sea water. Paper 682 presented at the NACE Corrosion/98 Conference, Houston, TX, USA

Chapter 6
Corrosion Phenomena

Familiar disease is safer than an incompetent doctor.
The author.

Abstract We can face all corrosion phenomena known and occurred in different industries and places. In this part, MIC, erosion-corrosion, cavitation, corrosion fatigue, thermal fatigue, and galling are described and case studies with remedies are given.

6.1 Microbiologically Induced Corrosion (MIC)

Bacteria are relatively complex, single-celled creatures. They were the first inhabitant on the Earth about 3.5 billion years ago, can reproduce on their own, and can survive in different environments. They are around us including water, air, soil, crude oil, and fuels. As all these media exist in oil refineries and petrochemical plants, bacteria are also present at different units and systems. Most bacteria are harmless, and some actually help in digestion food, destroying disease-causing microbes, and providing nutrients. Fewer than 1 % of bacteria cause diseases of people. Even less are the reason of corrosion of metals. Participation of bacteria in corrosion of metals and control methods in crude oil, fuels and water are described in [1–5]. Cooling water untreated by biocides can result in fast (during 15–30 days) biofouling and as a result to severe corrosion (Fig. 6.1). Four cases of MIC are described and analyzed in this section.

6.1.1 Case Studies

6.1.1.1 MIC of Heat Exchanger Tubes (Petrochemical Plant)

Conditions. The heat exchanger is intended for steam producing by heating of boiler feed water (BFW) by means of para-diethyl-benzene. BFW flowed outside

A. Groysman, *Corrosion Problems and Solutions in Oil Refining and Petrochemical Industry*, Topics in Safety, Risk, Reliability and Quality 32,
DOI 10.1007/978-3-319-45256-2_6

(a) (b)

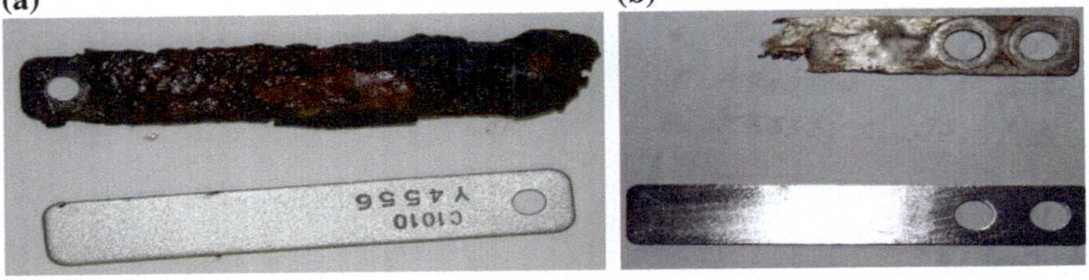

Fig. 6.1 Lower coupons are original. **a** Biofouling (15 days) on the coupon in cooling water untreated by biocides. **b** Half of the coupon corroded after cleaning from biofouling (30 days)

the tubes at 10 bar. The para-diethyl-benzene flowed inside the tubes at 25 bar, and temperature changes from 200 to 60 °C.

History. Outer space of tube bundle in the heat exchanger was filled by cooling water (contaminated with microorganisms) which was in stagnation during a year. Leak was detected after 2 months after start-up. Cooling water contained chloride (700–1500 ppm) and sulfate (700–1300 ppm) ions at pH = 7.5.

Material of construction. Carbon steel.

Service period before the failure: 2 months.

Visual examination and findings. Rusted tubercles, pits and holes were found on outer surface of tubes (Fig. 6.2).

Physicochemical analysis of rusted tubercles. Rusted tubercles contained: substances evaporated at 110 °C (60 wt%), organic substances (burning at 550 °C) (3.5 wt%), and ash (mostly inorganic substances) (36.5 wt%). EDS analysis showed the presence of the following elements in rusted tubercles (wt%): Fe (42.0–54.4), O (26.4–31.4), Si (5.6–8.6), Mg (4.9–7.8), C (4.8–5.7), Ca (2.1–2.3), P (1.1–1.2), Cl (0.6), S (0.5).

Microbiological analysis of corrosion products is shown in the Table 6.1.

Cause of failure and its explanation. Rust on carbon steel tubes in cooling water could be formed as a result of general corrosion by water and dissolved oxygen or due to activity of iron-oxidizing and/or iron-precipitating bacteria containing in

(a) (b) (c) (d)

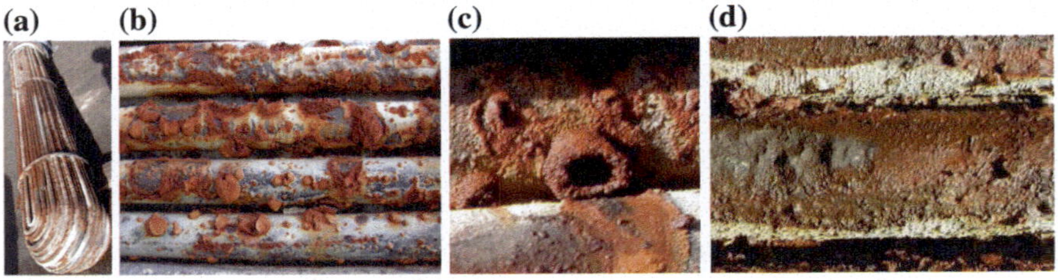

Fig. 6.2 **a** General view of failed heat exchanger. **b**, **c** Rusted tubercles. **d** Slimy deposits on outer surface of tubes and pits underneath

Table 6.1 Microbiological analysis of corrosion products

Bacteria	CFU[a]/g of rust
IOB[b]	$(0.9–2.2) \times 10^9$
HAB[c]	$(0.6–1.7) \times 10^9$
Fungi	$(0.1–1.5) \times 10^3$
SRB[d]	3.5×10^2

[a]*CFU* Colony Forming Units
[b]*IOB* Iron-Oxidizing Bacteria
[c]*HAB* Heterotrophic Aerobic Bacteria
[d]*SRB* Sulfate Reducing Bacteria

water. In the first case, rust has abiotic nature (non-living components are the cause of corrosion) while in the second case, it is biotic (living microorganisms are the reason of corrosion). Microbiological analysis demonstrates accumulation of Heterotrophic Aerobic Bacteria (HAB) (especially Iron-Oxidizing Bacteria, IOB) in corrosion products. The concentration of these bacteria in corrosion tubercles is very high ($\sim 10^9$ microbes in gram of rust) that could be related to biofilm formation by these bacteria. Physicochemical analysis of corrosion deposits shows that they consist mainly of iron and oxygen (main components of rust) but organic substances also are contained in relatively large amounts (5.7 wt% of carbon). Increased concentration of organic matter (metabolic products, viable and dead cells of IOB, HAB, SRB, fungi, and other microorganisms) detected in rust indicates active participation of these microorganisms in biofouling process. Abiotic rust, as a rule, free of carbon or contains its traces.

Thus, the corrosion failure (formation of rust tubercles, pits, and holes) occurred due to IOB activity. They proliferated intensively in cooling water producing large amounts of ferric hydroxide deposits in a very short period (several days–weeks). IOB generate energy by oxidation of ferrous (Fe^{2+}) to ferric (Fe^{3+}) ions. The morphology of rust tubercles and tube surface after their removing shows specific localized form of corrosion attack peculiar to MIC (see Fig. 6.2). Microorganisms which are present in cooling water under appropriate conditions (nutrients availability, water aeration, temperature 25–40 °C) can proliferate with enormous rates and create large biomass. Biofouling formation on carbon steel surface has irregular character which in its turn leads to localized corrosion during short period (2 months). Corrosion process is accelerated when cooling water contains chloride (700–1500 ppm) and sulfate (700–1300 ppm) ions. Anaerobic SRB could proliferate under biofilms formed by IOB inside of tubercles. Thus corrosion process with the participation of microorganisms occurred during a year of stagnation with enormously high rate which gave rise to failure during 2 months after the start-up of the heat exchanger service.

Failure phenomenon: *MIC*.

Solution and recommendations.

It is forbidden to use cooling water untreated by biocides, especially which does not flow. Water is becoming contaminated by numerous diverse microorganisms inducing corrosion. Only boiler feed water (not containing dissolved oxygen and inhibited by sodium nitrite, 1000 ppm $NaNO_2$) can be used for such purposes.

6.1.1.2 Corrosion of Heat Exchanger with Cooling Water (Ethylene Plant, Metathesis Unit)

Conditions. Heat exchanger is intended for cooling of hydrocarbons by means of cooling water. Cooling water flowed inside the tubes at 30–35 °C and 4–5 bar. Hydrocarbons flowed outside the tubes: inlet at 70 °C and outlet at 30–35 °C. Cooling water was treated with zinc-phosphate-phosphonate corrosion inhibitor, anti-scaling agent, and biocide. The volume of the cooling water system is 1000 m^3. Passivation of inner surface of the system (heat exchangers and pipes) was carried out before the beginning of chemical treatment of cooling water (injection of chemicals). The heat exchanger was shutdown several times during 3 years due to maintenance works in the cooling tower. Once sulfuric acid leaked into the cooling water system. At the time of the leak, the pH value of cooling water decreased from 7.5 to 2.2 and slightly increased to pH ~4 only after 2 h (Fig. 6.3d). The concentration of iron in water increased to 30 ppm as a result of this leak and severe corrosion of inner surface of carbon steel equipment. Many drains were performed, and the cooling water system returned to normal condition of service only after a month. During this period, the concentration of iron in water decreased from 30 to 1–5 ppm when the required value is less than 1 ppm. Another event of spilling of sodium hypochlorite (NaClO) into the cooling water occurred during the service, and the concentration of free chlorine increased to 200 ppm (when recommended value is 0.1 ppm). Microbiological examination was carried out once in 4–9 months. Note: Metathesis unit produces propylene from ethylene.

Material of construction. Carbon steel.

Service period before the failure: 3 years.

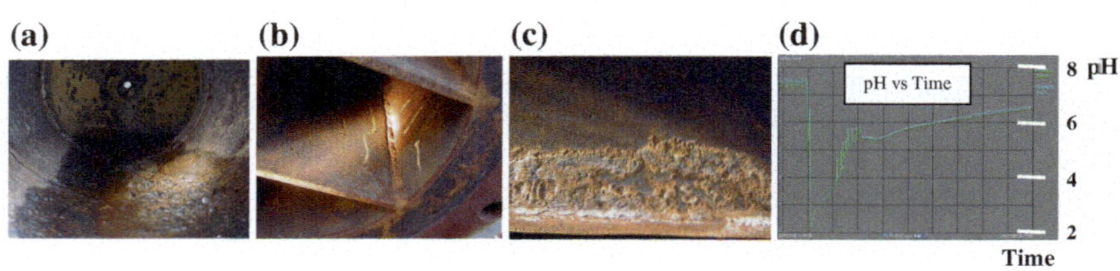

Fig. 6.3 **a** Deposits in the outlet pipe from failed heat exchanger. **b** Corrosion of welds and pits. **c** Shallow pits in the channel. **d** The changes of pH (from 7.5 to 2.2) during the leak of sulfuric acid into cooling water

Visual examination and findings. Deposits of black-brown-grey color and shallow pits underneath were found on the inner surface of floating heads, channels and tube-sheets including welds (see Fig. 6.3). Inner surface of heat exchanger severely corroded.

Failure phenomenon: *General and pitting corrosion; under deposit corrosion; MIC.*

Cause of failure and its explanation. The leak of sulfuric acid caused a decrease of pH to 2.2. Carbon steel corroded by sulfuric acid, and concentration of iron in cooling water increased from 1 to 30 ppm. In the system of the volume of 1000 m^3, 30 kg of iron were formed. The presence of dissolved oxygen in cooling water resulted in formation of Fe^{3+} ions which are the cathodic depolarizers and accelerated the corrosion of carbon steel [1]. When the value of pH of cooling water returned to recommended value 7–8, iron hydroxides were formed and deposited, predominantly in locations where water flow was slow. Frequent outages of the heat exchanger also resulted in deposition and accumulation of corrosion products, causing under deposit pitting corrosion. The performed microbiological tests have shown that the concentration of microorganisms (TBC) in cooling water ranged in 10^4–2×10^4 CFU/ml which is the borderline or even is higher than the permitted value 10^3 CFU/ml. When the concentration of microorganisms in the water (plankton bacteria) was above the permitted value, most likely that concentration of bacteria in corrosion deposits (sessile bacteria) was even higher, and this could cause MIC. Iron hydroxides in the form of tubercles were accumulated in certain locations inside the heat exchanger. These tubercles could be the consequence of activity of iron bacteria. The tubercles are highly structured, consist of a hard, brittle, outer shell of red corrosion products that encapsulates an inner core of soft, voluminous, dark corrosion products. As a result of the leak of sulfuric acid, sulfate ions appeared in large quantity in cooling water and they stimulated the growth of iron bacteria. Physicochemical examination of rusted tubercles showed the presence of substances evaporating at 110 °C (9.0 wt%), compounds burned at 550 °C (12.5 wt%) and ash (78.5 wt%). The relatively high amount of material burned at 550 °C shows that mostly organic substances (microorganisms) were present in tubercles' rust. Usually when rust consists only from inorganic substances (iron hydroxides and iron oxides), compounds burned at 550 °C constitute less than 1 wt%.

Solutions and recommendations.

(a) To fill shallow pits by means of welding to required thickness in respect that corrosion allowance was taken into account.
(b) To clean inner surface of heat exchangers by means of WOMA (cleaning machine) from deposits.
(c) To carry out passivation according to standard procedure before beginning (new start-up) of operating of cooling water system (Appendix I).
(d) To install CorrDATS (on-line corrosion monitoring system) [1] and carry out careful microbiological analysis of water (once a week) and deposits (once a month).

6.1.1.3 Corrosion of Carbon Steel Coils in Cooling Water Box

Conditions. Coils placing in the box (70 m^3) containing water condensate are intended for cooling of fuel flowing inside the coils. Water condensate is outside the coils in the cooling box which is open to the atmosphere. The box functioned only in the case of emergency (shutdown or unexpected outage of the unit), so the coils in the box were most time at ambient temperature. The water temperature can reach 80 °C during operation. Aluminumsacrificial anodes were installed on the surface of coils.

Material of construction. Coils and box—carbon steel.

Service period before the failure: 4 years.

Visual examination and findings. Severe corrosion of outer surface of coils was found during shutdown (Fig. 6.4). Rust tubercles and deep pits (~8 mm in dimensions) were found underneath. Sacrificial anodes from aluminum did not function.

Failure phenomenon: *MIC, water corrosion (general and pitting).*

Cause of failure and its explanation. Corrosion form and corrosion rate of carbon steel depend on the chemical composition of water and temperature. Water condensate is demineralized (deionized or pure) water, namely, water which does not contain ions and its electroconductivity is very-very low. Usually pitting corrosion of carbon steel occurs in pure water. Maximum general corrosion rate of carbon steel in pure water takes place at ~80 °C.

Microbiological analysis of water and corrosion products (biofouling in rust) showed that corrosion inducing microorganisms were present in large amounts and mostly were accumulated in biofilm in the rust (Table 6.2).

Solution and recommendations (Table 6.3).

(a) It is not recommended to use aluminum sacrificial anodes, because electroconductivity of water condensate is very-very low, aluminum does not work in non- or low-conductive media and cannot protect carbon steel in this weak electrolyte. Sacrificial anodes function only in electrolyte solutions when salt concentration is more than 0.3 wt%. In addition, aluminum and its alloys have passive film. Usually sacrificial anodes are not used in water condensate. It is possible to use magnesium (Mg) anodes at 80 °C but only in the case of addition small amounts of salts to increase electroconductivity of water.

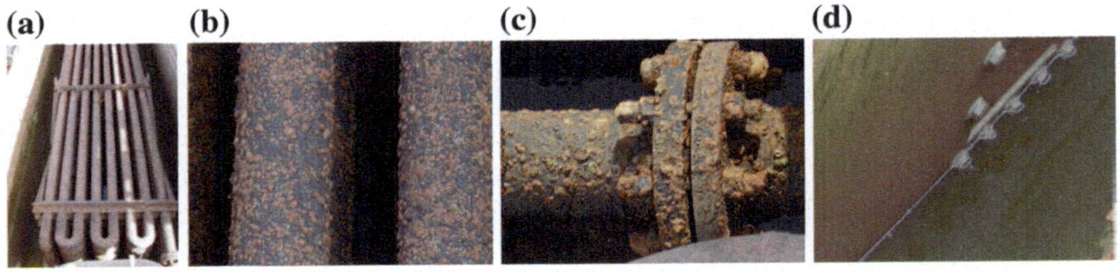

Fig. 6.4 **a** General view of corroded coils in water box. **b**, **c** Rust tubercles on coils and flange. **d** Biofilm on inner surface of carbon steel water box bottom

Table 6.2 Microbiological analysis of water and corrosion products (rusted biofilm)

Sample name	Unit	TPC[a]	Iron bacteria	Fungi + Yeasts	SRB[b]
Water	CFU[c]/1 ml	4.0×10^5	6.0×10^3	1.0×10^3	ND[d]
Biofilm (rust)	CFU[c]/1 g	4.62×10^8	4.35×10^6	3.3×10^5	ND[d]

[a]*TPC* Total Plate Count
[b]*SRB* Sulfate Reducing Bacteria
[c]*CFU* Colony Forming Units
[d]*ND* Non Detected

Table 6.3 Preventing methods of corrosion of carbon steel coils in cooling water box

Corrosion control method	Efficiency (%)	Requirements
Injection of sodium silicate (Na_2SiO_3)	~90	a. Keep Na_2SiO_3 concentration >300 ppm b. Check inhibitor concentration and pH once a week
Injection sodium chloride (NaCl)	~99	a. Keep concentration ≥ 26 wt% b. It is needed large amount of salt: 20 tons in 70 m^3 water c. Coating is needed on the surface of the border walls of water—air
Sealing the water box from the atmosphere and Injection sodium nitrite ($NaNO_2$)	~90	a. Sealing of cooling water box from the atmosphere b. Keep $NaNO_2$ concentration >400 ppm. It is desirable pH = 8.5–9 c. Check inhibitor concentration and pH once a week
Cooling water (treated by corrosion inhibitors at oil refinery) use at continuous circulation	~90	Pump installation
Use of water with salt concentration >0.3 % and zinc sacrificial anodes (cathodic protection)	~90	Cathodic protection design
Sealing and preserving inert atmosphere (nitrogen gas) above the water level to prevent air penetration	~90	a. Sealing of cooling water box from the atmosphere b. Injection of nitrogen gas above the water level
Using carbon steel coils with aluminum or zinc coating	100	Replacement existing rusted coils with that coated by aluminum or zinc
Periodical injection of biocide		Check concentration of microbes and biocide. If the concentration of bacteria in water is increased over 10^3 CFU/ml, it is necessary to use biocides

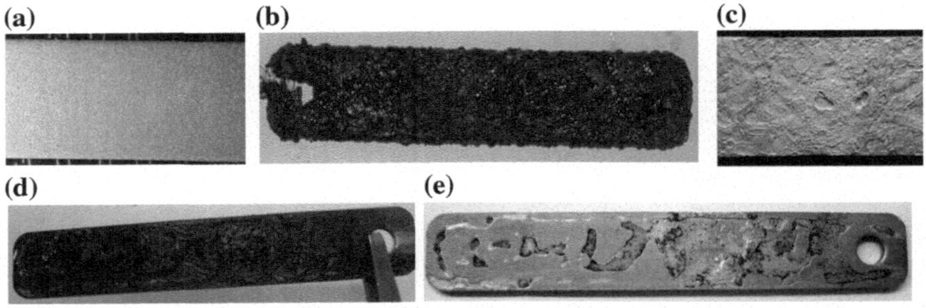

Fig. 6.5 **a** Original carbon steel (C1010) coupon. **b–e** Coupons after examination in cooling water condensate containing inhibitor Na_2SiO_3. **b** Coupon in cooling water in the box (stagnation, 628 days). Corrosion rate is 0.058 mm/year. Concentration of inhibitor varied from 0 to 470 ppm SiO_2. **c** Coupon's surface after chemical cleaning from corrosion products. **d** Coupon in cooling water in the box (stagnation, 123 days). Corrosion rate is 0.065 mm/year. Concentration of inhibitor varied from 13 to 190 ppm SiO_2. **e** Coupon's surface after chemical cleaning from corrosion products

(b) To inject corrosion inhibitor sodium silicate Na_2SiO_3 (>300 ppm) which can protect carbon steel in water condensate in stagnant conditions of the water box [1]. Sodium phosphate Na_3PO_4 (100 ppm) also can protect carbon steel at 80 °C but the presence of phosphor can stimulate proliferation of microorganisms in water.
(c) To use carbon steel coils with aluminum coating but to take into consideration that aluminum oxide which is present on aluminum surface is an insulator and can decrease heat exchange.
(d) It is not recommended to use organic coatings as they decrease heat exchange significantly.
(e) To inject biocides periodically and carry out monitoring.

PostScript. Corrosion inhibitor sodium silicate was injected into water and its efficiency according to coupons and chemical analysis of water was examined during ~2 years. Corrosion coupons are shown in Fig. 6.5.

Conclusion. Recommended concentration was more than 300 ppm SiO_2 under stagnation (see Table 6.3). It was impossible to protect by means of only sodium silicate because severe biofouling was developed.

6.1.2 Corrosion and Deposits in the Oil Cooler of Turbo-Generator (Power Station)

Conditions. The oil cooler of turbo-generator is intended for the cooling of oil (50–70 °C) by water (30–35 °C) (Fig. 6.6a). Cooling water (untreated by corrosion inhibitors, biocide and anti-scaling agent) flowed inside the tubes during 1 year and

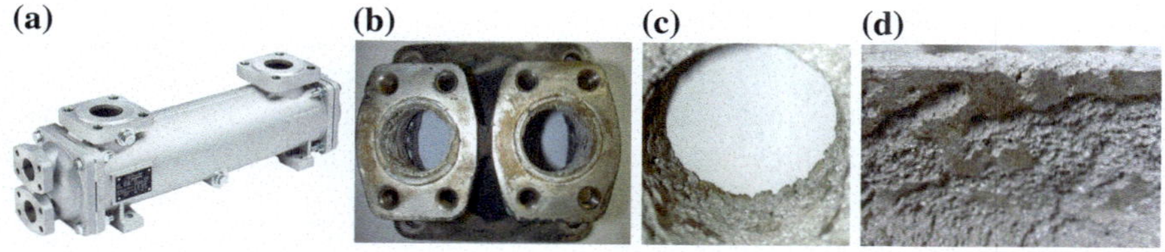

Fig. 6.6 **a** Oil cooler. **b–d** Corroded aluminum parts (water inlet)

10 months. This water was replaced with new cooling water treated by zinc-phosphate-phosphonate corrosion inhibitor, biocide and anti-scaling agent during 2 months before the failure.

Material of construction. Tubes—copper-nickel alloy CuNi10Fe1Mn (CDA 706); tube-sheets—copper alloy CuZn38Sn1(CDA 464); lids and parts for cooling water inlet—aluminum alloy AlSi9 Mg.

Service period before the failure: 2 years.

Visual examination and findings. White-grey deposits were found on the surface of the lid from the cooling water side and plugged water flow. Chemical analysis showed that the deposits consisted of phosphates and hydroxides of Al, Mg, Ca, and Zn.

Severe localized corrosion was detected on aluminum surface after removing of deposits (Fig. 6.6). Cooling water (untreated by corrosion inhibitors, biocide and anti-scaling agent) contained large amount (>10^8 microbes/ml) of diverse microorganisms (SRB, iron-bacteria, and other heterotrophic bacteria).

Failure phenomenon: *MIC and galvanic corrosion of aluminum parts.*

Cause of failure and its explanation. Cooling water (untreated by corrosion inhibitors, biocide and anti-scaling agent) which flowed through the oil cooler during ~2 years was particularly corrosive because of presence of large quantities of microorganisms inducing corrosion. Microbes deposited on aluminum surface. Aluminum alloy AlSi9 Mg containing magnesium is particularly vulnerable to MIC [3].

When corrosion inhibitors (e.g., azole) of copper alloys are not used in cooling water, copper corrodes and its ions accumulate on the surface of the aluminum, and work as cathode, causing severe galvanic corrosion of aluminum (see case Sect. 5.8.1.10). Usually copper ions serve as biocides but in this particular case in spite of this, microbes proliferated and aggravated corrosion of aluminum.

Solutions and recommendations.

(a) In no case do not combine copper and aluminum in one cooling water system.
(b) To use tubes, tube-sheets and lids from identical alloys CDA 706 or Admiralty brass CDA 443.
(c) It is necessarily to check the conformity of materials at design stage.

6.2 Erosion–Corrosion and Cavitation

Erosion is a destruction of materials by the abrasive action of moving fluids and gases, accelerated by the presence of solid particles or liquid drops in gases. The word *erosion* is translated from the Latin as *to gnaw*, or *to wear away*. When the substances flowing against metal surface are inert from the chemical point of view (sand or hydrocarbon fuels), erosion is a pure mechanical phenomenon resulting in the physical loss of a material. When chemical substances take part in corrosive processes (steam, liquid water or water vapors containing H_2S, HCl, CO_2, O_2, Cl_2), a conjoint action involving erosion (mechanical phenomenon) and corrosion (chemical/electrochemical phenomenon), is named erosion-corrosion [1]. It is very difficult to determine separately the contribution of erosion and corrosion in the general erosion-corrosion phenomenon context. Specific flow regime, or presence of aggressive components in moving media are the cause of mutual mechanical–chemical/electrochemical (erosion-corrosion) phenomena and is often named flow-assisted, flow-accelerated, or flow-induced corrosion, cavitation–erosion, liquid-impingement erosion, or mechanically assisted degradation. Impingement attack is a form of erosion-corrosion associated with the local hits of high velocity streams of fluids, gases, solid particles, or together (e.g., suspension) against a metallic surface. Cavitation is a destructive phenomenon occurring on metallic surface because of the formation and collapses of gaseous bubbles on it. The British engineer Froude R.E. suggested this term in 1895 to describe the clouds of bubbles produced by propeller blades.

Erosion-corrosion relates to localized corrosion because the results are pits, grooves, valleys and waves on the surface of metals. Usually erosion-corrosion results in failures in a very short period: from several weeks to several months. There are almost no metal or alloy which are not subjected to erosion-corrosion. Mild metals and alloys, such as aluminum, copper, and carbon steel, are especially susceptible to erosion-corrosion. However, titanium and alloys possessing large hardness also are susceptible to erosion-corrosion. It is very difficult to predict the failures if erosion-corrosion occurs because small changes in the flow regime, unexpected air ingestion in a circulating water, addition and operation of new pumps, appearance of new obstacles resulting in a velocity change and turbulence, the presence of aggressive gases (H_2S, HCl and Cl_2), temperature drop under dew point, and construction changes of equipment may result in the sudden development of erosion-corrosion. Therefore, failure may occur for a short period if hydrodynamic system has worked well for a long period before any changes in flow.

There are two main reasons for the occurrence of erosion-corrosion. The first reason is wear because of the impact of particulate matter or collisions of liquid drops on the surfaces of metals. In some cases there are no solids in liquids at all, but erosion may occur because of flow turbulence, namely, because of the severe mechanical action of two-phase mixture liquid–gas.

The second reason for the erosion-corrosion occurrence, namely, the cavitation, is the formation and collapse of the bubbles in the liquid, or condensation of molecules of vapors on a metallic surface during flow [1]. Cavitation of the first type is spread in centrifugal pumps, propellers, turbines, blades; the cavitation of the second type—in steam lines, and other equipment with a temperature drop under dew point. Chemical factors (if aggressive chemicals are present in liquids or in vapor phase) can accelerate cavitation.

Erosion-corrosion and cavitation often occur in the following equipment at oil refinery and petrochemical units: pumps and impellers, pipes (especially bends, elbows, and tees), valves, tube-sheets in heat exchangers. Nearly all liquid media (aromatics, water, amine solutions, fuel oil, gas oil, and gasoline) using in this equipment can cause flow-induced corrosion. We will describe and analyze such cases.

6.2.1 Case Studies

6.2.1.1 Erosion of Pump for Hydrocarbons (Petrochemical Plant)

Conditions. Aromatic hydrocarbons (reflux in the pre-fractionation BTX tower) were pumped out at 70–80 °C and 1 bar. The hydrocarbons contained ~80 ppm water.

Material of construction: Cast iron.

Service period before the failure: 5 years.

Visual examination and findings. The hole was detected inside the centrifugal pump (Fig. 6.7).

Failure phenomenon: *Erosion-corrosion.*

Cause of failure and its explanation. Boiling points of benzene is 80 °C and of toluene is 111 °C at 1 bar. Therefore, the vapors were formed and the hydrocarbon

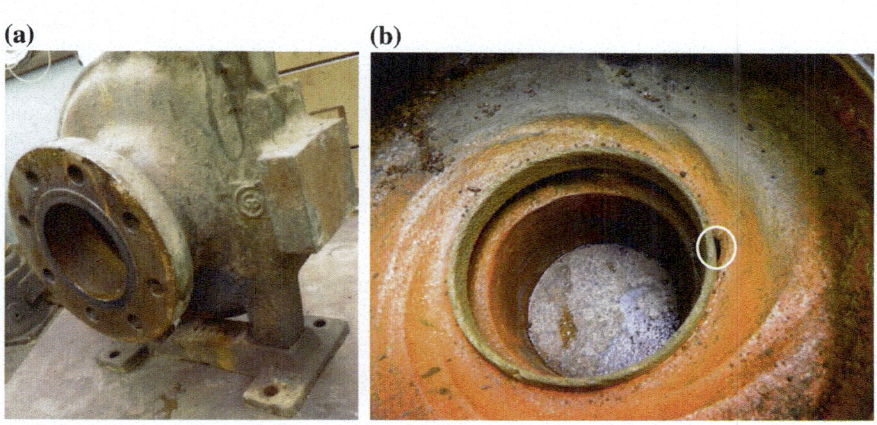

Fig. 6.7 **a** General view of failed pump. **b** Corroded inner surface of centrifugal pump with large corrosion hole (marked with *white circle*)

system contained two phases liquid and vapor. The hole is a result of erosion-corrosion by intensive turbulent flow of two-phase organic liquid-vapor system contained water.

Solution and recommendations.

(a) To coat with abrasive-resistant coating (epoxy-ceramic; novolac-ceramic; vinyl ester co-polymer containing glass flake and silicon carbide; thermoplast powder coating).
(b) To increase pressure in order to prevent formation of two phases.

6.2.1.2 General Corrosion of Inner Surface and Erosion of Impeller in the Water Pump

Conditions. The pump is intended for the pumping of the mixture of potable water with brackish water from wells. The pump worked under the following conditions: it was turned on about 6 months and it did not work subsequent 6 months (stand-by).

Materials of construction. The pump body—cast iron; the impeller-bronze.

Service period before the failure: 5 years.

Visual examination and findings. Severe corrosion was detected inside the centrifugal pump: thick rust in the form of tubercles and decrease of thickness of the wall impeller to zero (Fig. 6.8). The following things were detected: deep hollows in the surface of the pump body after removing the tubercles; large amounts of iron-bacteria (10^8–10^9 bacteria/g) both inside the tubercles and in brackish water from the wells; big quantities of sand in water. Deteriorated surface of the impeller was clean (free from corrosion products).

Failure phenomenon: *General corrosion, MIC, erosion.*

Cause of failure and its explanation. Three corrosion phenomena occurred with the pump: general corrosion, MIC of inner surface of the pump body and erosion of the impeller. Prolonged (during 6 months) stagnation of the mixture of potable and brackish (deteriorated with iron-bacteria) water inside the pump resulted in general corrosion, proliferation of iron-bacteria on inner surface of the pump body and formation of tubercles. The presence of hard particles of sand in water caused erosion of the impeller. Morphology of failed impeller surface showed typical erosion.

Solutions and recommendations.

(a) To repair and restore the inner surface of the pump body with epoxy-ceramic coating.
(b) It is impossible to repair the impeller and it should be replaced with new one.
(c) Good filtration of water from sand particles in order to prevent erosion of impeller.

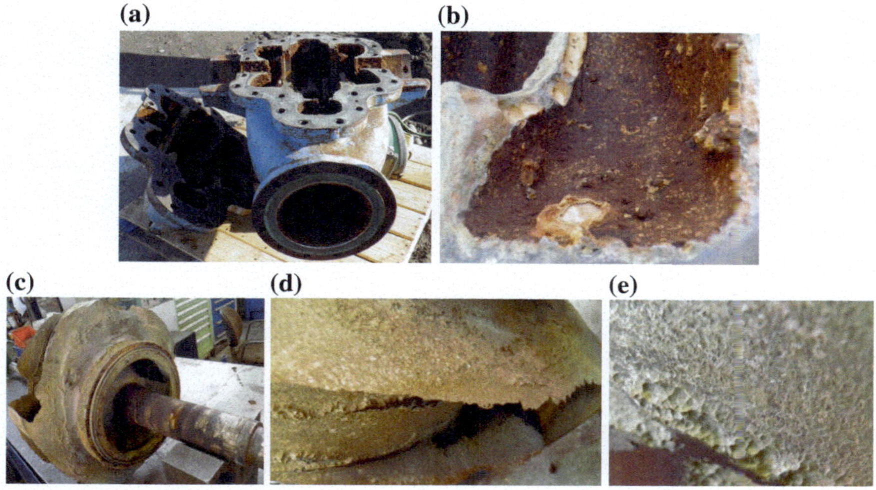

Fig. 6.8 Failed centrifugal pump for water service. **a** General view. **b** Rust with tubercles inside the pump. **c–e** Eroded impeller

(d) Prevention of stagnation of water in the pump: to drain water during the idle period and to keep inner surface dry. Alternatively, it is possible to keep water with added corrosion inhibitor sodium nitrite (1000 ppm) inside the pump.
(e) Biocide treatment of brackish water from wells.

6.2.1.3 Erosion of the Pump (FCCU)

Conditions. Clarified fuel oil (mixed with catalyst) at 365 °C. Catalyst contained oxides of aluminum and silica.

Material of construction: Stainless steel casting ASTM A217 Gr. CA15 (13 % Cr). Metallic coating based on tungsten was on the inner surface of the pump.

Service life before the failure: 10 months.

Visual examination and findings. Severe deterioration (up to holes) of the inner pump accessories was found (Fig. 6.9).

Failure phenomenon: *Erosion.*

Cause of failure and its explanation. Erosion by the catalyst—ceramic materials (oxides of aluminum and silica) possessing very high hardness.

Solution and recommendations.

(a) Unfortunately there is no material that can resist to erosion by catalyst for long time.
(b) To prevent penetration of catalyst into clarified oil by good filtration and settling of slurry oil in the settler. In other words, filtration of the fractionator

(a) (b) (c)

Fig. 6.9 **a** General view of the pump eroded by catalyst (FCCU). **b**, **c** Magnification of the picture **a**

bottom stream from catalyst particles which passed into main column (main fractionator) because of cyclone did not separate them in reactor.

(c) To replace coating with materials resistant to erosion: ceramics, carbides, oxides, alloys Colmonoy 6 or Stellite 720 (see Appendix F, Table F.9).

(d) To improve reactor cyclone efficiency in order to prevent catalyst entrainment to the main column.

6.2.1.4 Erosion-Corrosion of Lean Amine Pump (HDS Unit)

Conditions. The pump is intended for the pumping of aqueous solution of lean amine (DEA) from the tank at ambient temperature and pressure 5–8 bar.

Material of construction: Cast iron.

Service period before the failure: 4 years.

Visual examination and findings. Shallow pits are detected on the inner surface of the pump (Fig. 6.10).

Failure phenomenon: *Erosion-corrosion.*

Cause of failure and its explanation. The morphology of surface is typical to erosion phenomenon. When DEA solution is used long time, destructive substances can be formed which are corrosive to carbon steel and cast iron (see Sect. 4.4). In addition, turbulent flow of two-phase system (liquid + air) causes erosion. Thus, erosion-corrosion occurred inside the pump.

(a) (b) (c) (d)

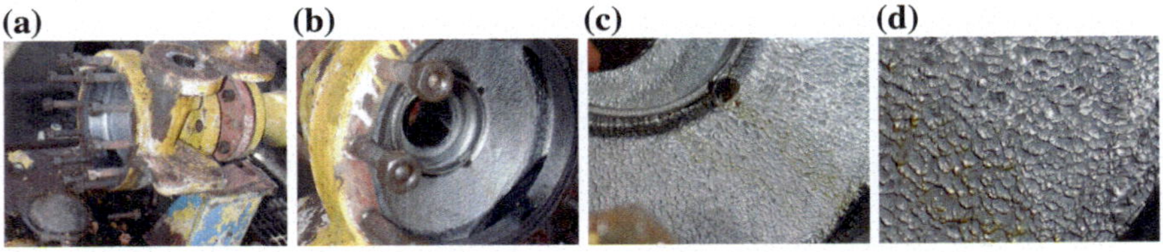

Fig. 6.10 **a** General view of failed pump casing made from cast iron pumping lean amine (DEA). **b**–**d** Inner surface of the pump house (erosion-corrosion)

Solutions and recommendations.

(a) To coat inner surface of the pump made from cast iron with SS 316 which is more resistant to erosion-corrosion by lean amine solution.
(b) To examine quality (chemical composition) and corrosiveness of lean amine solution periodically (once a week).

6.2.1.5 Erosion-Corrosion of Amine Pump (Amine Treating Unit)

Conditions. The pump is intended for the pumping of aqueous solution of lean amine (25–45 % MDEA) after stripping at 80–100 °C and 6–7 bar.

Service period before the failure: 4 years.

Material of construction: Cast iron (diffuser) and SS 316 (bushing).

Visual examination and findings. Shallow pits were detected on the inner surface of the pump (cast iron) (Fig. 6.11). Bushing (SS 316) was not damaged.

Failure phenomenon: *Erosion-cavitation.*

Cause of failure and its explanation. The morphology of damaged surface is typical to erosion-cavitation.

Solutions and recommendations.

(a) To fill shallow pits by welding with SS 316 which is more resistant to erosion-cavitation.
(b) To use the pump with inner coating from SS 316.

6.2.1.6 Erosion of Control Valve (Power Plant)

Conditions. Leak was detected from the control valve at the exit from the water storage tank. Usually the check valve is opened partially in order to allow water to flow.

Material of construction. Valve body—carbon steel; ball and bracket—SS 316.

Service period before the failure: 1 year.

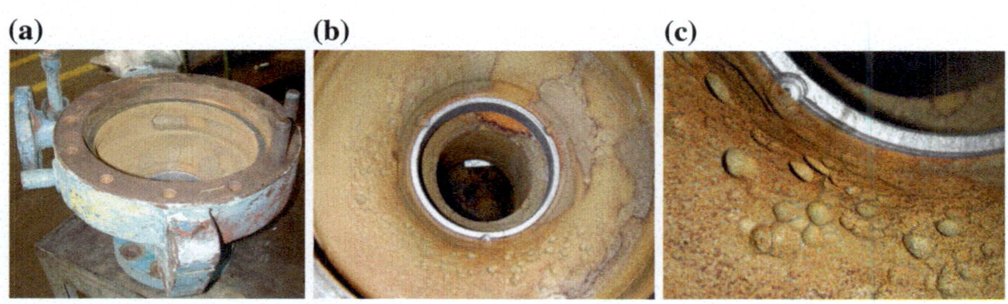

Fig. 6.11 **a** General view of corroded diffuser made from cast iron of amine pump. **b**, **c** Pitted inner surface (inner bushing made from SS316 is not damaged)

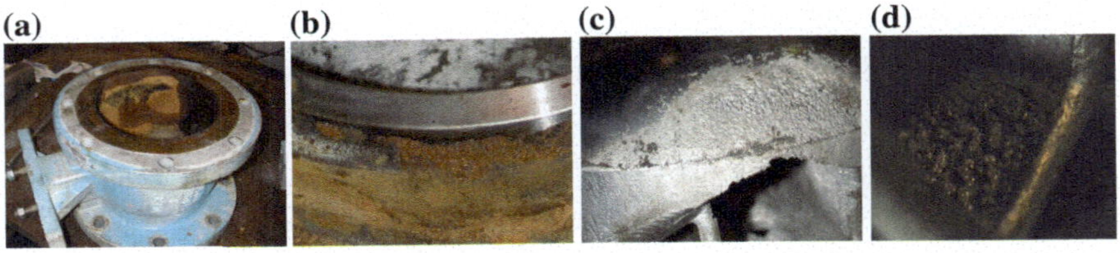

Fig. 6.12 **a** General view of failed control valve. **b** Severe erosion of carbon steel inner surface near the stainless steel bracket. **c** Erosion of inner part. **d** Morphology of eroded ball surface (magnification)

Visual examination and findings. The hole inside the valve body, severe corrosion of inner surface and localized corrosion of the ball was found (Fig. 6.12).

Failure phenomenon: *Erosion*.

Cause of failure and its explanation. The control valve was opened partly, and as a result, the difference of water pressure on both sides of the valve created severe flow rate which caused erosion of both carbon steel inner surface of the valve body and stainless steel ball. The formation of the hole shows that corrosion rate was 15 mm/year. This high value also shows the occurring of erosion.

Solution and recommendations.

(a) To change the water flow regime: in no way to open the valve partially, it should be opened completely.
(b) To repair by welding and to coat with abrasive resistant materials Colmonoy 6, or Stellite 720, or chromium oxide (see Appendix F, Table F.9).

6.2.1.7 Erosion of Valve at the Bottom of the Main Fractionator (FCCU)

Conditions. The valve is installed on the heating line of the pump which is pumping the bottom (clarified oil) of the main fractionator at the FCCU. The temperature of the clarified oil (fuel oil) is 367 °C and pressure is 9.7 bar.

Material of construction. Low-alloy 5 % Cr steel (ASTM A182 Grade F5).

Service period before the failure: 5 months.

Visual examination and findings. Severe erosion of the socket and inner surface of the valve was found (Fig. 6.13).

Failure phenomenon: *Erosion*.

Cause of failure and its explanation. The failure occurred as a result of erosion by catalyst which penetrated into the clarified oil (see case Sect. 6.2.1.3).

Solution and recommendations.

(a) Measures must be taken for preventing the penetration of catalyst into the clarified oil.

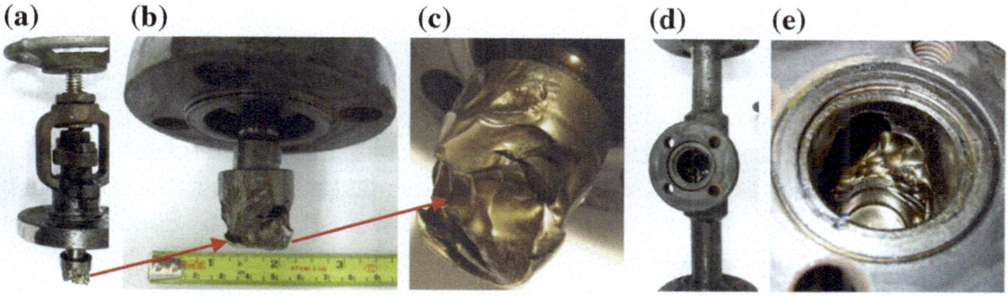

Fig. 6.13 **a–c** Erosion of the socket. **d**, **e** Erosion of inner surface of the valve

(b) To replace the material of the valve with Stellite 720 or Colmonoy 6 (see Appendix F, Table F.9).

6.2.1.8 Erosion of Tube-Sheet in Heat Exchanger (FCCU)

Conditions. The bottom (fuel oil) of the main fractionator at the FCCU flowed inside the tubes at 368 °C and 9.4 bar. HVGO flowed in the shell at 215–304 °C and 10 bar. The pumps with flowing fuel oil from the bottom of the fractionation column suffered several times from erosion during the last 3 years before the current failure of the tube-sheet.

Material of construction. Tubes and tube-sheets—SS 410 (martensitic stainless steel 12 % Cr).

Service period before the failure: 3 years.

Visual examination and findings. The tubes in heat exchanger were clogged by deposits. After removing and cleaning from the deposits, severe erosion was found in the central area of the tube-sheet (Fig. 6.14). Holes were formed at the ends of the tubes installing in the tube-sheet.

ED-XRF analysis. Deposits contained (wt%): Si (40–50), Al (35–40), and O (10–25). The catalyst used in the FCCU consisted from SiO_2 and Al_2O_3. This means that deposits are mostly contained catalyst.

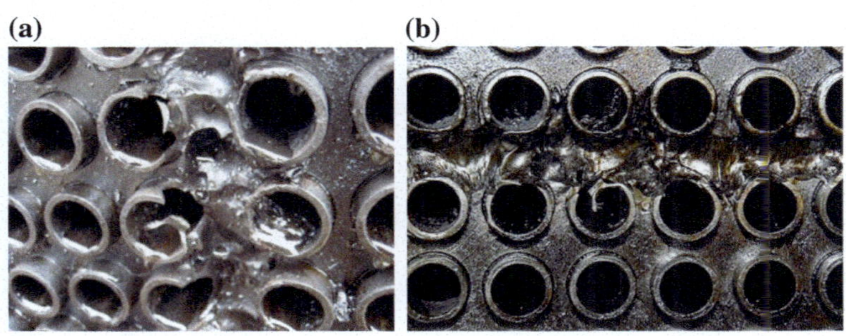

Fig. 6.14 **a**, **b** Eroded tube-sheet (after cleaning from deposits)

Failure phenomenon: *Erosion.*

Cause of failure and its explanation. The failure occurred as a result of erosion resulting from the attack by the catalyst escaping into the fuel oil coming from the bottom of the fractionation column (see case Sect. 6.2.1.3). High velocity with turbulence occurred at the entrance in the heat exchanger. Calculation showed that the velocity value was 3 m/s when the recommended value had to be less than 2 m/s. In addition, most of the tubes in the bundle were clogged by deposits which also gave rise to high velocity of the fuel oil containing catalyst. Thus, severe erosion of central area at the tube-sheet and tube endings occurred.

Solution and recommendations.

(a) To prevent penetration of catalyst into fuel oil (see case Sect. 6.2.1.3).
(b) To install pressure drop controllers (around heat exchangers) transmitted pressure changes in the tubes and monitor pressure drop.
(c) To increase frequency of cleaning of tubes and tube-sheets in heat exchangers.
(d) To coat with wear resistant alloy Stellite 720 or Colmonoy 6 (see Appendix F, Table F.9).

6.2.1.9 Erosion of Pumps with Light Atmospheric Gas Oil (Crude Distillation Unit)

Conditions. Two centrifugal pumps working in series were pumping Light Atmospheric Gas Oil (LAGO) reflux at 240 °C.

Material of construction. Carbon steel ASTM A 216 Gr. WCB.

Service period before the failure: 7 years.

Visual examination and findings. Pits and cracks were found on inner surface of the head and diffuser of two centrifugal pumps (Fig. 6.15).

Failure phenomenon: *Erosion.*

Cause of failure and its explanation. Turbulent flow of LAGO at 240 °C during long time resulted in erosion.

Solution and recommendations.

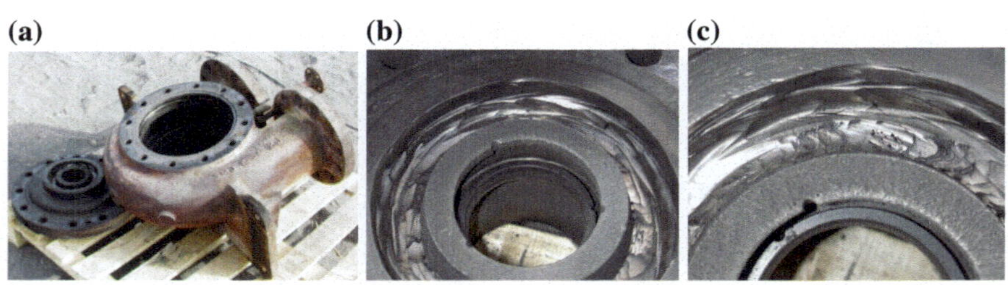

Fig. 6.15 **a** General view of failed centrifugal pump with LAGO reflux. **b** Erosion of the pump head. **c** Magnification of the picture **b** (erosion with cracks)

(a) To grind the eroded carbon steel surface, to weld by means of SS 304L and to polish to restored dimensions.
(b) To use wear resistant materials: Stellite 720 or Colmonoy 6 (see Appendix F, Table F.9).

6.2.1.10 Failure of Pipeline Outlet from the Controller of Condensate Water from the Flash Drum (Power Station)

Conditions. Water condensate (drain of the boiler drum) mixed with steam flowed in the pipeline. The pressure in the line dropped from 130 bar (before the controller) to 14 bar (after the controller) at 130 °C. The flange was welded to the pipe of smaller diameter (Fig. 6.16).

Material of construction. Carbon steel ASTM A105; diameter of the pipe—50 mm.

Service period before the failure: 5 years.

Visual examination and findings. Holes were revealed in welds of the flange and in bends of this pipeline with the frequency about 6 months. Localized corrosion in the form of pits and significant diminishing of wall thickness was found inside of weld area of the flange and in the bend (Fig. 6.17). Corroded inner surface of the flange and the pipe consisted of two oxides (red Fe_2O_3 and grey-black Fe_3O_4). Inner surface of pits were free from corrosion products.

Morphology examination (SEM photo). Signs of erosion caused by cavitation were found (Fig. 6.17c).

Metallographic examination. Smooth surface of the wall of the pit free from corrosion products manifests erosion caused by cavitation (Fig. 6.17d). This attack resulted in diminishing the wall thickness from original 3 to 0.6 mm.

Failure phenomenon: *Erosion-cavitation.*

Cause of failure and its explanation. Decrease of pressure from 130 to 14 bar from the controller to the flange gave rise to flash vaporization (evaporation) of

Fig. 6.16 The flash drum and failed pipes area after the controller

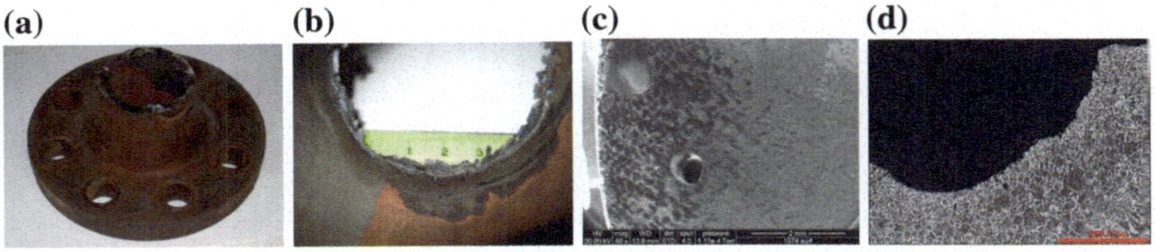

Fig. 6.17 **a** Failed flange. **b** Inner surface (eroded) of the flange with small holes in the region of welds. **c** Morphology of inner surface of the flange with the hole. **d** Metallographic cross-section through failed region: the pit with smooth walls without corrosion products

liquid water. As a result, two-phase mixture liquid water + vapor (steam) at 130 °C was formed, flowed and caused mechanical damage of the inner surface of the flange and the pipe located after the flange. Morphology of deteriorated inner surface of the flange (metal-colored valleys and shallow pits which did not contain corrosion products) and metallographic examination revealed erosion and cavitation. The holes in welds were formed as a result of cavitation: localized mechanical hurts of liquid drops of high energy during flash vaporization.

Solution and recommendations.

(a) To increase the diameter of the flange and the pipe after the controller with similar diameters.
(b) To replace the material of the flange and the pipe with SS 316L.

6.2.1.11 Failure of Disks of the Check Valve (Power Station)

Conditions. Mixture of several water sources (pH ~7.5) from the two large water storage tanks flowed within the check valve located after the sand filter. The pressure in submerged pump was 4–5 bar at ambient temperature. The check valve was open almost the whole service period of the submerged pump.

Material of construction. Butterfly discs—carbon steel ASTM A216 Gr. WCB. Pivots pins—SS 316.

Service period before the failure: 9 years.

Visual examination and findings. Severe corrosion with holes was found on inner surface of the check valve (Fig. 6.18). Serious corrosion damage (more severe than in other areas) was observed. This damage included deep pits and holes through the disc on the side of its connection to the pivot pin. Carbon steel bushings corroded severely (see Fig. 6.18c).

Metallographic examination. It was noted orientation of the pit which is reflected in the lack of symmetry between the sides.

EDS analysis. Iron, oxygen and silicon (~3 wt%) were found on the inner surface of the disc and inside the pit. This confirms the presence of sand in water.

Failure phenomenon: *Erosion-corrosion; galvanic corrosion.*

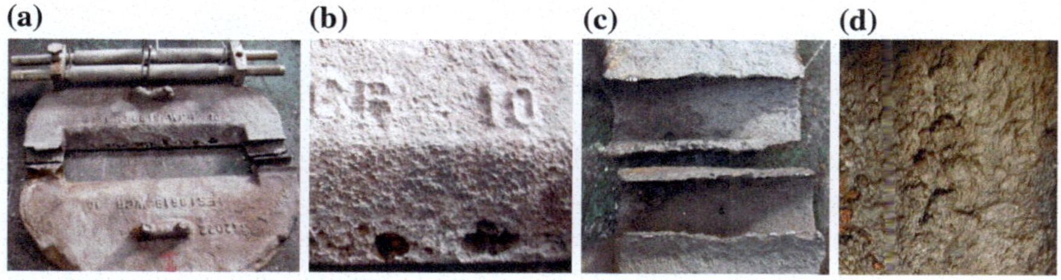

Fig. 6.18 **a** Failed butterfly check valve disc with two pivots. **b** Erosion area of the check valve. **c** Bushings after galvanic corrosion. **d** Inner surface of the butterfly check valve after erosion

Cause of failure and its explanation. Turbulent flow of water mixed with sand in the state of 'open' discs in the check valve caused mechanically-assisted damage. Morphology of deteriorated inner surface of the disc and the valve indicates that erosion occurred: valleys (shallow pits) with metallic color and without corrosion products.

Combination of pivots made from SS 316 and carbon steel bushings in water caused galvanic corrosion of inner surface of bushings.

Solutions and recommendations.

(a) It is impossible to repair failed butterfly discs.
(b) To improve the service of the sand filter (water should be well sand cleaned out).
(c) To paint the surface of the carbon steel butterfly discs with epoxy-ceramic coating resistant to erosion (e.g., epoxy phenol novolac resin filled with aluminum oxide, iron (III) oxide; siloxanes and silicones).
(d) To make the body and all accessories (including discs) in the check valve from similar material (SS 316).

6.3 Corrosion Fatigue and Thermal Fatigue

Fatigue is the weakening of a material caused by repeated or fluctuating stresses. We presume that the reader is familiar with phenomenon when in order to divide aluminum wire into two parts we had to bend and unbend it many times. Fatigue is the phenomenon leading to fracture under cyclic stresses having a maximum value less than the tensile strength of the material. If stresses are above a certain threshold, cracks are formed at macroscopic and microscopic discontinuities (pits, holes, keyways, sharp changes of direction), which cause stress concentrations. They are locations where fatigue begins. Eventually a crack will propagate and reach a critical size, and the structure can fracture. The shape of the structure significantly influences the fatigue life—the number of cycles of stresses that can be sustained prior to failure.

Corrosion fatigue is fatigue in a corrosive environment, namely, is the process in which a metal fractures prematurely under simultaneous corrosion and repeated cyclic loading at lower stress values or fewer cycles and less time than would be required in the absence of the corrosive environment [6]. In short, corrosion fatigue is degradation of a material under joint action of corrosion and cyclic loading.

Thermal fatigue can occur when heated metal is rapidly cooled repeatedly [7]. Rapid cooling of the metal surface establishes high triaxial stresses that can produce cross-hatched cracks.

Therefore, this phenomenon also is named *corrosion-fatigue cracking*.

Two sources of cyclic tensile stresses are cyclically fluctuating internal pressure, and constrained thermal expansion and contraction.

Corrosion fatigue like other corrosion processes is stochastic occurring with definite randomness having probabilistic nature. Damage is cumulative. Materials do not recover when rested.

The phenomenon should not be confused with SCC, where process leads to the development of brittle cracks, their growth under a sustained stress or residual loads and failure. SCC can occur in particular alloys in specific media, whereas corrosion fatigue can happen in nearly all alloys [8, 9].

Corrosion fatigue cracks develop according to the following mechanism [7]. The tube wall undergoes expansion during cyclic stresses. Iron oxide layer fractures and microscopic cracks appear through the oxide to the metal surface. Then the exposed metal surface at the root of the crack oxidizes, forming a microscopic notch in the metal surface. During the next expansion cycle, the oxide can fracture along this notch, causing it to deepen. As this cyclic process continues, a wedge-shaped crack propagates through the tube wall (see Fig. 6.20h). Cracks may originate on the external or the internal surface and are associated with pits. The pits act as stress concentrators for the initiation of corrosion fatigue cracks [7]. These cracks are usually straight, unbranched, needle- or wedge-shaped, and propagate perpendicularly to the metal surface. They often occur in families or parallel cracks and do not run long distances along the tube surface. Corrosion-fatigue cracks sometimes difficult to detect as they are often filled with iron oxides.

Corrosion fatigue failures most frequently occur in boilers that are in "peaking" service, used discontinuously, or otherwise operated cyclically. Rapid boiler startup or shutdown significantly increase the susceptibility to corrosion fatigue. The following parts of boiler are susceptible to corrosion fatigue: wall tubes, re-heater tubes, superheater tubes, economizer tubes, deaerators, the ends of the membrane on waterwall tubing, at points of attachment or rigid constraint (connections to inlet or outlet headers, tie bars, and buckstays) [7]. Cracks can be formed at grooves along the internal surfaces of boiler tubes that are only partly full of water, at points of intermittent steam blanketing within generating tubes, at oxygen pits in waterlines or feedwater lines, and blowdown lines.

Elimination or reduction of corrosion and thermal fatigue can be realized by controlling cyclic tensile stresses and environmental factors, and boiler redesign [7].

Reducing or eliminating cyclic operation of the boiler, extending startup and shutdown times are key ways to prevent corrosion-fatigue cracking. Four case studies of fatigue will be described in this section.

6.3.1 Case Studies

6.3.1.1 Fatigue of the Pipe Bend (HDS Unit)

Conditions. The pipe bend (elbow) connected tank (containing boiler feed water) with the pipe (5 m in length). Boiler feed water flowed inside the pipe at 50 °C.

Material of construction: Carbon steel ASTM A234 Grade WPB.

Service life: 2 years.

Visual examination and findings. Pipe bend was cracked (Fig. 6.19).

The crack starts from the inside of the bend and is perpendicular to the surface. The crack length is 85 mm outside and 100 mm inside the pipe bend center. Secondary crack comes out of the bottom of the main crack. At the distance of about 10 mm from the beginning of the main crack, a pit covered with the corrosion products and crack from this pit parallel to main crack were found (see Fig. 6.19e). The pipe (5 m length without supports) together with the bend suffered from vibrations in the horizontal plane.

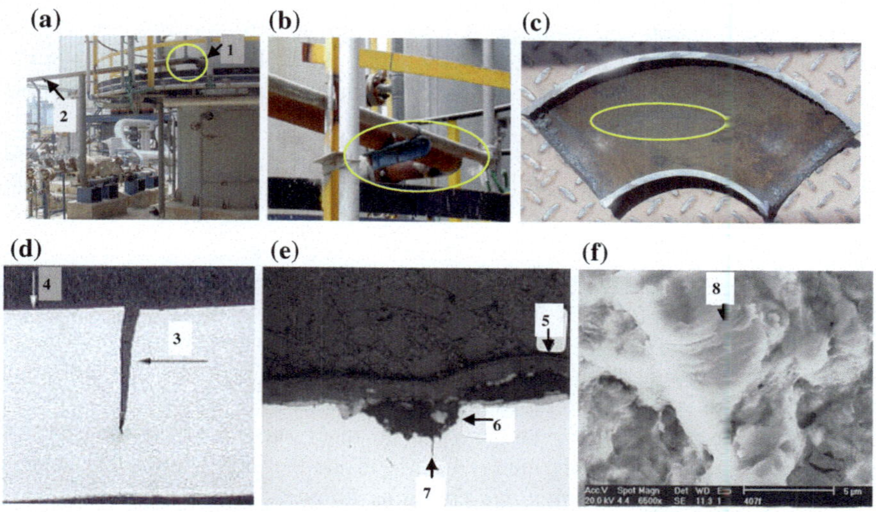

Fig. 6.19 **a** General view of failed pipe at the HDS unit: 1 location of failed bend, 2 the pipe. **b** Magnification of the failure area (pipe bend). **c** The inside of the bend with crack. **d** Macroscopic cross-section of the bend through the crack: 3 crack, 4 inner surface of the pipe. **e** Microscopic cross-section (×200): 5 corrosion products, 6 pit, 7 crack. **f** SEM photo of the picture **e** (×6500): 8 fatigue lines

Metallurgical examination. The chemical composition, mechanical properties and microstructure of the bend are in compliance with the requirements of the material specification.

SEM examination (Fractography). Typical lines like fatigue lines on the surface of fracture are detected (see Fig. 6.19f).

Failure phenomenon: *Fatigue.*

Cause of failure and its explanation. The pipe "moved" freely and suffered from horizontal vibrations which resulted in fatigue. The crack began from the pits inside of the bend. These pits were the source of stress concentration.

Solution and recommendations.

The pipe with elbow must be connected, strengthened and fixed correctly in order to eliminate the horizontal vibrations. The rules of ASME "Standards and Guides for Operation and Maintenance of Nuclear Power Plants" can be used for the guidance [10].

6.3.1.2 Failure of the Boiler Tube

Conditions. Boiler feed water and steam flow inside the tubes at 230 °C and 30 bars in the thermosyphon boiler. Corrosion inhibitor was injected into the water during the last 10 years of the service. Frequent shutdowns and startups (several times in a year) took place during the service of the boiler.

Material of construction. Carbon steel.

Service period before the failure: 15 years.

Visual examination and findings. A hole (4 mm length) was detected on the outer surface of the tube in the bended area (Fig. 6.20). Elongated defects in the form of cracks were found on the inner surface of the tube at the location of the hole and 180° opposite from it.

SEM examination of the inner surface of the tube at the location of the hole and close to it showed that pit was initiated from inner surface and propagated through the wall resulting in the hole (see Fig. 6.20f). Defects in the form of cracks and elongated pits were found close to the hole (see Fig. 6.20g).

Metallographic examination. Metallographic cross-sections were performed through an elongated pit close to the hole in the tube (see Fig. 6.20h, i). The pit filled with corrosion products and protective oxide layer (magnetite) can be seen on the surface and inside the pit. The layer inside the pit is broken at the edge, indicating the initiation of the pit from the inner surface.

EDS analysis revealed iron and oxygen on inner surface of the tube and in addition to these elements chlorine ($\sim$4 wt%) inside the pit.

Failure phenomenon: *Thermal and corrosion fatigue.*

Cause of damage and its explanation. Tensile stresses existed in bending area (especially with ovality). Frequent shutdowns and startups occurring during 15 years increased susceptibility to thermal fatigue. Usually cracking occurs when temperature fluctuations exceed 93 °C. Thermal fatigue cracks usually initiate on the surface discontinuities (pits, etc.). They are wide and often filled with iron

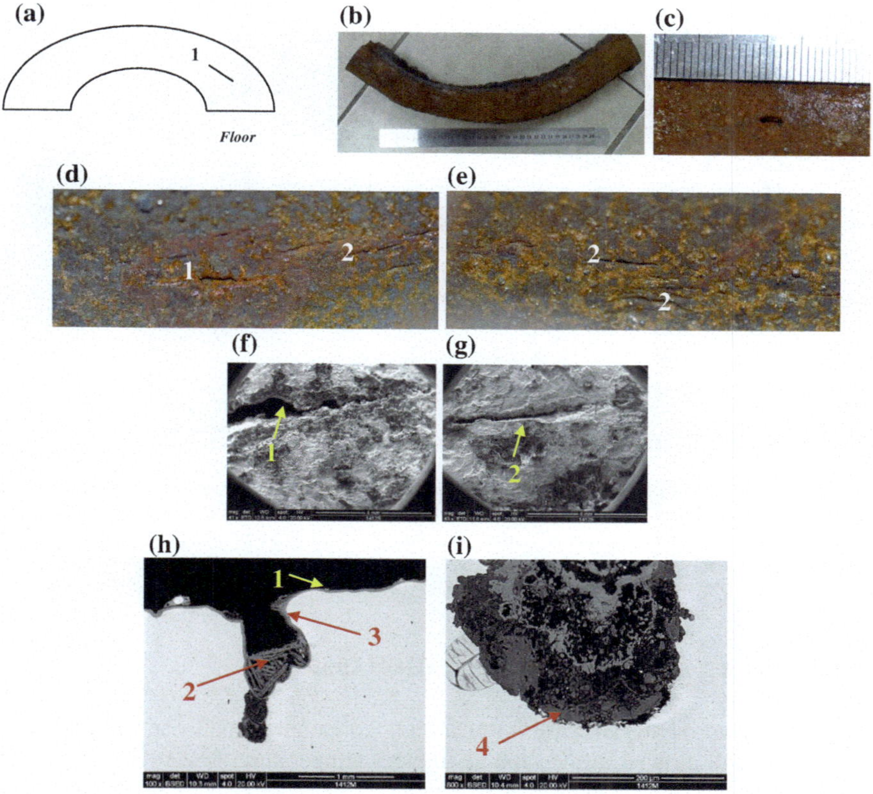

Fig. 6.20 **a** The sketch of the tube with the location of the hole (*1*). **b** General view of the tube with the hole. **c** Magnification of **b** at the location of the hole (outer side). Inner surface of the tube. **d** The side of the hole. **e** The opposite side. *1* the hole; *2* cracks. The SEM photo of outer surface of the tube. **f** The location of the hole (*1*). **g** The crack close to the hole (*2*). Metallographic cross-section through the defect (SEM images) of outer surface of the tube. **h** The pit. **i** The edge of the pit. *1* inner surface of the tube; *2* the pit; *3* oxide layer; *4* corrosion products at the edge of the pit

oxides due to high temperature (230 °C). Corrosion fatigue occurred under combined influences of corrosion and cyclic loading. Corrosion promoted failure at a lower stress value and not large number of cycles.

Solutions and recommendations.

(a) To minimize thermal cycles and thermal stresses.
(b) To control rates of heating and cooling during startup and shutdown in order to reduce stresses. To eliminate rapid cooling cycles.

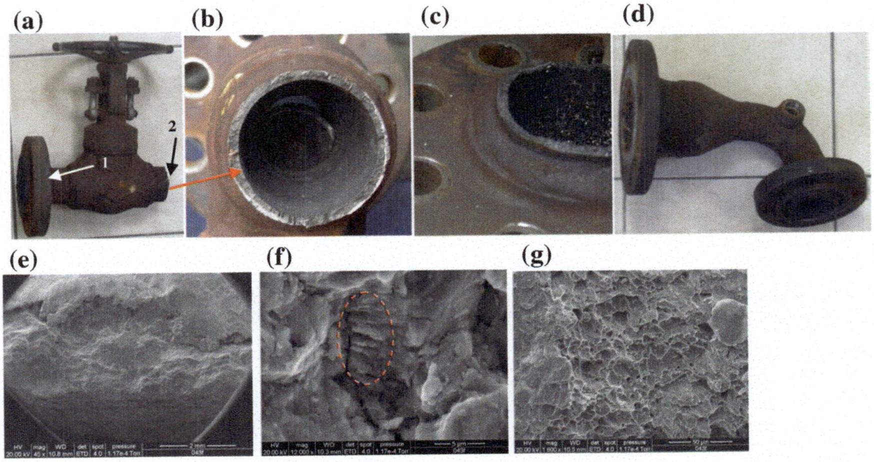

Fig. 6.21 **a** Failed part of the flange with pipe: *1* flange. *2* fracture of the pipe. **b** Enlargement of the fracture surface 2 on the picture **a**. **c** Enlargement of the fracture surface on the other side (near the welding flange). **d** The adjacent part to the flange (*1*). **e–g** SEM photo of the fracture of different magnifications. The *dashed red circle*—fatigue striations. **g** Dimples characterizing ductile fracture

6.3.1.3 Corrosion Fatigue of the Pipe (HDS Unit)

Conditions. Hydrogen gas with hydrocarbons containing small concentrations of hydrogen sulfide flowed in the pipe at ambient temperature. The two parts of the pipeline were connected by welds to flanges. The pipeline was in service only during outage (shutdown) of the unit. Cyclic vibrations and torque (twisting force) worked on these parts of the pipe.

Material of construction. Pipes—carbon steel ASTM A106 Gr. B. Flanges—carbon steel ASTM A105.

Service period before the failure: 30 years.

Visual examination and findings. One part of the pipe near the weld to the flange was broken (Fig. 6.21).

SEM examination. The morphology of fracture surface combines intergranular fracture with the fatigue striations. Such morphology is characteristic to *corrosion fatigue* (see Fig. 6.21f, g). The final fracture area near the outer side of failed part is characterized by dimples which are typical to ductile fracture. The crack started on the inner side and progressed through the wall thickness up to final failure.

Metallographic examination. Fracture occurred in the pipe, and not in the welding area or HAZ.

EDS examination. Fe, O and S (6–25 wt%) are defined in the fracture area.

Failure phenomenon: *Corrosion fatigue.*

Cause of failure and its explanation. The failure started on the inner surface of the pipe, and took place according to *corrosion fatigue* mechanism as a result of aggressive hydrogen sulfide that contained in hydrogen gas flowing in the pipe and cyclic vibrations (stresses) working on this part. No failure was observed on similar element where the same media flowed inside but it was not under cyclic vibrations (stresses). Thus, the failure would not occur without cyclic stresses.

Solution and recommendation.

To exclude cyclic vibrations (stresses) on the pipe.

6.3.1.4 Corrosion Fatigue of Aluminum Fins in the Fin-Fan Air Cooler (CCR Unit)

Conditions. Air cooler composing from carbon steel tubes and aluminum fins is intended for cooling of gas oil by fans located under the tubes at the HDS unit. The temperature of aluminum fins reaches 140 °C.

Material of construction. Fins—aluminum EN AW-1050A (Al 99.5; commercially pure aluminum); 0.4 mm in thickness.

Service period before the failure: 2 years.

Visual examination and findings. Damages of aluminum fins were detected in several appearances (Fig. 6.22): broken edges of the aluminum fins; the fins acquired the shape of waves which is typical form of damage due to vibration; surface of fins has grey color (typical to aluminum oxides). Screws in some places have untwisted. Thus, certain groups of tubes were not connected tightly to the base.

SEM examination. Fatigue lines covered by corrosion deposits were detected. These lines are typical to corrosion fatigue (see Fig. 6.23).

Metallographic examination. The fracture is brittle.

EDS analysis of fracture surface: Al, O, Fe, Cu, and Cl.

Failure phenomenon: *Corrosion fatigue.*

Cause of failure and its explanation. The failure was caused by improper connection of tubes in a group. The presence of Cu and Cl on the surface of the fracture shows that the atmosphere was aggressive to aluminum. Vibrations of tubes in the presence of corrosive atmosphere resulted in corrosion fatigue of aluminum fins.

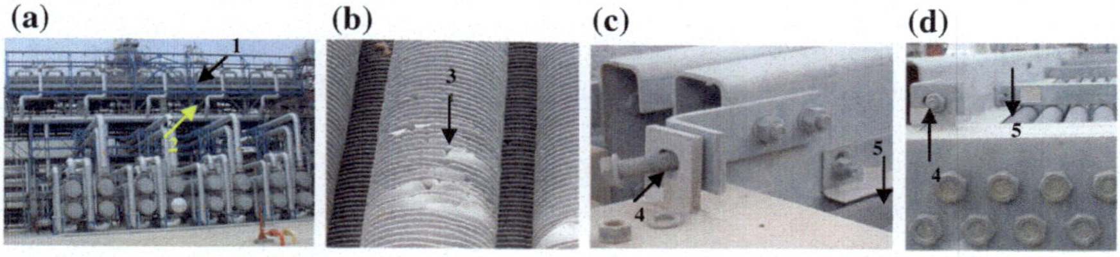

Fig. 6.22 **a** General view of air cooler (upper part of the picture): *1* locality of failed tubes, *2* locality of fans. **b** Failed aluminum fins (*3* the location of damaged fins). **c** Improper connection with the bolt (*4*). **d** Bolt (*4*) connection and failed aluminum fins (*5*)

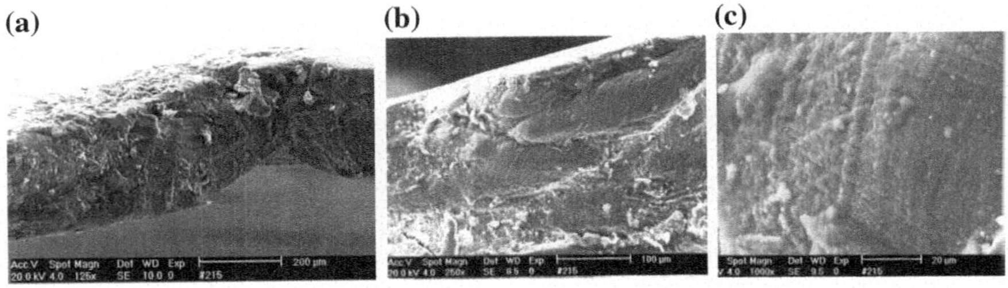

Fig. 6.23 SEM photos of aluminum fin fracture with different magnification: **a** ×125, **b** ×250, **c** ×1000

Solution and recommendation.

To ensure strict and correct connections of tubes in groups in order to prevent vibrations.

6.4 Galling

One mechanical engineer at the oil refinery called me to observe several bolts and nuts made from SS 316. He told me that during installation the bolt was twisting off and the bolt's threads were seizing to the nut's threads. It was impossible to disassemble this system. All mechanical properties of stainless steel bolts and nuts were checked and they were consistent with required technical data. The mechanical engineer was frustrated. I explained him that this phenomenon was known as *thread galling*, or *cold welding*, or simply *galling*'.

Galling is a form of wear caused by adhesion between sliding surfaces. Galling of mating surfaces can be a result of fretting corrosion. Galling is caused by a combination of friction and adhesion between the surfaces, followed by slipping and tearing of crystal structure beneath the surface and localized welding. This will generally leave some material stuck or even friction welded to the adjacent surface. These self-mated coupled become "frozen" fasteners. Galling can happen with fasteners made from stainless steel, aluminum, titanium, nickel-chromium alloys which self-generate an oxide surface film for corrosion protection. During fastener tightening, as pressure builds between the contacting and sliding thread surfaces, protective oxide layers are broken, wiped off, and interface shear and lock together. This cumulative shearing-clogging-locking action causes increasing adhesion. As a result, galling leads to seizing—the actual "freezing" together of the threads. This can occur not only with threads but when two surfaces contact under pressure, for instance, pin and bearing. Roughness of mate surfaces also affects galling. The rougher the flanks, the greater the likelihood galling will occur. In a wear system, a galling or seizure failure occurs first, followed by dimensional loss due to wear, and finally corrosion. Galling and wear must be the first concerns of the engineer during designing of valve stems, seats, trim, fastening systems, screening, pins, pump shafts, bushings, and roller bearings.

There are several solutions for this problem [11–15]:

(a) Using different stainless steel alloy grades (i.e., SS 316 and SS 304) for two surfaces reduces galling. The key here is the mating of materials of different hardnessesbecause different alloys work-harden at different rates.
(b) Slowing down the installation RPM (revolutions per minute) speed frequently reduce, or sometimes solve completely the problem. As the installation RPM rate increases, the heat generated during the tightening increases, which in its turn the tendency for thread galling also increases.
(c) Lubricating the mating surfaces can eliminate galling. Such lubricating materials are molybdenum disulphide (MoS_2), mica, talc, or waxes. Graphite is not recommended because it can cause severe galvanic corrosion in the presence of electrolyte because the graphite has the most positive electric potential comparing to that of metals. Some companies produce proprietary anti-galling or anti-seize lubricants (pastes).
(d) Use of alloys resistant to galling (Appendix F, Table F.10).

The case of galling is described in this section.

6.4.1 Case Study Galling of Blower Axis (CCR Unit)

Conditions. In order to replace bearings in blower axis it was tried to remove the axis, but without success. The axis was "sticked" to the bearing without the possibility of separation between them. Service temperature was 500 °C.

Material of construction. Axis and bearing—Incoloy 800H.

Service period before the failure: 3 years.

Visual examination and findings. The "sticking" of the axis to the bearing.

Cause of failure and its explanation. The phenomenon happened is galling. Galling occurs when the two surfaces are "glued" with no possibility of separation. It happens with specific types of stainless steels, Ni-Cr–Fe, and aluminum alloys.

Failure phenomenon: *Galling.*

Solutions and recommendations.

(a) To replace Incoloy 800H with Nitronic 60 stainless steel (Alloy 218) which is more resistant to galling (see Appendix F, Table F.10).
(b) To coat the axis with Cr_2O_3.

References

1. Groysman A (2010) Corrosion for everybody. Springer, Dordrecht
2. Groysman A (2014) Corrosion in systems for transportation and storage of petroleum products and biofuels. Springer, Dordrecht
3. Starosvetsky J, Armon R, Groysman A, Starosvetsky D (2002) MIC of storage tank aluminum floating roofs during hydrotest. Mater Perform 41(4):50–54

4. Javaherdashti R (2008) Microbiologically influenced corrosion: an engineering insight. Springer, London
5. Little BJ, Wagner P (1997) Myths related to microbiologically influenced corrosion. Mater Perform 36(6):40–44
6. ASM Handbook (1987) vol 13: Corrosion. ASM International, USA
7. Port RD, Herro HM (1991) The Nalco Guide to Boiler Failure Analysis. McGraw-Hill Inc., New York
8. Schijve J (ed) (2009) Fatigue of structures and materials. Springer, Dordrecht
9. Gangloff RP (1995) Environmental cracking—corrosion fatigue. In: Baboian R (ed) Corrosion tests and standards: application and interpretation. ASTM, Philadelphia, pp 253–271
10. ASME (2007) Standards and guides for operation and maintenance of nuclear power plants OM-S/G, part 3. ASME, USA
11. IFI Inch Fastener Standards Book (2014) Industrial Fasteners Institute, 9th edn. Ohio, USA
12. Davis JR (2001) Surface engineering for corrosion and wear resistance. ASM International, USA
13. Speck JA (1997) Mechanical fastening joining and assembly. Marcel Dekker, New York
14. ASTM G98-02 (2009) Standard test method for galling resistance of materials. ASTM International, West Conshohocken, PA, USA
15. ASTM G40-13 (2013) Standard terminology relating to wear and erosion. ASTM International, West Conshohocken, PA, USA

Chapter 7
Fouling, Corrosion, and Cleaning

Scientists study the world as it is. Engineers create the world that has never been.

Theodore von Kármán (1881–1963),
the Jewish-Hungarian-American aerospace engineer.

Abstract Fouling is severely connected to corrosion. Organic, inorganic, preoperational, process-side, water-side, and fire-side fouling are described. The reasons and problems arising from fouling, its mechanism, control methods of fouling in organic and cooling water systems are analyzed. Mechanical and chemical cleaning procedures, preoperational cleaning and passivation, safety, environmental and general recommendations for implementation of cleaning are presented.

Fouling is the accumulation of unwanted material on solid surface of equipment and structures.

Deposits, sludge, scaling, sediments, precipitates, unused residue are related to fouling. Heat exchangers, pipelines, tubes, tanks, pumps, filters, compressors, air coolers, columns, boilers, reactors, and furnaces are prone to fouling. Its nature, appearance and extent depend on the flowing media, equipment design and geometry location, metallurgy, surface conditions (i.e. roughness, presence of scale, corrosion products), and process conditions (temperature and its fluctuations, pressure, flow regime and velocity, number of phases, injection of chemicals, efficiency of desalter operation, the frequency of outages).

We can safely say that total efficiency of oil refinery and petrochemical units depends on uninterrupted (trouble-free) operation of all equipment. In this perspective, fouling problems occupy paramount attention. Sometimes these problems associate only with heat exchange equipment and really this network is very significant. However, other equipment not related to heat exchange (tanks, pipelines, pumps, filters, compressors, columns, ejectors, reactors) also are severely prone to fouling (Fig. 7.1). For instance, sludge in tanks, wax deposition and corrosion products in crude oil pipelines, biofouling in cooling water pipelines, corrosion products in filters, to name a few.

A. Groysman, *Corrosion Problems and Solutions in Oil Refining and Petrochemical Industry*, Topics in Safety, Risk, Reliability and Quality 32,
DOI 10.1007/978-3-319-45256-2_7

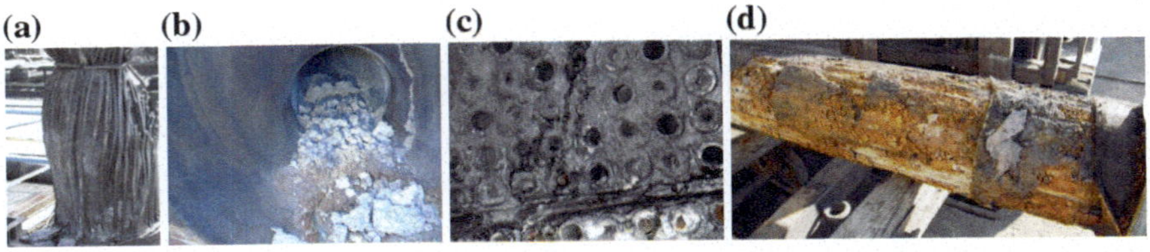

Fig. 7.1 **a** Solidification (fouling) of phthalic anhydride led to distortion of SS316 tubes of the heater (4 years, Petrochemical Plant). **b** Inorganic deposits in ejector (hot acid gases reacted with aqueous soda solution at 45 °C, 9 months, CCR unit). **c** Amine solution fouling (4 years, Amine Treating Unit). **d** Cooling water: corrosion products (5 years, Power Station)

Although physicochemical phenomena underlying the fouling are complex, the understanding of mechanism in most cases is known. For instance, fouling in the hot end of the preheat exchangers is originated by a series of chemical reactions triggered by the high operating temperatures [1]. Critical velocities exist above which fouling does not occur due to deposit removal by shear forces from the fluid flow. The 'fouling threshold' model was developed which allowed users to estimate operating conditions where the fouling rate formation would be close to zero [2–4].

Fouling can be caused by various mechanisms, and they are different in organic, aqueous, and gaseous media. The most important characteristics of a fouling are its thermal resistance (two orders more than that of metals) and the thickness of the layer (from several mm to several meters, e.g., at the bottoms of tanks).

The reasons of fouling in organic streams (crudes, petroleum products, fuels, organic solvents) are instability and incompatibility of components of crudes after blending, decomposition of organic substances resulting in formation of coke and tar, polymerization and oxidation of hydrocarbons, large concentration of metals in media, precipitation of sludge and corrosion products. Extensive use of opportunity crudes and oil sands increases likelihood of fouling. Typical problem areas include heat exchangers and furnaces at crude distillation units, thermal cracking process and visbreaker units, hydrotreater exchangers and reactor beds, FCCU slurry exchangers, hydrogen recycle compressors at catalytic reforming units, pumps in some applications (aromatics) at petrochemical plants, and heat exchangers in cooling water systems.

The reasons of fouling in cooling water systems are crystallization (scale formation of calcium and magnesium carbonates, sulfates, silicates, and phosphates); deposition of microorganisms, their metabolic products; and corrosion products (hydroxides, oxides, and salts of corroded metals).

The reasons of fouling in gaseous media (oxygen and flue gases in furnaces, or process gases in refining and petrochemical units) are hot corrosion, fuel ash corrosion, or metal dusting. However, the fouling in gaseous phase is considered as only corrosion products and not fouling alone, namely, corrosion fouling.

Fouling may be organic, inorganic, or a mixture of the two. Fouling mechanism is different not only for formation of organic and inorganic materials but also differs at various equipment and units.

Organic fouling in a crude unit results from the precipitation of organic compounds which become insoluble in the media, such as asphaltenes, and high molecular weight hydrocarbons (i.e. paraffins). Asphaltenes are the fraction of crude oil that is soluble in toluene (aromatic solvent) and insoluble in the n-alkane (usually heptane) [5]. The asphaltenes can become unstable because of the blending of incompatible crudes or heating. The precipitated organic compounds can be dehydrogenated forming coke. The coke can be formed also as a result of thermal degradation of hydrocarbons due to long heating periods. Organic fouling on the feed side of the hydrotreater feed/effluent exchangers and reactors systems and in a FCCU slurry system is usually caused by polymerization of olefines and diolefines, condensation reactions between carboxylic acids and N-organic compounds, and degradation of heavy molecular weight polynuclear aromatics [6].

Organic fouling may be formed not only in organic streams but also in water media as a result of bad filtration of water from organic substances or the presence and proliferation of microorganisms (biofouling).

Inorganic fouling can result from several sources. Corrosion products (mostly metal oxides, hydroxides, and sulfides) and solid inorganic contaminants (sand, silt and salts) can deposit mainly in areas with low velocity of liquids or outside tubes in furnaces as a result of hot corrosion, fuel ash corrosion, or metal dusting. Inorganic fouling can be formed in organic, aqueous and gaseous streams. Despite of removing most of inorganic salts in desalters, some salts can remain and deposit in the preheat exchangers. Aqueous caustic soda solution is injected into the desalted crude to neutralize HCl forming as a result of hydrolysis of calcium and magnesium chloride salts. Excess of caustic soda (usually under uncontrolled injection) can cause fouling in downstream exchangers and coking in furnaces. Inappropriate water washing in the overhead at the distillation unit can cause fouling by ammonium chloride and amine chloride salts (see Sect. 8.2.1). Catalyst fines can contribute to fouling in the FCC slurry system. Ammonium chloride fouling (with severe corrosion consequences) forms in hydrogen recycle compressors at catalytic reforming, catalytic cracking, hydrotreating and hydrocracking units. Inorganic fouling (mostly iron sulfide as corrosion product) is formed in pumps and columns at TPX and BTX units at petrochemical plants. However, the biggest problem with inorganic fouling occurs in cooling water systems. It is possible to classify fouling according to sources: preoperational, process-side, water-side, and fire-side [7].

Preoperational fouling is formed during the fabrication and erection of equipment. In addition to mill scale (formed during metallurgical processes and heat treatment operations), soil, dust, dirt, weld spatters, corrosion products, oil, grease, lubricants, fat, and shop primers (temporary coatings) can be attached on the equipment surface. Surface of stainless steel, copper alloys and titanium alloys may become contaminated with iron particles from tooling or welding of carbon steel parts and cause galvanic corrosion of iron dust with further rusted spots (Fig. 7.2). In most cases such metal corrosion fouling causes unaesthetic appearance of outer surface of equipment and constructions, but sometimes personnel accepts this situation as failed stainless steel or copper alloys. Really, if not to remove these

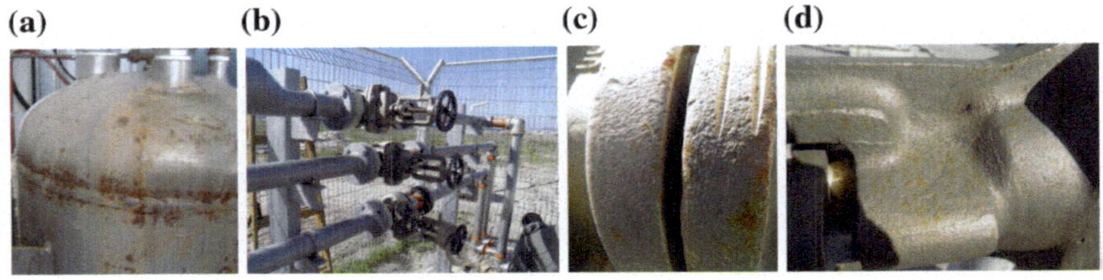

Fig. 7.2 Preoperational fouling of outer surface. **a** Stainless steel (SS 316) vessel after erection (3 months). This surface needs cleaning and passivation according to standards [8, 9]. **b–d** Al-Bronze (UNS C95800) valves in the fire-fighting system. **c**, **d** Rust on the surface of Al-Bronze valves (1 month in winter during harsh rains). It is easy to remove rusted spots by means of aqueous solution of nitric acid (25 wt%)

corrosion spots (rust) from the surface, possible crevice corrosion can occur. Special cleaning procedures are needed to remove this fouling from stainless steel, copper and titanium alloys [8–10]. In order to prevent a buildup of small iron particles and formation of rusted spots, the equipment should be covered with tarpaulin for the period of welding.

In order to decide what type of cleaning should be carried out for specific equipment, physicochemical identification of fouling should be done.

Process-side fouling at oil refinery and petrochemical units is very diverse and represents organic substances and coke alone or in mixture with corrosion products (sulfides, oxides and hydroxides) (Appendix J). Sometimes deposits (iron sulfides) are pyrophoric when exposed to air (see Sect. 4.1.1). Preheat exchanger fouling is defined by crude oil type, caustic soda and slop oil injection and can include following components: organic compounds (polymers, asphaltenes, gums, waxes, amine neutralizers, and amine corrosion inhibitors), inorganic substances (chlorides, sulfates, carbonates, silicates, elemental sulfur, caustic), chloride salts of amines, sediments (sand, silt, dirt, particulates), and corrosion products (iron sulfides, hydroxides, oxides). Numerous organic and inorganic substances and inappropriate conditions (disturbance of processing parameters) can cause fouling on both shell and tube sides of heat exchangers, in tanks, pipelines, reactors, pumps, compressors and other equipment. Fouling on the shell side is often associated with low flow velocity 'dead' zones close to the baffles. Various hydrocarbon streams, injection of ammonia, amine neutralizers and corrosion inhibitors can result in fouling by organic substances and chloride salt deposits on shell side of heat exchangers in atmospheric and vacuum overhead systems (see Sect. 8.1.1, Fig. 8.1a). Chloride salts may be inorganic (ammonium chloride NH_4Cl) and organic (amine chloride RNH_3Cl) ones.

Bottom (residue) from crude distillation unit flowing to visbreaker unit is always highly prone to fouling, therefore, this is a huge problem at visbreaker.

Decomposition and oxidized products of sulfolane and amine solutions (HSAS) may result both in severe fouling and corrosion in BTX units (petrochemical plants) and amine treating units respectively.

In order to understand the importance of fouling problem in oil refining and petrochemical industry, we will give only one very impressive figure. Fouling in crude oil distillation units costs the world ~4.5 billion US $ per annum and results in excess fuel being burnt in refinery furnaces [11]. This value is greater than direct corrosion cost (~3.7 billion US $ annually) in oil refining industry in the USA [12]. The cost of fouling (economic penalties) includes extra energy consumption, throughput loss, maintenance costs, and cost of antifoulants.

Water-side fouling also may be of many chemical types. The deposits may consist of carbonate hardness salts alone, or only biofouling inside cooling water pipelines or in heat exchangers. In addition, sulfates, silicates, phosphates, hydroxides, oxides, sulfides, silica, mud, and debris also can be present in deposits. Petroleum products and other organic media can leak into cooling water and foul inner surface of heat exchangers, pipelines and pumps.

Fire-side fouling is formed from the burning of fuel oil and diesel fuel in furnaces and boilers. Because sulfur compounds and metals are present in fuel oil, combustion products (as fouling) are severely corrosive to carbon steel and fuel ash corrosion can occur. Acid dew point corrosion of outer surface of carbon steel and low alloy steel pipes in furnaces by condensed sulfuric acid (forming by absorption SO_2 and SO_3 gases by H_2O vapor at 100–180 °C) can result in fouling by sulfate salts. These salts during shutdown are hydrolyzed by condensed water vapor from the atmosphere, result in acidic environment (up to pH ~1) and cause severe pitting corrosion of outer surface of pipes in furnaces. Therefore, these sulfate acidic salts should be immediately neutralized by aqueous soda ash solution (2–3 wt% Na_2CO_3) and removed during shutdown.

Switch onto natural gas (namely, methane, as clean fuel) instead fuel oil and diesel fuel significantly improves situation in furnaces and boilers, and as a result fouling does not occur.

Thus, this complexity and diversity of fouling formation makes prediction of fouling, its monitoring and anti-fouling treatment complicated. First of all, we will describe fouling mechanism.

7.1 Fouling Mechanism

It was suggested to differentiate the five major categories into which fouling can be divided according to the mechanism that produces it [13]:

A. Crystallization fouling—solidification or precipitation of salts from supersaturated solutions.
B. Particulate fouling—deposition of suspended solid particles (dirt, silt, clay, rust).
C. Chemical reaction fouling—deposition resulting from chemical reactions between reactants.

D. Corrosion fouling—deposition produced by reactions between reactants and the metal surface.
E. Biofouling—formation of microorganisms' layers.

Proportion of each of these five mechanisms is different in organic and aqueous media. For instance, crude oil and petroleum product fouling at high temperatures (>300 °C) can happen via any of the above mechanisms, except biofouling. In cooling water systems, crystallization, corrosion and biofouling may play leading role. Establishing which is the overarching mechanism is paramount in order to take remedial action at both design and operation stages. In most cases, there is the combination of different fouling mechanisms. Prolonged experience in oil refining and petrochemical industry shows that two mechanisms for organic fouling are more likely to occur in heat exchangers:

a. A combination of corrosion and chemical reaction fouling via the auto-oxidation mechanism.
b. Precipitation of asphaltenes.

Detailed description of these five fouling mechanisms is given in Appendix J.

Generally, several fouling mechanisms occur at the same time, nearly always being mutually reinforcing. Each of five generic types of fouling comprise five distinguishable steps: initiation, transport to surface, attachment, removal, and ageing [11]. As many of these steps can operate in parallel, fouling is a complicated physicochemical process. Transport to surface, attachment, and removal are understood well, while initiation and ageing till now remain poorly understood.

There is induction period for beginning of fouling formation lasting from few seconds to several hours, days, weeks, and even months. For instance, this period is few hours at T > 275 °C in HVGO [14]. Induction period decreases significantly with increasing temperature, roughness of surface, and when the surface is not clean initially [15–17]. Like in kinetics of localized corrosion (pitting, crevice, SCC, MIC), after the initiation stage, the process of accumulation of fouling can develop quickly up to full plugging of heat exchangers, filters in pumps, and pipes. Fouling rates increase moderately with surface temperature, i.e. an increase of 25 °C from 260 to 285 °C results in a doubling of the initial fouling rate [5].

In all reaction mechanisms, the fouling precursors (foulants) must be transported by diffusion mechanisms from the bulk of the fluid toward the zone in which they are converted into deposits by adhesion to the surface [1]. In a similar manner, the foulant or deposit, if mobile enough, could be transported back toward the fluid bulk. The transport of foulants is governed by the difference between their buck concentration and their concentration at the surface.

Ageing is a phenomenon when deposits become more cohesive and harden their structure. Ageing may increase the strength of the deposit by polymerization, re-crystallization, and de-hydration [17]. For instance, biofouling where iron-oxidizing microorganisms take part in the process, may be hardened by the presence of ferric ions (Fe^{3+}) and creation of solid shell composed from ferric

Fig. 7.3 Biofouling created by iron-oxidizing microorganisms

hydroxides (FeOOH) strongly adherent to metal surface (Fig. 7.3). Every fouling is subjected to ageing.

Corrosion with formation of iron sulfide can play an important role in the induction period and in later steps of crude oil fouling. Iron sulfide can be formed by different ways (see Appendix D and Sect. 4.1). Iron sulfide can promote fouling by increasing the surface area and decreasing the interfacial tension for stronger adhesion to insoluble asphaltenes [18]. Thus, fouling from crude oil heat exchangers can contain >35 wt% of iron sulfide (see Appendix J). Iron sulfide is black and, without chemical analysis, could be mistaken for organic carbonaceous material (also black). It is very easy to define if it is iron sulfide (see Appendix D).

Corrosion products provide resistance to heat transfer, create areas with roughen surface and thus promote the accumulation of fouling. In addition, corrosion products become particulate fouling for downstream section of the equipment.

Increasing of temperature promotes chemical reactions of fouling formation, corrosion, shortens induction period and severely influences the ageing rate of deposits. They may be hardened and become difficult to remove or, alternatively, become weaker and tend to spall under the influence of temperature and time [1].

7.2 Problems Arising from Fouling

Fouling leads to the following problems.

A. Under deposit corrosion (see Sect. 2.4). Corrosion products in their turn contribute to the fouling. As a result, a vicious circle takes place constantly.
B. Decreasing of heat transfer and outlet temperature of streams. Fouling material has a thermal conductivity 0.2–2.9 $W \cdot m^{-1} \cdot K^{-1}$, up to two orders of magnitude lower than that of the tube metal wall ($\sim$20–40 $W \cdot m^{-1} \cdot K^{-1}$, depending on the metal use and temperature) [1, 17]. As a consequence, heat transfer is impaired and the energy recovered reduced. Refinery energy consumption is 10–30 % greater because of fouling.

C. Increasing temperature on the hot side and decreasing temperature on the cold side. For instance, when water is heated to 250 °C in tubes of the boiler the average temperature of the tube wall is 400 °C. However, when the layer of scale gets thicker than 50 μm, the temperature of the tube wall should be increased to 510 °C in order to keep the water temperature 250 °C inside. This temperature is close to the value 550 °C at which thermal oxidation of carbon steel becomes excessive [19].
D. Extra cooling capacity may have to be provided to properly cool product streams on the hot sides of heat exchangers [20].
E. Increasing pressure drop and, as a result, decreasing flow rate, and the need of energy costs.
F. Blockage of inner surfaces of pipelines, tubes, heat exchangers, pumps, and compressors.
G. Increasing fuel gas burning in furnaces in order to provide required temperature of the stream.
H. Higher firing temperatures in furnaces shorten heater tubes' life.
I. Environmental impact occurs because of increased release of green-house gas emissions (CO_2). Another ecological aspect includes the necessity of the disposal of the carbonaceous and hydrocarbon deposits containing sulfur, nitrogen, and metals in an adequate place.
J. Decreasing of capacity of storage tanks when large volumes of fouling are accumulated at the bottoms and intensive uncontrolled localized under deposit corrosion.
K. Fouling can cause deterioration of fuels' quality because of partly dissolution in finished products or releasing of such contaminants as H_2S by SRB (biofouling) at the bottom of kerosene storage tanks.
L. Unscheduled shutdowns for periodic cleaning from fouling. During shutdowns, carbon steel surfaces of equipment and constructions are exposed to air. As a result, rust is formed, and then after start-up this rust flakes off and fouls downstream equipment.
M. Safety hazards and environmental problems (see Sect. 7.6).

7.3 Control Methods of Fouling in Organic Media

The following control methods of fouling in organic streams (crude oils, petroleum products, fuels, organic solvents) are used:

A. Adequate preparation of crude oil. Water, aqueous contaminants (salts, H_2S), and sediments are separated from crude by settling, draining and sometimes filtration in the field. Unfortunately some clay minerals (micron-sized) are not filterable. Some substances (surfactants) are not separated from crude oil and form stable emulsions. Then at refinery, demulsifiers are injected into the crude to separate salts and water from the crude in desalters.

B. Crude oil and other more fouling and more corroding fluids should be carried on tube side to facilitate cleaning. The inside of tubes can be cleaned much easier than the outside.
C. Bypass piping or parallel, spare, heat exchanger and pump should be provided.
D. The concentration of injected caustic solution should be not greater than 1.5 wt% NaOH to improve caustic dispersion in the crude.
E. Maintaining a high fluid velocity prevents particles from settling onto the equipment surface. Eliminating stagnant and low velocity regions. Heat exchangers, pumps, tanks, columns and other equipment should be drained, washed and cleaned from fouling during outages and shutdown.
F. Chemical treatment, or use of antifoulants. Antifoulants are chemical substances which are injected into desalted crude oil, or heavy petroleum products (atmospheric bottom, coker feed), and in hydroprocessing stream to diminish or eliminate fouling. They can present one or several chemicals. For instance, amine phosphate ester detergents, inorganic phosphorus-containing acids and salts thereof are used as antifoulants in crude oils. Usually antifoulants contain complicated mixture of dispersants, corrosion inhibitors, metal deactivators, and inhibitors of polymerization. Each component has its purpose. Dispersants prevent smaller particles from agglomerating to form larger particles which deposit more easily. They also prevent the small particles from being attracted to already existing deposits. Corrosion inhibitors prevent the contact between the metal surface and the corrosive fluid. Metal deactivators "neutralize" metals by complexing, thus reducing the catalytic activity of the metal (namely, Fe and Cu) to initiate polymerization. Polymerization inhibitors react with radical (R·) to form stable molecules (R–R) and thus interrupting the chain reaction which leads to the formation of polymers. It is not easy to select antifoulant and its dosage because its efficiency depends on composition of fouling and hydrocarbon stream, temperature, conditions of its formation, flow regime, traits of unit operation and its history. Before the choice of antifoulant type, careful physicochemical analysis of feedstock and deposits should be made. This analysis can give an indication of deposit constituents (organic, inorganic or a combination) and possible mechanism of fouling formation. Fouling often occurs on both sides of heat exchanger tubes, but at different rates. Antifoulants can delay the onset of fouling, but not prevent it. They require clean equipment to start with in order to be effective. Fouling intensity can vary with time. Therefore, it is often impossible to keep the right concentration of antifoulant for maximum effectiveness.

7.4 Fouling and Its Control in Cooling Water Systems

Cooling water usually results in fouling by scale and deposits (see 'crystallization fouling') inside or outside of tubes, tube-sheets and in channel boxes of heat exchangers, and even can fully clog tubes and piping (Fig. 7.4).

Carbon steel, low-alloy steel, cast iron, copper alloys, copper-nickel alloys, stainless steel, titanium alloys, in short, all alloys using in cooling water systems are prone to fouling (Figs. 7.5, 7.6). Usually slow flow velocity and high temperature (>80 °C) result in deposition of sediments. It is impossible to remove this fouling from inner surface by means of acid cleaning. Only mechanical cleaning can be applied in such cases.

The pressure outside of tubes generally is higher (12–25 bar, hydrocarbons stream) than inside (4–5 bar, cooling water). Therefore, in the case of corrosion and

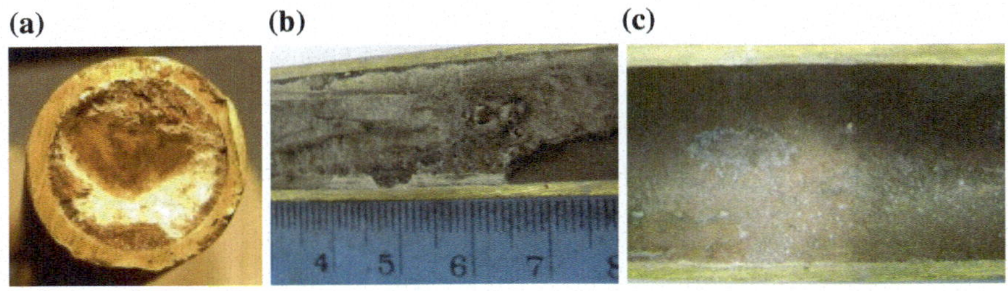

Fig. 7.4 **a**, **b** Full blockage of inner space of Admiralty brass tubes after 3 years of service at Visbreaker unit. Inside of tubes—cooling water, outside—diesel fuel (240 °C). Deposits contain phosphates and sulfates of calcium and zinc with small amounts of silicates and chlorides. **c** Under deposit corrosion on brass (after removing of sediments)

Fig. 7.5 Three years of service. **a** Blockage of inner space of Ti Gr.2 tubes at crude distillation unit. Inside of the tubes—cooling water (30–35 °C); outside—hydrocarbons (120 °C). Deposits contain phosphates and sulfates of calcium and zinc (T > 80 °C). **b** Inside Naval brass tubes—cooling water (30–35 °C); outside—LVGO (120 °C). Clogging by debris because of bad water filtration. **c** Inside of carbon steel tubes—'tempered water' (~50 °C); outside—LVGO (120 °C). The tubes are clogging by *black* polymeric substances forming as a result of penetration of LVGO into 'tempered water'

(a) (b) (c) (d)

Fig. 7.6 **a–c** Full blockage of inner space of 90–10 Copper-Nickel tubes after 3 years of service at the Isomerization unit. Inside tubes—cooling water (treated by corrosion inhibitors), outside—LPG (had to be 60–40 °C). Deposits contain zinc, calcium, and iron phosphates with chlorides. High temperature (≫80 °C, Fig. 7.7) resulted in deposition of sediments. Incorrect design was in this case: air cooler did not decrease temperature of LPG to required value 60 °C. **d** *Green-blue* corroded surface of copper-nickel tubes

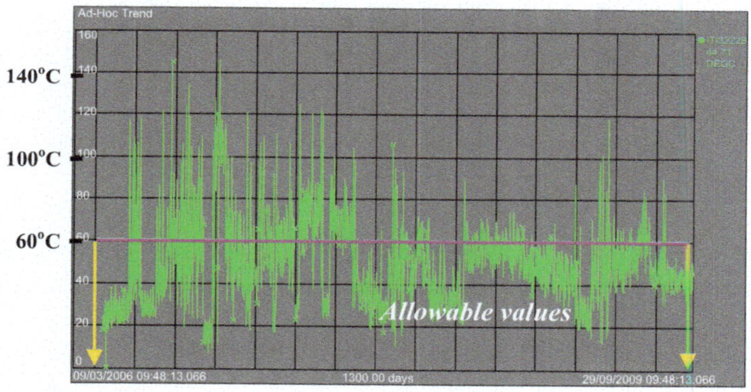

Fig. 7.7 Temperature fluctuations of LPG at the inlet to the heat exchanger during 3 years, PI data

formation of holes in tubes, hydrocarbons penetrate into cooling water and can cause formation of foam and exacerbate organic deposits inside of tubes. These phenomena decrease flow velocity of cooling water up to water stagnation and further corrosion development.

In spite of different thermal conductivity of various alloys, their nature does not influence significantly chemical composition of scale and deposits formed on their surfaces. Fouling formation on water side is defined by the following factors:

A. The source of cooling water (its physicochemical properties) and make-up water quality, its treatment: use of precipitation tanks for removing Ca^{2+} and Mg^{2+} ions—hardness salts; inadequate and insufficient filtration and sedimentation of suspended solids and particulates.
B. Chemical treatment of cooling water: injection of anti-scaling agents, keeping recommended pH and ions' composition. For instance, $CaCO_3$ and $MgCO_3$ scale can be formed if anti-scaling agents (scale dispersants) are not used. Sometimes even application of anti-scaling agents in cooling water cannot

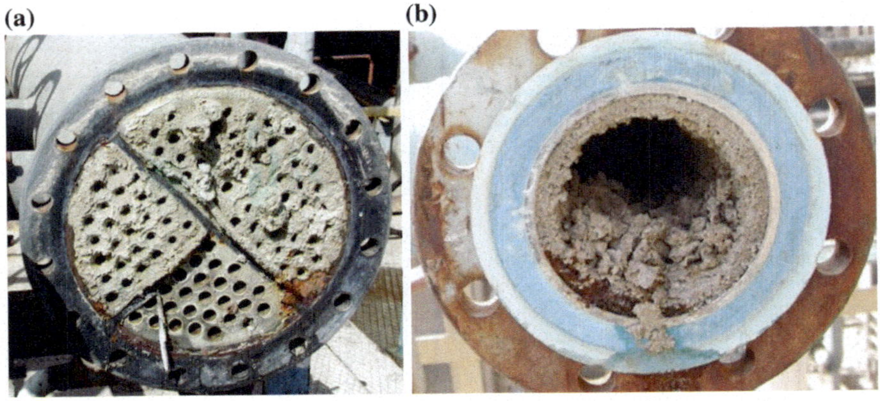

Fig. 7.8 **a** Heat exchanger clogged with the scale $CaCO_3$ (9 months of service). Cooling water (30–35 °C) flowed inside of carbon steel tubes and steam (200 °C) flowed outside of tubes. **b** Pipe close to the heat exchanger also was clogged. Design was wrong in this case. It is forbidden to inject steam at high temperature (~200 °C) in heat exchanger with cooling water containing hardness salts. This heat exchanger was intended for cooling steam in vacuum system. Two solutions exist: to cool steam (200 °C) by means of pure (deionized) water or to cool steam not by cooling water but by air (air cooler)

prevent scale formation if concentrations of Ca^{2+} and Mg^{2+} ions are higher than recommended values, pH > 8, and temperature exceeds 60–80 °C (Fig. 7.8).

C. Operating conditions: temperature, flow velocity, number of cycles of concentration, correct blowdown, realization of reversal flow (back flushing) in heat exchangers during their service. Therefore, it is very important to maintain physicochemical parameters of circulated cooling water and temperature below maximum allowable values. Besides the most common and often formed insoluble in cooling water carbonate salts, other compounds (sulfates, silicates, hydroxides and oxides) also can form scale.
D. Uncontrolled corrosion in cooling water can cause fouling by corrosion products. Usually iron hydroxides and oxides, but aluminum hydroxides and oxides (when aluminum sacrificial anodes are used in heat exchangers) also may be present in corrosion products.
E. Injection of corrosion inhibitors (i.e. zinc-phosphate-phosphonate), if not properly controlled (wrong dosage and conditions, such as unsuitable pH, off-limits concentrations of ions, temperature, frequent outages), zinc and calcium phosphates also can be important constituents of fouling.
F. Low water velocity (usually below 1.2 m/s) can result in settling of organic and inorganic suspended solids. Recommended designed flow velocities of water for some alloys and metals are shown in Appendix J. If 'half-moon' baffles are used, the baffle cut should not exceed 20 % of the shell inside diameter [17].
G. Inappropriate use of biocides and biodispersants can result in proliferation of microorganisms and biofouling. Attached microorganisms trap suspended

solids, corrosion products, silt, mud, debris, and thus very complicated mixture of living and nonliving materials appears on metal surface. As a result, under deposit corrosion can progress.

Chemical composition of fouling depends on the abovementioned factors, and any sudden changes can result in fouling formation. In order to control, diminish or prevent fouling in cooling water systems, following measures must be carried out:

(a) To remove hardness ions from the water source of cooling water and its make-up.
(b) To monitor all physicochemical parameters of cooling water, namely, to maintain recommended pH, chemical composition, and temperature of cooling water in order to prevent scale formation.
(c) To maintain operational conditions: water flow velocity above 1.2 m/s; immediate drainage of water during outage or shutdown; elimination of 'dead' zones; realization of periodical reversal flow of cooling water in heat exchangers.
(d) To monitor corrosion by means of analytical methods, installation of WL coupons, on-line LPR-probes (including measurements of pitting index) and heat transfer resistance index [21].

Heat exchangers are mostly prone to fouling. To reduce fouling, heat exchanger performance should be monitored and cleaned periodically. Despite that fouling at different process units is unique, control methods of fouling include pretreatment with correctly designed and applied chemical treatment, bypassing, and cleaning procedures which will be described in the next section.

7.5 Cleaning Procedure

The physicochemical nature of fouling and alloy influences adherence extent of scale to metal surface and, as a result, the procedure of removing and cleaning. It is impossible to decide about the cleaning procedure without opening equipment (heat exchanger, condenser, pipe, pump, compressor) and identification of chemical composition of fouling/scale/deposits. For each kind of fouling, there are specific methods of its removing and cleaning. Sometimes it is enough only hydrojetting to remove organic deposits, or light carbonate-sulfate scale, or rust. In most cases it is necessary to tuck up suitable acid and its concentration to dissolve scale/deposits. If new chemical compounds in deposits are found, it is needed to examine cleaning solutions under laboratory conditions with variations of their concentrations, temperature, time, and flow regime (circulation or stagnation). Organic substances can form biofouling which includes microorganisms and products of their metabolism. Carbonates, sulfates, phosphates and hydroxides have different solubility in cleaning acid solutions. We should take into consideration that after removing the scale and deposits, pure bare metallic surface will be exposed to cleaning acid solutions. Therefore, their corrosiveness should be defined and estimated possible corrosion loss. Acidic solutions should be used with corrosion inhibitors (usually organic amines). These acidic solutions do not dissolve scale/deposits if organic substances are present in deposits too. Therefore, the first step is alkaline cleaning (degreasing) which is carried out by means of alkaline solutions with detergents to remove organic substances containing in deposits. If fouling occurred only mechanical and chemical cleaning should be carried out. Operational measures include control all parameters influencing fouling formation (flow velocity, temperature, chemical composition of streams, injection of antifoulants and dispersants) and compliance with recommended values. Mechanical and chemical cleaning methods will be described separately.

7.5.1 Mechanical Cleaning

Mechanicalcleaning includes hydraulic, abrasive, and thermal methods [7]. Hydraulic cleaning is carried out at pressures 20–2200 bar and includes water cleaning, high pressure and ultra-high pressure water jet (hydrojetting), cleaning by steam, ultrasonic cleaning, and explosive methods. Abrasive cleaning includes pigging and scraping, turbining, drilling, rodding, sandblasting, and explosive removal. Abrasive cleaning is implemented where hydraulic cleaning may not be practical. Following forms are used for abrasive cleaning: moulded plastic cleaners (pigs) and brushes for removing soft deposits, light silt and biofouling (inside tubes); metal cleaners for harder types of deposits (inside tubes): small cylindrical stainless steel wire brush or spring coil. The construction of these cleaners is diverse. A water pressure of 20 bar is very effective for propelling the cleaning tools through the tubes. Usually heat exchanger bundles contain between 500 and 1100 tubes, with 6 m in length and 25.4 mm in diameter. Therefore, use of small cleaners for each tube is very laborious and time consuming operation.

Thermal cleaning includes the use of steam for removing debris, mill scale, coke, and polymers. Periodical reversal flow of cooling water in heat exchangers also relates to mechanical cleaning by water. Application of particular method depends on fouling properties, specific equipment and medium. For instance, for cleaning of inner surface of pumps intended for kerosene, LAGO, HAGO, LVGO, cracked diesel fuel and tempered water, is enough to wash by steam. For cleaning of inner surface of pumps intended for light and heavy naphtha, or water condensate, ultra-high pressure water jet is needed.

Shell-and-tube heat exchangers usually require disassembly, removal of the tube bundle, transportation to a cleaning facility and cleaning, reassembly and leak testing. Mechanical cleaning is a multistage process, especially complicated and dangerous when heat exchangers are installed high above and access is difficult. Special equipment is needed in most cases to extract the tube bundle. Mechanical cleaning can take between 3 and 14 days, depending on heat exchanger size and weight, and severity of fouling. Cleaning facility is also dangerous for people and environment as pollution of organic and inorganic mist occurs during washing with a high pressure water jet. The cost of mechanical cleaning of average heat exchanger is ~50,000 $. Large penalties of up to ten-times that cost can also be incurred, depending on the duration of outage. Therefore, chemical cleaning procedures were developed which are cheaper than mechanical cleaning and which can be carried out in short period, usually 1 day or even less (excluding time for preparation, generally 8 h).

7.5.2 Chemical Cleaning

Chemical cleaning is a technique of use of liquid chemicals to dissolve and remove solid deposits. Chemical cleaning is performed by circulating the solution through the equipment or soaking without its dismantling. This method has several advantages over mechanical cleaning: cheaper and quicker, no need to dismantle equipment, more uniform removal of deposits (chemical solutions reach inaccessible areas), and longer intervals between cleanings. Chemical cleaning is used for removing both organic and inorganic deposits and, if needed, for passivation. In general chemical cleaning procedures consist of five stages, each being monitored for results (quality) before proceeding to the next.

A. Alkaline cleaning (degreasing). The aim is to remove the organic deposits or the organic part (grease, oil, fat) of the fouling in order to render the inorganic part hydrophilic and thus to make the following acid cleaning effective.
B. Rinsing. The aim is to remove physically softened material formed during degreasing.
C. Acid cleaning. The aim is to remove inorganic part (salts, hydroxides, and oxides) of fouling.
D. Neutralization and rinsing. The aim is to neutralize and remove residual acid, dissolved inorganic compounds, and sludge formed during acid cleaning.
E. Passivation. The aim is to create protective anti-corrosive passive film on the surface of metallic equipment. 'Passivate' essentially means 'de-activate'.

However, not all five stages are realized in all cases of fouling removing. It depends on type of the fouling and equipment needed to clean. These five stages are described in Appendix K.

7.5.3 Preoperational Cleaning and Passivation

In order to decide about preoperational cleaning procedure, we should take into consideration the material of construction, particular equipment service, and the degree of cleanliness required. Inner surfaces of the equipment first should be inspected and decision about cleaning procedure should be taken. Usually preoperational cleaning alone (only degreasing) is used for process equipment (distillation and other columns, reactors) start-up, steam and water condensate pipelines. For instance, for low pressure steam pipelines (3–14 bar, 145–225 °C), degreasing and blow out with steam is enough. For water condensate piping, only degreasing is enough. Cleaning with passivation is used for boilers, steam-generating (high pressure 35 bar, 250 °C) and compression systems, cooling water systems (piping, heat exchangers). Passivating solutions for steam piping and cooling water systems are different (see Appendixes K, L, I). Preoperational cleaning procedures are different for carbon steel, stainless steel, and copper alloys. Distillation and stripper columns and reactors are cleaned inside to remove protective grease, organic deposits, sand, dirt, and sometimes mill scale. They are cleaned by fill and soak.

Procedure for boilers is called *boil-out* and it provides degreasing, removing mill scale and formation protective film of magnetite (Fe_3O_4). This is the only iron oxide (among three FeO, Fe_2O_3, and Fe_3O_4) which possesses protective anti-corrosion properties in contact with steam at high temperatures. Therefore, magnetite is produced in the final step (passivation) of cleaning procedure in boilers and compressors with piping intended for contact with steam (see Appendix L). Magnetite has no protective properties in contact with atmosphere (condensed water vapor) and water. Therefore, it is not necessary to produce magnetite layer inside equipment (heat exchangers, pipes, and pumps), which is intended for contact with cooling water. For this, special cleaning procedure and formation of phosphate protective passive layer is carried out (see Appendix I).

Piping and heat exchangers (inside, especially after erection of new unit) may contain dirt, sand, grit, grease, loose paint, weld spatter, rust, and mill scale. Dirt, sand, grit, and loose paint can be removed by flushing with water, or blowing with steam, or compressed air. Grease and other hydrocarbon contaminants (oil, fat) are removed by alkaline aqueous solution containing detergents at 50–60 °C. Sometimes mechanical cleaning is required, depending on the amount of weld spatter, rust, and mill scale. In some cases, these steps are enough (i.e. piping in amine treating units) but piping intended for cooling water is cleaned by degreasing and passivation (see Appendix I).

7.6 Safety and Environment

Frequent equipment outages and openings (usually pressurized processing hydrocarbons) for cleaning increase the risk of leakages, specifically during start-up and cleaning procedures. Although equipments are flushed with steam, flammable atmospheres and pyrophoric material can be present. Extraction of bundles with cranes in confined spaces, transportation on trucks to the facility of mechanical cleaning, after the cleaning returning process delivery and reinsertion are needed. High pressure ($\sim$1000 bar and can reach 2000 bar) water jet cleaning is provided with horrible noise and pollution of toxic materials into the atmosphere.

Works with acids, alkalis, some salts ($NH_4F{\cdot}HF$) and organic corrosion inhibitors (using with acid solutions) are dangerous for human health and the environment. Special safety precautions should be applied during cleaning from sulfide deposits because hazardous and toxic H_2S is generating during reaction with acidic solution. Organic solvents using for removing organic viscous deposits are flammable and special safety measures must be applied.

There are difficulties associated with a crowded work space and pumping corrosive solutions of acids through temporary connections. There is the need to neutralize acid solutions and dispose of wastes. In this connection, appropriate safety measures and precautions must be carried out during transportation, storage, handling, application, and disposal of chemical substances using in alkaline and acid cleaning.

In literature, sometimes chromic acid (H_2CrO_4) is recommended for chemical cleaning from iron sulfides. Chromic acid (like other chromium (VI) compounds, i.e. chromates) is considered as human carcinogen and in any case should not be used.

All area at oil refineries and petrochemical plants where cleaning proceeds should be fenced and signed. It is recommended to carry out cleaning at night when less personnel is present and endangered. Disposal requirements for waste materials should be carried out according to local regulations.

7.7 General Recommendations for Implementation of Cleaning

Every oil refinery should have its proprietary standard or procedure which should be applied for each case of cleaning of special equipment having specific fouling. If a contractor is used for chemical cleaning, it can have its own particular procedure. This procedure should be written, considered, documented and approved by following departments at oil refinery: production, maintenance, materials and corrosion, laboratory, ecological, and safety. A scope of work may varied with changes in materials, operational conditions at units, physicochemical content of fouling, and objectives of cleaning.

During planning stage, following actions should be documented: listing of equipment assigned for cleaning, its volume, materials of construction, disconnections from operating equipment, isolation, installing fittings and connections to equipment destined for chemical cleaning, listing of all chemicals, contact times for each stage, temperature, pressure, circulation rate, corrosion control and monitoring methods, safety requirements and procedures, disposal requirements, temporary conservation (if equipment is not to be used immediately).

During preparation, when system is ready for cleaning, following actions should be carried out: testing for leak, establishment and control of circulation rate,

Fig. 7.9 **a** Rusted control metal parts before cleaning and passivation. **b** The same parts after the passivation. **c** Visual examination (inspection) of inner surface of the pipe (its end) after cleaning and passivation (well formed passive oxide Fe_3O_4 layer of *grey color* is formed)

temperature and pressure, control of correct choice of substances and their concentrations.

Cleaning stage. Circulation and monitoring, which include control and measuring of all parameters: temperature, concentration of oil in solution (degreasing and rinsing steps), concentration of acid, pH, hardness and metal ions' concentration (see Appendixes K and L).

After finishing of chemical cleaning, system is drained, flushed with water, neutralized, and passivated (if required). The effectiveness of cleaning and passivation may be evaluated according to control metal parts which are immersed in solutions (in bath) which are used during all stages of cleaning (Fig. 7.9). In the case of unsatisfactory cleaning of the equipment, the cleaning procedure should be repeated. Solutions after draining should be disposed at appropriate facilities with the permission of ecological department and local authorities (if required)

Corrosion monitoring during chemical cleaning. Corrosion coupons and spool parts containing a sample of deposits should be installed in vessels containing acidic solutions. Corrosion rates should be calculated after finishing of chemical cleaning procedure [21]. The corrosion rates of WL coupons are usually greater than the actual corrosion rate of the equipment during the cleaning procedure. We can evaluate possible damage from corrosiveness of cleaning solutions. For instance, if the corrosion rate of carbon steel coupon was measured 10 mm/year, this means that during 6 h of circulation the wall of the equipment was diminished on 0.007 mm. As a rule, corrosion rate of carbon steel coupons is less than 5 mm/year during chemical cleaning. Therefore, corrosion loss 0.0035 mm of the equipment wall is negligible.

Some notes about correct decisions. It is important to make the right decision regarding the chemical cleaning with acidic solution of inner surface of large carbon steel pipes (diameter 100 mm and more) containing 'heavy' corrosion products accumulated inside (Fig. 7.10).

Generally these corrosion products represent rusted tubercles (formed as a result of the activity of iron bacteria) with severe underneath localized corrosion which can penetrate through whole wall thickness of the pipe. Thus, chemical cleaning by acidic solutions can dissolve rusted tubercles and uncover corrosion through-holes

Fig. 7.10 **a**, **b** Biofouling (slime and rusted tubercles) in cooling water system after 4 years of service (before complex biocide treatment). **c** Inner surface after treatment with NaOCl, biodispersant and organic biocide

in pipes. This situation may cause perforation of piping and unexpected leaks of acidic solution (pH ~2.8!) representing a hazard to working personal and corrosion risk to nearby equipment. When severe corrosion with rusted tubercles are revealed inside piping, it is not recommended to clean inner surface of such pipes but completely to replace them. Unfortunately, not always there is economic justification to renovate such system immediately, and refinery should remain with this pipes for several years. For instance, refinery is planning to replace cooling towers in 2–3 years. In this case, it is possible to avoid acid cleaning because of possible perforation and to apply treatment with mixture of NaOCl, biodispersant and organic biocide. After several days of such treatment, rust becomes more dense, and this piping can be used for several years (see Fig. 7.10c).

Inspection and documentation. Careful inspection and reporting should be carried out during all steps of preparation, cleaning, rinsing, and disposal of wastes. Physicochemical analytical measurements should be carried out systematically and results should be analyzed in order to decide about the quality of each stage and continuing the next step. When cleaning is finished, the equipment must be opened and inspected visually or by means of boroscope. All data must be documented and estimated.

7.8 Conclusions

(a) All alloys which are used at oil refinery and petrochemical units are susceptible to fouling.
(b) Fouling can happen not only on heat transfer surfaces but also at any surface at ambient temperature. Fouling rates increase with increasing temperature and decreasing flow velocity.
(c) Petroleum products, fuels, organic solvents and cooling water containing suspended solids are fouled at much higher rates than filtered products.

(d) It is recommended to keep dissolved oxygen concentrations in petroleum products to minimum by good design and operating practices. Equipment in vacuum systems should be sealed unfailingly at all times.

(e) In order to determine optimal procedure for cleaning, coordination between refinery operators, researchers, suppliers of chemicals and contractors for carrying out cleaning are needed. Individual approach should be done in each case because of different physicochemical content of fouling. As a result, individual procedures should be applied to each equipment.

References

1. Coletti F, Hewitt JF (eds) (2015) Crude oil fouling: deposit characterization, measurements, and modeling. Gulf Professional Publishing, Elsevier, USA
2. Ebert WA, Panchal CB (1995) Analysis of Exxon crude-oil slip stream coking data. In: Panchal CB et al (eds) Fouling mitigation of industrial heat exchange equipment. Begell House, NY, pp 451–460
3. Deshannavar UB, Rafeen MS, Ramasamy M, Subbarao D (2010) Crude oil fouling: a review. J Appl Sci 10:3167–3174
4. Wilson DI, Polley GT, Pugh SJ (2005) Ten years of Ebert, Panchal and the 'threshold fouling' concept. In: Müller-Steinhagen H, Malayeri MR, Watkinson AP (eds) Proceedings of the 6th international conference on heat exchanger fouling and cleaning—challenges and opportunities. Engineering conferences international, Kloster Irsee, Germany, June 5–10, 2005, ECI symposium series, vol RP2, pp 25–36
5. Hong E, Watkinson AP (2007) Precipitation and fouling in heavy-oil diluent blends. In: Müller-Steinhagen H, Malayeri MR, Watkinson AP (eds) Proceedings of 7th international conference on heat exchanger fouling and cleaning—challenges and opportunities, Engineering conference international, Tomar, Portugal, July 1–6, 2007, ECI symposium series, vol RP5, pp 23–31
6. Refinery Process Fouling Control (2012) Technical paper TP1194EN, General Electric Company
7. Jones L, Haff JD, Moniz BJ (1987) Chemical cleaning of process equipment. Corrosion, vol 13. ASM Handbook, ASM International, USA, pp 1137–1143
8. ASTM A380-13 (2015) Standard practice for cleaning, descaling, and passivation of stainless steel parts, equipment, and systems. ASTM International, West Conshohocken, PA, USA
9. ASTM A967M-13 (2013) Standard specification for chemical passivation treatments for stainless steel parts. ASTM International, West Conshohocken, PA, USA
10. ASTM B600-11 (2011) Standard guide for descaling and cleaning titanium and titanium alloy surfaces. ASTM International, West Conshohocken, PA, USA
11. Epstein N (1983) Thinking about heat transfer fouling: a 5 × 5 matrix. Heat Transfer Eng 14 (1):25–36
12. Flemming HC, Murthy PS, Venkatesan R, Cooksey KE (eds) (2009) Marine and industrial biofouling. Springer, Berlin
13. Asomaning S, Watkinson AP (1999) Deposit formation by asphaltene-rich heavy oil mixtures on heat transfer surfaces. In: Bott TR (ed) Understanding heat exchanger fouling and its mitigation. Begell House, NY, pp 283–290
14. Watkinson AP, Li YH (2009) Fouling Characteristics of a heavy vacuum gas oil in the presence of dissolved oxygen. In: Müller-Steinhagen H, Malayeri MR, Watkinson AP

(eds) Proceedings of international conference on heat exchanger fouling and cleaning VIII-2009, June 14–19, 2009, Schladming, Austria, pp 27–32
15. Yang M, Crittenden B (2012) Fouling thresholds in bare tubes and tubes fitted with inserts. Appl Energy 89:67–73
16. Yang M, Young A, Niyetkaliyev A, Crittenden B (2012) Modeling fouling induction periods. Int J Thermal Sci 51:175–183
17. Müller-Steinhagen H (2000) Heat exchanger fouling: mitigation and cleaning technologies. PUBLICO Publications, Germany
18. Engineering Sciences Data Unit (ESDU) 3-00016 (2005) Heat exchanger fouling in the pre-heat train of a crude oil distillation unit. London, UK. https://www.esdu.com/cgi-bin/ps.pl?sess=unlicensed_1151215113504mzp&t=doc&p=esdu_00016b
19. Port RD, Herro HM (1991) The Nalco Guide to Boiler Failure Analysis. McGraw-Hill, NY
20. Gutzeit J (2004) Crude unit corrosion guide. PCC Process Corrosion Consultants, USA
21. Groysman A (2010) Corrosion for everybody. Springer, Dordrecht

Recommended Literature

22. Joshi MH, Brons G (2003) Chemical cleaning of oil refinery heat exchangers—the need for a joint effort. In: Proceedings heat exchanger fouling and cleaning: fundamentals and applications. http://dc.engconfintl.org/heatexchanger/29/
23. Speight JG (2015) Fouling in refineries. Gulf Professional Publishing, Elsevier, USA
24. Putman RE, Optimizing the cleaning of heat exchangers. https://www.concosystems.com/sites/default/files/userfiles/files/techical-papers/optimizing-cleaning-heat-exchangers.pdf
25. TEMA (1988) Standards of the tubular exchanger manufacturers association, 7th edn. Tarrytown, NY
26. Hewitt GF, Shires GL, Bott TR (1994) Process heat transfer. CRC Press, Boca Raton
27. Koch GH et al (2002) Corrosion costs and preventive strategies in the United States. FHWA-RD-01-156. Federal Highway Administration (FHWA), Washington
28. Rafeen MS et al (2007) Crude oil fouling: PETRONAS refineries experience. In: Müller-Steinhagen H, Malayeri MR, Watkinson AP (eds) Proceedings of the 7th international conference heat exchanger fouling and cleaning—challenges and opportunities. Engineering conference international, Tomar, Portugal, July 1–6, 2007, ECI symposium series, vol RP5, pp 8–12
29. Melo LF, Bott TR, Bernardo CA (eds) (1988) Fouling Science and Technology. NATO ASI (Advanced Study Institute) Series, Series E: Applied Sciences, vol 145. Kluwer Academic Publishers, Dordrecht
30. Joshi HM (2013) Crude oil fouling field data and a model for pilot-plant scale data. In: Malayeri MR, Müller-Steinhagen H, Watkinson AP (eds) Proceedings of the international conference heat exchanger fouling and cleaning, June 09–14, 2013, Budapest, Hungary, pp 22–26
31. Watkinson AP (2005) Deposition from crude oil. In: Müller-Steinhagen H, Malayeri MR, Watkinson AP (eds) Proceedings of the 6th international conference heat exchanger fouling and cleaning—challenges and opportunities. Engineering conferences international, Kloster Irsee, Germany, June 5–10, 2005, ECI symposium series, vol RP2, pp 7–16
32. Klemeš JJ, Arsenyeva OP, Kapustenko PO, Tovazhnyanskyy LL (2016) Compact heat exchangers for energy transfer intensification: low grade heat and fouling mitigation. CRC Press, Taylor & Francis Group, Boca Raton
33. Bories M, Patureaux T (2003) Preheat train crude distillation fouling propensity evaluation by the Ebert and Panchal model. In: Proceedings of conference heat exchanger fouling and cleaning: fundamentals and applications, engineering conferences international, Art. 27, ECI digital archives http://dc.engconfintl.org/heatexchanger/27

34. Wilson DI et al (2009) Ageing: looking back and looking forward. In: Müller-Steinhagen H, Malayeri MR, Watkinson AP (eds) Proceedings of international conference heat exchanger fouling and cleaning VIII-2009, June 14–19, 2009, Schladming, Austria, pp 221–230
35. Paikert P (1983) Verschmutzung von Kondensatoren und Kühltürmen. GVC Weihnachtstagung, pp 371–390
36. Process Design (1997) P-001 Rev. 3, Norsok Standard

Chapter 8
Corrosion Control at Oil Refinery Units

If a problem can be solved with money, it is not a problem, it is an expense.

Jewish proverb.

Abstract The following methods of corrosion control at oil refineries' and petrochemical units are described: use of materials compatible with corrosive media (the experience of use of titanium is presented in detail); treatment and change the environment (diminishing the concentration of corrosive substances in streams, e.g., neutralizing of acidic corrosives in overhead; wash by water and soda ash; using of corrosion inhibitors). Corrosion inhibitors in the overhead of crude distillation system and use of polysulfides as corrosion inhibitor at the FCCU are analyzed. Examples of corrosion monitoring at the overhead of the crude distillation system and in the petrochemical plant are given.

Corrosion control must be started at the stage of equipment and operation design and correct use of standards, codes of practice (engineering materials guidelines), specifications, and recommended practices. We can show many examples of wrong selection of materials, their combination, and mistakes in processing/operation design (e.g., see Figs. 7.6 and 7.8). Unfortunately there is no methodology taking into account the effect of process upsets on design. In other words, we cannot predict all possible faults (including outages) during design. For instance, many corrosion problems appear during startup or shutdown. In addition, we cannot take into consideration influencing of human factor in wrong or correct decision of designing and further steps of planning corrosion management which includes also corrosion control measures (see Chap. 9).

The following methods are used to lessen and control corrosion at oil refineries' and petrochemical units: use of corrosion resistant alloys (suitable materials); treatment and change the environment (diminishing the concentration of corrosive substances in streams; using of corrosion inhibitors; wash by water and soda ash); technological measures (keeping temperature under dew point; elimination of stagnation and high velocities of streams); use of anti-corrosion coatings; and cathodic protection. These corrosion control actions also should be planned at the

A. Groysman, *Corrosion Problems and Solutions in Oil Refining and Petrochemical Industry*, Topics in Safety, Risk, Reliability and Quality 32,
DOI 10.1007/978-3-319-45256-2_8

design stage. Corrosion allowance and appropriate heat treating during fabrication also should be considered at this stage. It is important to differentiate between forging and casting components when ordering them. Forging is preparing by heating an object in a furnace and beating or hammering it (like a cutlet prepared by your mother or wife). Casting is an object made by pouring molten metal into a mold and then allowed to solidify. As a result, casting contains unpredictable defects (pores, cracks on surface) while forging has significantly better structure and less flaws. Thus, casting components susceptible to fatigue failures. This is the reason that steel castings are not used for critical parts, i.e. pressure components in contact with hydrocarbons.

Then corrosion monitoring and inspection methods of behavior of materials and tracking that anti-corrosion measures are correctly applied must be also provided at the design stage. Vessels, piping, reactors, heat exchangers, pumps, compressors, columns, tanks, furnaces, stacks, valves, ejectors, in short, all equipment and structures (small and large) should be analyzed from the view of corrosion risk during design.

Using of coatings and cathodic protection were described previously [1, 2]. Use of alloys and treatment of environment in refining streams will be described in this chapter.

8.1 Materials Compatible with Corrosive Media

We should remember that there is no ideal material which is resistant to all media under all conditions. Therefore, using the term '*corrosion resistant material*' is problematic and conditional. All materials have advantages and disadvantages and can be used only under appropriate conditions. Diversity of physicochemical properties of media (organic and inorganic, their mixtures; from acids to alkalis) and conditions (from −196 to 1400 °C; from vacuum to 1000 bar; the presence of several phases; from stagnation to severe flow) requires using different materials possessing suitable properties. There are limits for use of each material. It is better to use the term '*materials compatible with corrosive media*'. We will describe alloys using for equipment and structures at oil refineries and petrochemical plants.

Carbon steel is used as the main material of most (>80 %) refinery equipment and structures at temperatures below 230 °C. Carbon steel is resistant to alkaline (pH ~9–13) solutions but in the presence of chlorides severely corrodes. Carbon steel is passivated in strong H_2SO_4 (>90 wt%) and HNO_3 (>70 wt%) acids, and there is no need to use stainless steel vessels containing these concentrated acids. Special killed steels (>0.1 wt% Si) and steels with symbol HIC (e.g., A516 Gr.70 HIC) are resistant to sulfidic corrosion (see Sect. 4.1 and Appendix F). Carbon steel is not always resistant to crude oils, petroleum products and fuels because of corrosive impurities, such as H_2S, H_2O, dissolved O_2, salts, acids, and microorganisms. Carbon steel also has low resistance to cooling water, atmosphere, soil, acid (pH < 4) and very alkali solutions (pH > 13.5 at T > 90 °C), and anti-corrosion measures should be applied: injection of corrosion inhibitors, biocides, and neutralizers, use of coatings, and cathodic protection. Other alloys are used instead of carbon steel in most corrosive environment of oil refineries and petrochemical plants: low-alloy steels, stainless steels, copper alloys, nickel alloys, and titanium alloys [3–5].

Low-alloy Cr–Mo steels (5Cr–0.5Mo to 9Cr–1Mo) are used instead of carbon steel for improved resistance to sulfidic corrosion (see Sect. 4.1).

Stainless steels include five groups (ferritic, martensitic, austenitic, duplex, and precipitation hardening) and the first four of them are used for equipment at refineries. They have improved resistance to sulfidic corrosion comparing with low-alloy steels. Austenitic stainless steels containing above 3 wt% Mo are resistant to NAC. Stainless steels are not resistant to HCl, HBr, HF, and aqueous solutions containing Cl^-, Br^- and F^- anions; sea water (under stagnant conditions); and organic acids (formic, oxalic, acetic, and lactic). Pitting, crevice, SCC (including polythionic acid SCC), MIC, and intergranular corrosion of stainless steels can occur under appropriate conditions.

Copper alloys include three groups: brasses, bronzes, and copper-nickel alloys. Brasses are used instead of carbon steel for improved resistance against cooling water corrosion. However, if cooling water is treated with corrosion inhibitors and biocides, there is no necessity to use brass, and carbon steel is suitable material. When copper alloys are used in cooling water, corrosion inhibitors of copper (azoles' derivatives) should be injected into water. Brasses are susceptible to corrosion by H_2S and NH_3, to pitting corrosion, SCC, and dezincification under particular conditions. Bronzes are somewhat stronger than the brasses, yet they still have a high degree of corrosion resistance to many environments. Copper-nickel alloys have improved resistance against SCC and water corrosion than brasses. All copper alloys are not resistant to media containing H_2S and NH_3.

Selection of specific material is based on potential corrosion problems and risk which are known in most cases at refineries' units. Titanium alloys acquired good reputation and will be described in the next section.

8.1.1 Use of Titanium in the Oil Refining and Petrochemical Industry

When I began working as a corrosion engineer at the oil refinery in 1990, the chief chemist once called me and gave me a shiny thin tube about 10 cm length and 1 cm in diameter, and asked me to determine which material the tube was made of. He also said that I should not use chemical analysis. I remembered a story how Archimedes uncovered a fraud in the production of a golden crown commissioned by Hiero II, a king of Syracuse. He tested whether it was pure gold or a mixture of gold with silver, according to the volume and mass of the crown (density). Therefore, I decided to determine the density of the shiny small tube. Weighing it and measuring its volume, I obtained the density 4.5 g/cm^3. This value is nearly half that of stainless steel which was my first guess when I saw the tube—so a light metal! There was no internet in those times. I took the encyclopedia and found that only one metal had this density (4.5 g/cm^3)—titanium. Thus, I began my acquaintance with titanium at the oil refinery. In 10 years more than 45 heat exchangers in different applications were functioning at oil refinery units.

Titanium has found wide application in different industries and spheres of life: from spacecraft, aircraft and missiles to laptops, body implants, architecture, and in jewelry and decorations as the beautiful yellow coating titanium nitride (instead of gold). Although titanium alloys were originally developed in the early 1950s for aerospace applications where their high strength-to-density ratios were especially attractive, the chemical and petrochemical industries were the first where titanium began to be used because of its high corrosion resistance to diverse media [6–11]. When the price of titanium tubing fell during the 1960s, it became possible commercially to exploit the resistance of the metal to attack by brackish, saline and cooling water, and other media (hydrocarbons containing HCl, H_2S, NH_3, and organic acids) of oil refineries. Titanium heat exchangers, piping, vessels, tanks, valves and other equipment in different applications have been used since 1967 in oil refining and petrochemical plants [12]. It is the combination of high corrosion resistance, high strength and low density that has led to the increased use of titanium in the oil refining and petrochemical industry from 1970 onwards. Titanium is no longer regarded as a new metal or an unusual one used solely for aerospace applications, but simply as another high-performance material available to the design engineer. Nowadays there is nearly 50 years' experience of the use of titanium as a construction material in the oil refining and petrochemical industry [12–18]. Only a few cases of failure have occurred.

In physical appearance, titanium is a beautiful lustrous white metal with a silver color like steel. It is easy to confuse with stainless steel on the basis of its appearance alone. One important physical property of titanium, its lightness, can easily identify it. The physicochemical properties and chemical resistance of titanium alloys to different media, advantages, corrosion phenomena and limitations, experience of applications at oil refinery and petrochemical units, problems and solutions are described in this section.

8.1.1.1 Properties of Titanium

The density (4.5 g/cm^3) of titanium is midway between those of the light alloys of aluminum (2.7 g/cm^3), and magnesium (1.74 g/cm^3) and those of steel (7.87 g/cm^3) and nickel alloys (8.4–8.9 g/cm^3). In alloy form, titanium retains useful strength at temperatures substantially higher than those considered safe for the more conventional light alloys and is an attractive metal for applications demanding high specific strength at temperatures ranging from sub-zero to ~250 °C. When titanium is alloyed with Al, Sn, Zr, Nb, Mo, and Si, it can be used up to ~600 °C. Titanium loses strength when heated and at 300 °C it is half as strong as at 20 °C. This feature must be taken into account when using it at elevated temperatures. Titanium is as strong as steel but 45 % lighter at ambient temperatures. It is twice as strong as aluminum, but 60 % heavier. Titanium is not as hard as some grades of heat-treated steels, it is non-magnetic and a poor conductor of heat and electricity.

Titanium at high temperatures (600–1000 °C) greedily absorbs nitrogen, oxygen, hydrogen and carbon from the atmosphere and becomes brittle and unsuitable for practical purposes. Therefore, it has been impossible to obtain pure titanium and so it had not been used for many years. Titanium became a commercial reality only in the 1950s. All high temperature processes with titanium (including welding) must be done in vacuum or in an atmosphere of inert gas (argon or helium).

Another important characteristic of titanium is its corrosion resistance in a wide range of chemical and natural media, particularly with respect to general corrosion, pitting corrosion, SCC, MIC, erosion, and cavitation. In many applications, especially where the material is used in the presence of chlorides, its corrosion resistance is far superior to those of stainless steels. We should remember these properties of titanium when using it at refinery units. There are about 40 grades of Ti alloys differing in composition and, as a result, in mechanical properties and chemical resistance. Five of these grades (Gr.1, Gr.2, Gr.7, Gr.12, and Gr.16) are used in the oil refining industry (Appendix F, Table F.13).

8.1.1.2 Chemical Resistance of Titanium Alloys

Titanium is a reactive metal: $E^{o}_{Ti^{2+}/Ti} = -1.63\ V_{SHE}$ (SHE—Standard Hydrogen Electrode) at 25 °C [19]. For comparison, $E^{o}_{Al^{3+}/Al} = -1.66\ V_{SHE}$ at 25 °C. However, the chemical resistance of titanium and its alloys is due to the fast formation of a tenacious, hard, tightly adherent protective titanium dioxide surface layer (TiO_2). Therefore, if titanium is immersed in an aqueous solution of electrolytes the electrode potential of the surface (not pure Ti but TiO_2) is more positive and can reach −0.7 V_{SCE} (−0.383 V_{SHE}) (SCE—Saturated Calomel Electrode) in HCl and H_2SO_4 to −0.1 V_{SCE} (+0.217 V_{SHE}) in sea water at 25 °C. Titanium dioxide as white pigment has a wide range of applications, from paints to polymeric materials, sunscreen, toothpaste, paper, and food colouring. Actually, three titanium oxides (similar to the three iron oxides formed on an iron surface at high temperature) are formed: TiO (close to the titanium surface), Ti_2O_3 (intermediate layer), and TiO_2 (outer layer close to the atmosphere or other medium). Thus, titanium alloys have a tendency to be spontaneously passivated at ~20 °C and this behavior is retained at 250 °C in the presence of an oxygen source. Oxidizing conditions are heating in air (thermal oxidation), anodic polarization in an electrolyte, and exposure to HNO_3; the thickness of the oxide layer on titanium increases with time: from 1–2 nm to 6 nm during 70 days, and then to 11 nm after 1.5 year, and even to 25 nm after 4 years (in natural conditions). Chemical resistance of titanium can be enhanced by increasing the surface oxide film thickness, alloying titanium with certain elements, and adding inhibitors (oxidizing compounds) to the media. Anodization thickens the oxide layer to give a beautiful colorful spectrum.

Usually naturally formed TiO_2 films are invisible to the eye. This oxide layer is highly chemically resistant and is attacked by only hot concentrated HCl, H_2SO_4, NaOH, and HF. Without the protective dioxide layer TiO_2, titanium would dissolve rapidly in most solutions.

8.1.1.3 Corrosion of Titanium

The following corrosion phenomena can occur with titanium: general corrosion, pitting corrosion, crevice corrosion, SCC, hydrogen embrittlement, fatigue, erosion, and galvanic corrosion.

General corrosion. Titanium offers moderate resistance to some mineral (inorganic) (HCl, H_2SO_4, H_3PO_4) and organic acids. Titanium oxides are not soluble in most inorganic acids. The corrosion rate depends on acid type, its concentration, and temperature. Amine corrosion inhibitors are not effective for protecting titanium. Titanium is very resistant to HNO_3 up to its boiling point. The corrosion rate of titanium in diluted HCl (up to 5 wt%) is very low at 24 °C. Significant corrosion of Ti Gr.2 (>0.2 mm/y) occurs at concentrations of HCl greater than 8 wt% at 24 °C and greater than 1.5 wt% at 52–66 °C (Table 8.1). For Ti Gr.12, considerable corrosion begins at concentrations of HCl greater than 8 wt% at 24 °C, and greater than 4 wt% at 52–66 °C (Table 8.2). For Ti Gr.7, severe corrosion begins at concentrations of HCl greater than 26.5 wt% at 24 °C, greater than 15 wt% at 52 °C and greater than 11 wt% at 66 °C (Table 8.3). Titanium is not resistant to HF solutions at all.

Titanium is resistant to most organic acids over a wide range of concentrations. However, it severely corrodes in formic acid (>10 wt%), oxalic acid (even in dilute solutions), and in citric acid (50 wt% at 100 °C). Alloying additions of Mo and Ni (Ti Gr.12) reduce the susceptibility of Ti to anodic dissolution. The latter additions increase the alloy strength at the expense of ductility. Palladium is added in small concentrations (0.04–0.25 wt%) (Ti Gr.7, Ti Gr.16—see Appendix, Table F.13) to increase the corrosion resistance of titanium in reducing acid environments. Titanium exhibits low corrosion rates in alkaline solutions (NaOH, KOH) at low and moderate concentrations and temperatures. Titanium is not resistant in strong caustic solutions (pH > 12, T > 80 °C) (8.1).

$$2Ti_{(s)} + 2NaOH_{conc.} + 2H_2O_{(l)} \rightarrow 3H_{2(g)} + 2NaTiO_{2(aq)} \qquad (8.1)$$

However, titanium is resistant to aqueous ammonia (NH_4OH) at all concentrations and temperatures and to anhydrous ammonia. Thus, titanium can be considered as an amphoteric metal, which means that it dissolves both in concentrated acids and alkalis. Titanium is resistant to solutions of alcohols, aldehydes, esters, ketones, and hydrocarbons. Aeration may be required to maintain its passivity. Titanium alloys are highly resistant to air, sea water, fresh water, steam (up to 400 °C), H_2S and CO_2, aqueous solutions of neutral salts (up to 100 °C, even more resistant than stainless steel), dilute HCl (up to 5 wt% at 24 °C), hypochlorites

Table 8.1 Corrosion of Ti Gr.2 in naturally aerated HCl solutions at different temperatures [6]

T (°C)	24						52					66					
Conc. of HCl, wt%	5	6	8	9	17.3	26	1.4	5.8	6	7	11.5	1	1.5	1.7	2	3	4.7
Corr. rate, mm/y	0	0.07	0.2	0.25	0.51	2.59	0.02	0.51	0.68	1.27	3.07	0.01	0.02	0.13	0.61	1	7.08

Table 8.2 Corrosion of Ti Gr.12 in naturally aerated HCl solutions at different temperatures [6]

T (°C)	24					52				66			
Conc. of HCl, wt%	6	8	10	12	28.5	3	4	5.9	7	2.4	3.6	5.9	7
Corr. rate, mm/y	0.008	0.008	1.4	2.54	5.58	0	0.001	0.518	5.30	0.01	0.03	0.51	8.98

Table 8.3 Corrosion of Ti Gr.7 in naturally aerated HCl solutions at different temperatures [6]

T (°C)	24					52						66			
Conc. of HCl, wt%	9	18	20	26.5	27	9	11.5	14.7	16.8	19	21.9	6	9.6	11.5	16.8
Corr. rate, mm/y	0	0	0.01	0.02	0.70	0.008	0.02	0.03	0.06	0.08	0.41	0.01	0.03	0.04	0.13

(HOCl, NaOCl, $Ca(OCl)_2$), ammonium chloride (NH_4Cl), amine chloride, and ammonium bisulphide (NH_4HS). They also are resistant to concentrated NaCl, $CaCl_2$ and $MgCl_2$ aqueous solutions containing sulphur (120 g/l) and at pressure 7 MPa H_2S up to 180 °C [10]. Above this temperature, the titanium alloys are susceptible to localized attack.

Allowable corrosion rate of titanium. The allowable corrosion rate is the maximum (acceptable, permissible) value for which the equipment or construction can serve without failure during a defined period (design life). This value is defined from experimental data about general (uniform) corrosion rate and depends on the type of metal, requirements of a specific industry, and the type of equipment (the type of the material and its thickness). Allowable corrosion rates are very important parameters for designers and engineering personnel at oil refineries. These values allow decisions to be made about the correct and proper application of corrosion control measures. For carbon steel heat exchanger tubes in oil refining and petrochemical industry, the allowable corrosion rate is defined as 0.11 mm/year [1]. For titanium heat exchanger tubes this value should be defined. The design life of heat exchangers is 15 years in the oil refining and petrochemical industry. The widespread wall thickness of heat exchanger titanium tubes is 0.9 mm (instead of 2.336 mm for carbon steel). The allowable corrosion rate of titanium tubes in this particular case is

$$\frac{0.9\,\text{mm}\cdot 0.7}{15\,\text{years}} = 0.04\frac{\text{mm}}{\text{year}} \tag{8.2}$$

where 0.7 is the safety factor. This means that we may only use heat exchangers up to a corrosion wear (from a mechanical point of view—strength of remaining tube material) of 70 % [1]. Thus, the allowable corrosion rate (0.04 mm/year) of titanium tubes is one third of that of carbon steel (0.11 mm/year). We can emphasize that corrosion rates of titanium in most media are much lower than the allowable value 0.04 mm/year because of the presence of the protective film TiO_2 on the titanium surface and its growth during service. As a result, titanium equipment is often designed with a zero corrosion allowance in passive conditions in different media of the oil refining and petrochemical industry.

Pitting corrosion. Titanium is highly resistant to pitting attack in chloride media and failures are very rare. This is explained by the protective oxide film which exhibits very high anodic pitting potentials in chloride solutions: +5 to +10 V (versus SCE).

Crevice corrosion. Titanium with a passive film is susceptible to crevice corrosion, similar to other metals and alloys with passive films (stainless steels, aluminum, and nickel alloys). Crevice corrosion on titanium can occur under metal joints, non-metal gaskets, or under deposits. Temperature and pH are critical factors for the occurrence of crevice corrosion on titanium. Crevice corrosion does not occur at pH > 10. Initiation can occur rapidly at pH < 7, titanium chlorides formed within the crevice are hydrolized, forming HCl (pH < 1) and titanium oxide/hydroxide corrosion products. At 7 < pH < 10, crevice attack can also occur.

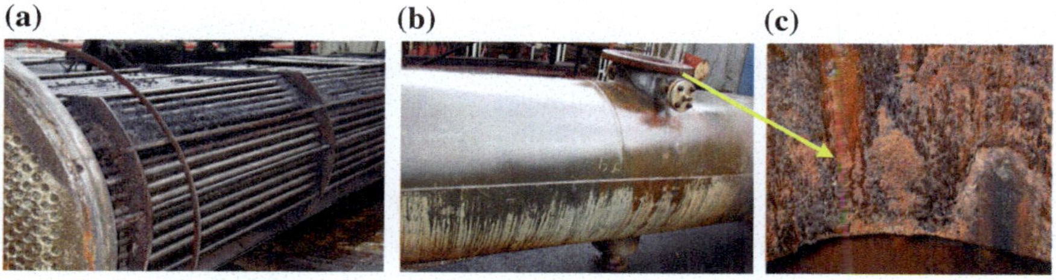

Fig. 8.1 **a** Titanium Gr.12 tube bundle of the condenser in the overhead at the vacuum unit. **b** Carbon steel inlet of hydrocarbons containing small concentrations of HCl and H_2O (T = 50 °C). **c** Corroded carbon steel inlet after 4 years. As opposed to this, titanium tubes have good appearance (**a**)

At pH < 7 and T > 80 °C, the resistance of titanium alloys to crevice corrosion depends on the presence of noble metal alloying additions of Mo, Pd, and Ru. Titanium alloys (Gr.7, Gr.16, Gr.12) are susceptible to crevice corrosion when subject to "dry" amine salts. Addition of liquid water ($\sim$1 wt%) in the overhead at the crude distillation column is required to passivate the titanium alloys and thus to prevent under deposit "dry" amine salt crevice corrosion. Ti Gr.7 contains 0.12–0.25 wt% Pd and its resistance to crevice corrosion is very high. Ti Gr.16 contains 0.04–0.08 wt% Pd and its resistance to crevice corrosion is significantly less than that of Ti Gr.7.

Ti Gr.12 contains 0.3 wt% Mo and 0.8 wt% Ni, has improved strength compared to unalloyed Ti Gr. 2, and is used when crevice corrosion under salt deposits is expected in the overhead of atmospheric and vacuum columns (Fig. 8.1). In spite of formation salt deposits on Ti Gr.12 tubes, they can serve with these deposits (in the case of insufficient water wash) without failure for more than 20 years (current experience). Removing and cleaning titanium tubes from salt deposits should be carried out every 4 years. Titanium tubes after 20 years of service in the overhead of atmospheric and vacuum columns have good appearance while carbon steel undergoes severe corrosion (see Fig. 8.1c).

Specific case of crevice corrosion. Once, during the assembling of heat exchanger tube bundle made from Ti Gr.2, carbon steel parts were welded in the vicinity of titanium tubes. Nobody noticed that iron particles were smeared on the titanium tubes. Several months after the beginning of service of the titanium heat exchanger with cooling water (aqueous solution of electrolytes), leaks from the titanium tubes were detected. Visual examination of titanium tubes revealed corrosion holes formed on the outer surface. This phenomenon is called 'smeared iron pitting' occurring because of galvanic corrosion of iron particles adhered on titanium. The resulting corrosion product is acidic ferric chloride ($FeCl_3$) which attacks the titanium surface under iron particles where they breach the passive titanium oxide film TiO_2. Usually 'smeared iron pitting' occurs at T > 80 °C but sometimes at lower temperatures in electrolyte solutions. Ti Gr. 7, 12, and 16 are more resistant to this attack. The general rule is that in no case should work with carbon steel and low-alloy steels (i.e. welding, tooling, cutting) be carried out when titanium parts are fabricated in the vicinity. Stainless steel tools are recommended for use on titanium. If, however, smeared iron particles are present on titanium, they should be removed by immersion in aqueous solution of mixture HNO_3 and HF for several minutes.

Stress corrosion cracking (SCC). The majority of titanium alloys are resistant to SCC which can occur in pure methanol and aqueous halide (Cl^-, Br^-, I^-) solutions. The addition of 2 wt% H_2O to methanol inhibits the SCC of titanium.

Hydrogen damage (hydriding, hydride embrittlement, hydrogen-induced cracking).

The oxide film on titanium is an effective barrier to hydrogen atom intrusion. Disruption of this film allows easy absorption of hydrogen atoms. However, the presence of ~2 wt% of H_2O or O_2 passivates the titanium surface, forming again an effective barrier for hydrogen atoms up to T = 300 °C and P = 55 bar. At T < 80 °C, hydrogen atom diffusion is very slow and atoms remain on the surface. At T > 80 °C atomic hydrogen penetrates into the titanium lattice leading to the formation of titanium hydrides (TiH_2) in the alloy microstructure. Brittle titanium hydrides (dark, acicular needle-like structured platelets) can lead to a loss of mechanical properties (ductility, toughness) and then embrittlement (crack) under applied stresses. Hydriding occurs when titanium corrodes actively (when protective dioxide film is absent or disturbed) or titanium is coupled to a less noble corroding metal/alloy (i.e. carbon steel). The most favorable conditions for hydriding are: pH < 3 or pH > 12; generating nascent hydrogen on a titanium surface (which can occur from a galvanic couple; an impressed cathodic current; damage to the protective titanium dioxide layer, i.e. abrasion or smeared iron); T > 80 °C (where the diffusion rate of hydrogen atoms into titanium is significant). In order to prevent hydrogen damage, these conditions must be avoided. Therefore, galvanic contacts of titanium tubes with carbon steel tube sheets, baffles and shells must be excluded. Like arsenic, antimony and cyanide species, the sulfides act as a hydrogen recombination poison (that is, prevent the recombination of atomic hydrogen) and enhance hydrogen uptake by titanium. No hydrogen absorption and embrittlement occur when titanium contacts fully passive alloys (stainless steels and nickel alloys).

Impressed cathodic potentials below the hydrogen evolution line should be avoided. High-temperature (>80 °C) alkaline conditions should be avoided as they may cause hydrogen uptake (as a result of general corrosion) and embrittlement of titanium alloys.

It is interesting to emphasize that the same deleterious titanium hydride (TiH_2) which is formed inside the microstructure when atomic hydrogen penetrates, and then can cause environmentally stress cracking of titanium alloys, can play the positive role of catalyst of hydrogenation of some organic compounds and as a reducing agent in powder metallurgy.

Pyrophoricity. When titanium severely corrodes (oxidizes) forming a high surface area (sponge-like structure), it can be pyrophoric (see Sect. 4.1.1). Burning titanium produces little smoke. In order to prevent this phenomenon, the corroded titanium equipment must be wetted.

Biofouling and MIC. Since titanium is nontoxic, it is susceptible to biofouling ($\sim$0.5 kg·m^{-2}·year^{-1}) when immersed in seawater [19]. However, microorganisms do not attack dioxide film and titanium remains resistant to localized corrosion. Water velocities greater than 2 m/s reduce the extent of biofouling. Ozone, UV (ultraviolet light) and chlorination help in controlling biofouling on titanium.

Galvanic corrosion. Titanium is corrosion resistant in many media and is often the more noble metal (the cathode) in a galvanic couple. Rapid corrosion of the less noble alloys (carbon steel, copper alloys) occur when these materials contact titanium (Fig. 8.2). Therefore, these contacts must be avoided.

Titanium may be safely coupled to stainless steel (in their passive condition), nickel alloys (i.e. Hastelloy C 276, Inconel 625), and nickel-copper Alloy 400 (Monel) which have similar electric potentials (Table 8.4).

Erosion, erosion-corrosion, and cavitaion. Titanium is highly resistant to erosion and cavitation as a result of its hard adhered dioxide layer TiO_2 (ceramic material). For instance, Ti Gr.2 gave a minimal flow enhanced corrosion rate of 0.01 mm/year (comparing with stainless steel) at sea water velocity of 35 m/s (in the absence of suspended particles) at 25 °C. High local turbulence, impingement and cavitation favor erosion of titanium. Inlet turbulence in shell and tube heat exchanger, entrained gas bubble and mist-laden vapor impingement, and the presence of corrosives (i.e. HCl drops) can result in erosion-corrosion of titanium in very rare cases (Fig. 8.3).

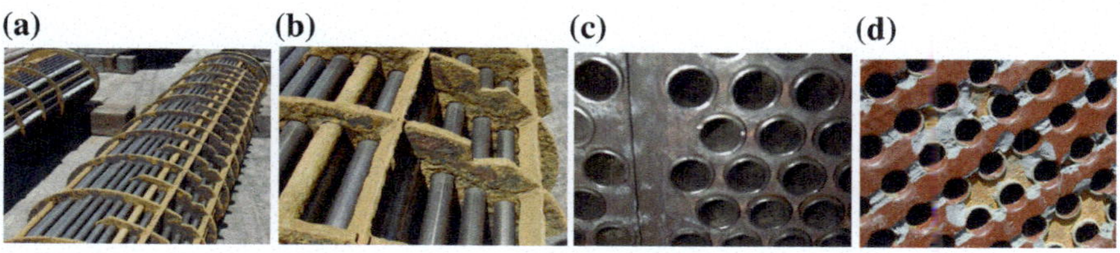

Fig. 8.2 Examples of galvanic corrosion after 4 years of service. T = 30–35 °C. **a**, **b** Ti Gr.2 tubes. Baffles and tie rods—carbon steel. Medium—hydrocarbons + water condensate after separation tank (drum) at the crude oil atmospheric unit. Solution—baffles and tie rods must be made from SS 304. **c** Tubes—Ti Gr.2. Tube-sheet—Admiralty brass (copper-zinc alloy). Medium—cooling water. Electric potentials are −0.10 V of Ti Gr.2 and −0.29 V (vs. SCE) of Admiralty brass (see Table 8.4). The latter worked as anode relatively to titanium cathode and corroded. Solution—the tube-sheet must be made from Ti Gr.2 or carbon steel cladded with Ti Gr.2. **d** Tubes—Ti Gr.2. Tube-sheet—carbon steel coated by epoxy-ceramic coating system. Medium—cooling water. Solution: **a** In no case to paint anode (carbon steel tube-sheet); **b** The tube-sheet must be made from Ti Gr.2, or SS 316, or carbon steel cladded with Ti Gr.2

Table 8.4 Galvanic series in flowing sea water[a] [6]

Metal/alloy	Pt	SS 316[c]	SS 304[c]	Monel[d]	Hastelloy[e]	Ti	SS 316[f]	Adm. brass[g]	CS[h]
E, V versus SCE[b]	+0.15	−0.05	−0.08	−0.08	−0.08	−0.10	−0.18	−0.29	−0.61

[a]Some metals and alloys are shown. Velocity: 4 m/s. T = 24 °C
[b]*E, V* versus SCE—electrode potential in Volt versus saturated calomel electrode
[c]Austenitic stainless steels in passive state
[d]Monel—alloy 400, copper-nickel alloy
[e]Hastelloy—'C-type' alloys (C-276, C-22, C-2000), nickel alloys
[f]SS 316—austenitic stainless steel in active state
[g]Adm. brass—admiralty brass (copper-zinc alloy)
[h]CS—carbon steel

(a) **(b)**

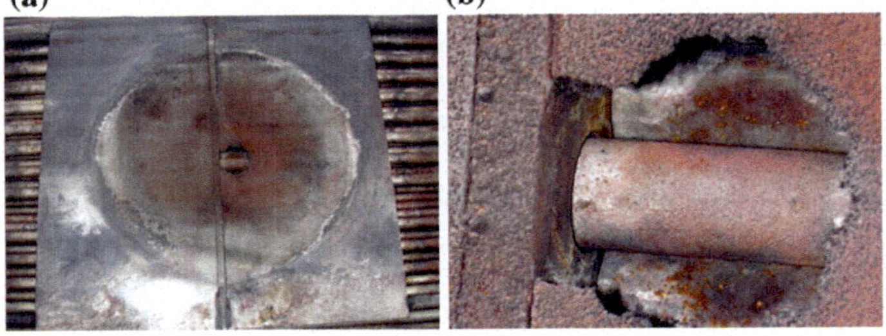

Fig. 8.3 Overhead condenser at the vacuum unit. Medium: hydrocarbons with contaminants HCl and H_2O (T = 50 °C, 4 years). **a** Erosion–corrosion of the impingement plate (Ti Gr.2) **b** Magnification of the picture (**a**)

Table 8.5 Hardness of some titanium alloys

Titanium grade	2	4	5
Hardness Brinell (HB)	160	265	379

The harder higher-strength titanium alloys (Ti Gr.4 and Ti Gr.5) are more resistant to erosion than Ti Gr.2 (Table 8.5).

8.1.1.4 Experience of Use of Titanium Alloys at Oil Refinery Units

Two-phase mixtures of hydrocarbons and water condensate containing HCl, H_2S and NH_3 forming in the overhead in crude distillation and FCC fractionation columns are aggressive to carbon steel, copper alloys and copper-nickel alloys. Usually heat exchangers and condensers made from these materials serve 2–4 years, when economical design life is 15 years. The service life of this equipment was extended to 20 and more years after replacing them with titanium. Five types of titanium alloys (Ti Gr.1, Ti Gr.2, Ti Gr.12, Ti Gr.7, and Ti Gr.16) are used as construction material in equipment of the oil refining and petrochemical industry. Titanium heat exchangers and condensers (shells, tubes, baffles, tie rods, and spacers), air coolers (tubes and header boxes), piping, and pressure vessels can be used in:

- the overhead at crude oil distillation atmospheric and vacuum columns (the medium: two-phase mixtures of hydrocarbons and aqueous solutions of H_2S, HCl, NH_3, amine neutralizers, and amine corrosion inhibitors);
- the overhead at FCC fractionation column;
- delayed coking units (DCU);
- the overhead condensers in sour water strippers (SWS);
- hydrodesulfurizer (HDS) (to 170 °C and 50 bar);
- cooling water systems.

The FCCU and DCU fractionation overhead streams in addition to hydrocarbons and water condensate contain large concentrations of NH_3 and H_2S along with small amounts of HCN, HCl, SO_2, and CO_2. Ammonium polysulfide or proprietary amine corrosion inhibitors are sometimes injected in these overhead condensing systems to control corrosion. Sour waters are stripped in SWS to remove H_2S and NH_3. The condensed overhead water in SWS can contain up to 40 wt% NH_4HS along with cyanides and chlorides ($\sim$ 100 ppm). Cooling water systems can contain different salts and their concentrations at about neutral pH: brackish water, sea water, and potable water.

In spite of successful use of titanium as construction material in the oil refining and petrochemical industry, some limitations exist.

8.1.1.5 Limitations of Use of Titanium Alloys

Certainly there is no material with only desired properties. Titanium has also some disadvantages which limit its use. Misapplication of titanium or operation of titanium equipment beyond established process limits can result in failures which are described below.

Titanium is prone to crevice corrosion under deposits (dry amine salts at $\sim$ 121 °C or smeared iron particles), hydrogen embrittlement, and galvanic corrosion of other alloys. Temperature limitation of use of Ti Gr.2 in the overhead at crude oil distillation units, FCCU, DCU and SWS is 121 °C. Ti Gr.12 in the overhead condensers can be used up to 171 °C where formation of solid, dry amine salt is avoided. Proper water wash or selection of amine neutralizers is important in ensuring the successful use of titanium.

Contacting titanium with more active alloys (carbon steel, copper alloys) can cause their galvanic corrosion. Titanium may contact austenitic stainless steels without fear of galvanic corrosion. The factors that influence the breakdown of passivity on titanium alloys are temperature, pH, halides (F^-, Cl^-, Br^-, I^-), and elemental sulfur. Titanium alloys are resistant to dilute HCl (up to 5 wt%) at ambient temperatures, formic and acetic acids (up to 10 wt%) up to 100 °C. Titanium alloys are not resistant to concentrated HCl. Addition of corrosion inhibitor sodium molybdate (Na_2MoO_4) to HCl aqueous solutions significantly reduces corrosion rate and hydrogen pick up. Titanium is not resistant to concentrated NaOH and KOH (pH > 12) and diluted NaOH (2 wt%) with addition of H_2O_2 (1 wt%). Injection of corrosion inhibitors sodium silicate (1 wt% Na_2SiO_3) and sodium hexametaphosphate (0.05 wt% $(NaPO_3)_6$) to the latter alkaline-peroxide solution significantly decreases corrosion of titanium [20].

8.1.1.6 Fouling in Titanium Heat Exchangers

One of serious problems is a formation of fouling(scale and deposits) on titanium tubes, tube-sheets and plates in heat exchangers from cooling water-side (Figs. 8.4, 8.5). Usually this is similar to the formation of scale and deposits on other materials with passive films, such as stainless steel and copper alloy heat exchange surface (see Sect. 7.4).

In most cases, scale formed in cooling water is composed of $CaCO_3$ and $MgCO_3$. Sometimes $CaSO_4$ is also included in deposits.

If cooling water is treated by corrosion inhibitors (zinc-phosphate-phosphonate), deposits can contain calcium and zinc phosphates and calcium sulfates, and can even clog the inner surface of titanium tubes after 4 years of service (see Fig. 8.5a). Plate heat exchangers (because of close position of plates) can be clogged even after a year of service (see Fig. 8.5b, c). Sometimes corrosion products (iron hydroxides/oxides) may be present in deposits. Carbonates, sulfates, phosphates and hydroxides have different solubility in cleaning acid solutions.

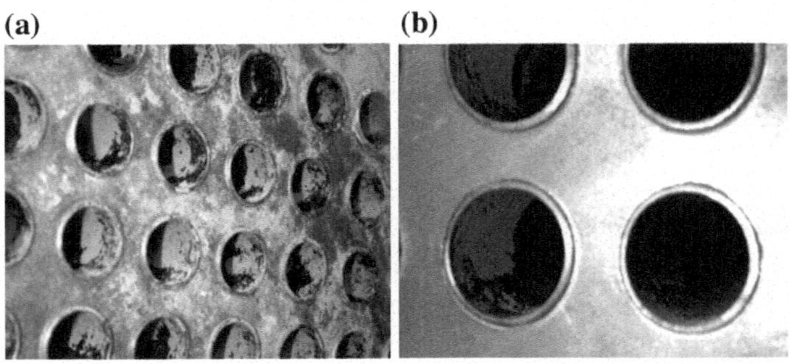

Fig. 8.4 **a** Scale ($CaCO_3$) inside titanium (Ti Gr.2) tubes (cooling water-side, T = 30–35 °C. P = 4–5 bar, 4 years). **b** Titanium (Ti Gr.2) tube-sheet and tubes after chemical cleaning by citric acid (5–8 wt%) from the scale

Fig. 8.5 Clogged heat exchanger (cooling water-side) with titanium tubes (Ti Gr.2) after 4 years of service (**a**), and titanium plates (Ti Gr.1) after 1 year of service (**b**, **c**). White-grey deposits: calcium phosphate, zinc phosphates, and calcium sulfates. T = 30–35 °C. P = 4–5 bar. (The picture **c** is the magnification of the picture **b**)

Slow water velocity and absence of sufficient concentration of anti-scaling agent in cooling water results in the precipitation of deposits. In order to diminish and prevent fouling, the following solutions and recommendations should be carried out:

(a) To keep cooling water velocity more than 1.2 m/s.
(b) To optimize the concentration of anti-scaling agent.
(c) To perform periodical reversal flow (flushing) in heat exchangers (once in 3–4 months).
(d) To clean inner surface of titanium tubes by means of citric acid (5–8 wt%). Corrosion rate of titanium in solutions of citric acid (5-8 wt%) is less than 0.01 mm/y (when allowable value is 0.04 mm/y), therefore not necessary to inject corrosion inhibitor during the cleaning by citric acid. Carbonate, sulfate and phosphate scale may be removed from titanium surface also by aqueous solution HCl (2–5 wt%). This dilute HCl solution does not require corrosion inhibitor as sometimes is recommended.

Unusual extraordinary failure of titanium tubes during assembling in the tube-sheet of new heat exchanger is described below.

8.1.1.7 Unique Failure of Titanium Tubes

Conditions. A mandrel with 50–60 mm length was used while assembling titanium tubes in the tube-sheet during rolling. Titanium tubes were cracked during rolling operation by means of mandrel.

Service period before the failure. Zero time. The assembling of tube bundle before the service.

Material of construction. Ti Gr.12.

Visual examination and findings. A longitudinal crack (30 mm length) initiating perpendicular to the edge of the tube was found (Fig. 8.6).

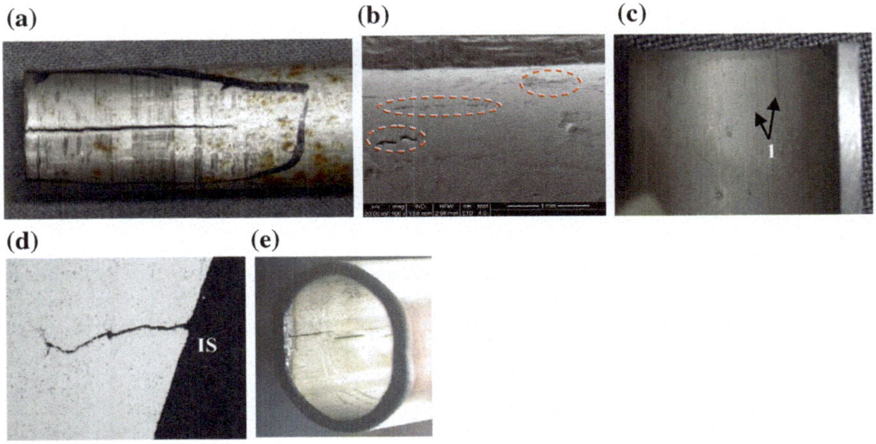

Fig. 8.6 **a** The titanium tube edge with crack. **b** SEM photo of the fracture inner surface of the tube (marked with red circles—cracks). **c** Macroscopic section through the tube (200 mm from the edge, inner surface): 1—cracks. **d** Microscopic cross-section through the tube (200 mm from the edge—crack tip). IS—inner surface. **e** New titanium tube after flattening test (cracks were formed inside)

Metallographic examination. Macroscopic examination of the tube (200 mm from the edge) in the area which did not contact mandrel showed longitudinal cracks up to 3.5 mm length on the inner surface (see Fig. 8.6c). Microscopic examination of this surface showed branched cracks with up to 0.3 mm depth initiated perpendicular to the inner surface of the tube.

Flattening test was performed according to ASTM B338 [21] on two samples that were taken from the tube far from the edge, until the distance between the load platens was 14 mm. Cracks were revealed on the inner surface after the flattening test relatively far from the expansion area—the mandrel was 50–60 mm in the length (see Fig. 8.6e). They were developed from the tube manufacturing cracks, the defect during the production of the tube at the plant. Thus, these original cracks were the reason of the flattening test failure.

Failure phenomenon: *Stress cracking*.

Cause of failure and its explanation. Overload mechanism while using the mandrel during rolling caused the failure because of the presence of longitudinal cracks in the inner surface of the tube. These cracks originated in the manufacturing of tubes and propagated during their installation. Flattening test showed that these titanium tubes were delivered with small cracks on the inner surface.

Solutions and recommendations.

(a) Correct manufacturing of titanium tubes according to the ASTM B338 Gr.12 [21].
(b) Incoming examination and control of tubes' quality.

8.1.2 Conclusions

A. Experience of use of titanium alloys (five grades) as construction material for heat exchangers, condensers and other equipment for different services and applications at oil refinery and petrochemical units during 20–40 years proves its high durability and benefits.

B. Titanium alloys Ti Gr.2 and Ti Gr.12 can be used in contact with hydrocarbons, H_2O, HCl (diluted), H_2S and NH_3 up to 121 and 171 °C at 1 bar respectively. It is necessary to prevent formation of solid ammonium or amine chloride salts in titanium heat exchangers. Water wash and suitable amine injection should be used. Ti Gr.12 in some cases is resistant to solid ammonium or amine chloride salts.

C. Titanium alloy Ti Gr.2 can replace carbon steel, copper alloys and copper-nickel alloys in the overhead service at FCCU fractionation column where H_2S and NH_3 are the corrosion problem. Copper and nickel are not resistant both to H_2S and NH_3.

D. Titanium alloys Ti Gr.1 and Ti Gr.2 are recommended for use in cooling water containing large concentrations of chlorides and sulfates (i.e. sea water, brackish water or potable water) at neutral pH at 10–40 °C and 5 bar.

E. Titanium alloys are susceptible to crevice corrosion, therefore, special attention must be paid to tube-to-tube sheet joints, gasket-to-metal flanges and other seal joints, and weld joints.

F. In no case should titanium be used in contact with less noble alloys (carbon steel, copper alloys) in electrolyte solutions, otherwise galvanic corrosion of the latter can occur. Tube-sheets and tube bundle parts (baffles, tie rods, bolts) should be made from suitable materials (usually SS 304 or SS 316), not carbon steel, to prevent possible galvanic corrosion and to optimize bundle life.

G. Use of the thinner wall titanium tubes (0.9 mm wall thickness instead 2.336 mm for carbon steel tubes), requires appropriate handling of bundles. This situation also requires closer baffle spacing to prevent excessive tube vibration that can result in premature fatigue failures. It is recommended to install extra vibration dampening bars between the baffles to provide additional support for the titanium tubes. It is not a good practice to retube an existing carbon steel bundle with titanium tubes.

H. It is forbidden to weld carbon steel in the vicinity of unalloyed titanium (Ti Gr.1, Ti Gr.2) equipment or to use iron instruments when titanium parts are fabricated. Iron particles can be smeared on titanium surface and then cause crevice corrosion of titanium. Smeared iron particles on titanium should be removed by means of aqueous solution HNO_3 and HF. Stainless steel tools should be used on titanium.

I. Scale and deposits can be formed on titanium in cooling water service. Cleaning with dilute aqueous HCl solution (not more than 5 wt% HCl) without corrosion inhibitor can be used for removing carbonates, sulfates and phosphates.

8.2 Treatment and Change the Environment

Removal of the aggressive components, such as desalting of crudes, deaeration, removal of H_2S, neutralization, use of biocides, drying the media, the use of corrosion inhibitors—all these measures relate to treatment and change the environment. We will consider some of them.

8.2.1 Diminishing the Concentration of Corrosive Substances in Streams

The most common way to diminish corrosive compounds in atmospheric and then in vacuum distillation units is desalting and then caustic injection. Caustic injection prevents hydrolysis of calcium and magnesium chlorides in desalting crude oil and possible formation of hydrochloric acid. It is impossible to reach absolute removing of corrosive substances. In any case, they (H_2O, H_2S and HCl) move to the overhead of distillation column.

Injection of water (water wash) in the place where gases exit from an overhead of the distillation column can dilute the corrosives when they condense or dissolve in liquid aqueous solution. In addition, correct and good water wash will dissolve any water soluble salts (ammonium and amine chlorides) and thus will prevent under deposit corrosion. Water wash program must be carefully designed and applied. The quality of wash water (its source) is very important. Boiler feed water is the best choice. Because of its high cost, the most commonly used wash water sources are atmospheric and vacuum column accumulators and sour water strippers. Neutralization is one of the old chemical procedure to diminish corrosiveness and environmental hazard of acidic solutions.

8.2.1.1 Neutralizing of Acidic Corrosives in Overhead

In order to neutralize acidic substances in the overhead and increase pH to nearly neutral (the target is pH = 5.5–6.5), alkalis ammonia and amines are injected (see Appendix E). Acidic compounds are mainly HCl and H_2S, but H_2SO_3 and H_2SO_4 (in the case of appearing dissolved oxygen) and light organic acids (formic and acetic) can be present. Neutralizing amines vary by type of amine, solvent, and concentration of active ingredients. Some proprietary neutralizers consist of mixtures of several amines. Other proprietary neutralizers consist of amine derivatives. Most neutralizers are water soluble. Some neutralizers are both water soluble and hydrocarbon solvent miscible solutions (in naphtha or kerosene). The following physicochemical properties of neutralizing amines should be considered before selecting for overhead application: strength and basicity; boiling point; solubility in naphtha and in water; melting point of the amine hydrochloride salt; solubility of this salt in naphtha and water [3, 22].

The acids react with injected alkalis, pH increases, and salts are formed (see Appendix E). The hydrochloride amine salts formed after neutralizing usually are soluble in water and insoluble in hydrocarbons. One of the commonly used neutralizers, the ethylene diamine, forms the salt which has the highest thermal decomposition temperature. This salt is the least soluble in water ($\sim$30 wt% at 80–90 °C). Therefore, even effective water wash cannot completely prevent deposit formation and corrosion underneath.

Sometimes neutralizing amine is injected into the naphtha reflux line that is amine falls directly into the distillation column where it can react with hydrogen chloride to form salt and deposit as fouling on the upper trays in the column. This fouling can plug trays and cause under deposit corrosion. There are two other ways of appearance of amine salts in the distillation column. The first is the use of the atmospheric or vacuum column accumulator water in the desalter. A small amount of the desalter water enters the column with the crude oil feed. The amine salt dissolved in this water is heated to $\sim$350 °C with the crude. At this temperature the amine and ammonium salts either decomposes or sublime. If they sublime they can deposit in upper parts of the column.

The decision of whether to use separately ammonia, or a neutralizing amine, or both, and the type of neutralizing amine must be made carefully. Ammonia was the first neutralizer used in the overhead systems. It was readily available, cheap and provided acceptable protection of the overhead condensers. The use of ammonia, especially gaseous ammonia, sometimes has the problems to control the feed rate because of possible pressure changes on the unit.

Another drawback of using ammonia is that it easily vaporizes and it is poorly soluble in hot water ($\sim$100 °C) and so does not enter the water when it first condenses. Opposite, hydrogen chloride is strongly attracted to the hot water and can significantly diminish the pH to acid values in such areas. As a result, ammonia as a neutralizer does not protect the area where water begins to condense. However, in systems with low amounts of acid (good and sufficient desalter) and large amounts of water present ammonia can be used successfully. Ammonium chloride salt is insoluble in naphtha. Ammonia easily increases the pH of condensed water to 5.5–6.5. If too much ammonia is injected in the presence of H_2S, bisulfide corrosion can occur (see Sect. 3.1 and Appendix D). What is more important, ammonia is considered as a hazard compound and poses a threat to humans and environment.

Nowadays there is the tendency to inject a neutralizing amine alone because of safety reasons. Amines are easier and safer to ship, store, and handle. They are less concentrated than ammonia and injection rates are usually easier to regulate and control. Amines have boiling point temperatures in the range of 115–175 °C and remain liquids after injection (except diethylamine which vaporizes) [3]. They absorb readily HCl vapor and neutralize HCl droplets during the acid condensation. All amine neutralizers (except methoxypropylamine—MOPA) tend to form hydrochloride salt deposits. They are well soluble in water and not soluble in naphtha and many other hydrocarbons. Sometimes these salts can contaminate petroleum products and cause fouling in downstream operating units (e.g., in naphtha desulfurizer). Another possible problem of using ammonia and amines in the overhead is their aggressiveness to copper alloys. Fortunately copper alloys do not corrode in the overhead system because the concentrations of ammonia and amines are low.

The choice of the target pH in the accumulator water is important. Usually it is recommended to keep pH in the range 5.5–6.5 (see Sect. 3.1). If pH < 5.5 corrosion can occur. If pH > 7.0, ammonium and amine salts can be formed which can result in under deposit corrosion. Another problem is that at higher pH equilibrium between sulfide (S^{2-}) and bisulfide (HS^-) ions shifts in the formation of the last ions. Bisulfide ions form less protective iron sulfide films and also can react with existing protective iron sulfide films to partially convert them to bisulfide less protective films.

Usually pH measurements of water condensate are carried out in the atmospheric and vacuum accumulators (boots). These pH values do not inevitably reflect the pH in the overhead and throughout the condensing system. As ammonia does not dissolve in hot acidic water, the pH in the area of the first condensation of water is lower than the pH in the accumulator. Most neutralizing amines are more strongly attracted to the first condensing hot water than is ammonia and thus provide much improved pH control in this region. The second reason of the tendency to use neutralizing amines alone is that they are usually easier to handle, provide much better neutralization in the area where water first condenses, have more water soluble hydrochloride salts, and are usually less aggressive toward copper alloys. If the neutralizer is not atomized well, large drops or even a stream of the neutralizer can cause erosion-corrosion of the pipe wall. The following methods are used for injection of neutralizers: directly into the hydrocarbon vapors exiting the top of a distillation column; into naphtha reflux line; and into wash water.

8.2.1.2 Overhead Wash Water Injection

Wash water is carried out in the overhead at distillation crude unit, FCCU, hydrotreater, hydrocracker, HDS, amine treating units, and catalytic reforming.

Distillation Crude Unit. If crude oil is desalted well and salt content is less than 1 ptb (3.5 ppm NaCl) in crude after desalting, expected chloride concentration in the overhead is less than 20 ppm HCl, and ammonium chloride deposits cannot form. In such cases wash water is not necessary in the overhead. However, this is realized only in the case of trouble-free desalter functioning and proper injection of caustic soda solution into the crude oil downstream of the desalter. Really unforeseen desalter upsets can occur and caustic soda injection pumps can unexpectedly fail. Therefore, it is necessary to provide using intermittent or continuous water injection in the overhead. Important indicator to use water wash is deterioration of heat transfer because of fouling, which usually means that crudes are no longer properly desalted. Since fouling of overhead heat exchangers is usually caused by salt and corrosion product deposits, intermittent or continuous water wash injection often is necessary to remove these deposits and thus minimize or even prevent under deposit corrosion. Thus, wash water injection into crude unit overhead at distillation column is applied with the following aims:

(a) Dissolving and removing ammonium chloride and amine hydrochloride salts, and loosening and removing chloride-containing corrosion product deposits in order to prevent corrosion underneath.
(b) Restoring heat transfer of fouled overhead condensers by removing salt and corrosion product deposits.
(c) Dilution hydrogen chloride in condensed water and thus to diminish its corrosiveness.
(d) Better mixing of neutralizer and corrosion inhibitor with the process stream.

Following sources of wash water exist at refineries (they are listed in order of decreasing desirability and likely increasing cost) [3]:

(a) Recirculated crude oil column overhead water condensate.
(b) Vacuum column overhead water condensate.
(c) Stripped sour water.
(d) Boiler feed water.
(e) Steam condensate.

It is not recommended to use following sources of water (even how readily available or cheap): cooling tower blowdown; water condensate from vapor recovery sections of hydrotreating or FCCU; potable water; brackish or sea water. The final choice of wash water type is often defined not only by quality of water, but also by the quantity of available water at refinery. Wash water injection can influence on dew point and this fact is very important for possible corrosion to occur [3].

Insufficient wash water amounts in the overhead can be counterproductive, because dissolved salts and suspended corrosion products can be remained on metal surface as water evaporates.

Intermittent wash water injection requires relatively small amounts of water, therefore it is sometimes more convenient to use boiler feed water, or steam condensate, instead of recycled overhead water condensate. Continuous water addition is spread more often. Wash water rate should be monitored continuously together with analytical measurements of overhead water condensate collected in the accumulator (pH, iron and chloride content), corrosion on-line monitoring and corrosion coupons, and of course water hardness (calcium and magnesium ions) should be checked (once a week at least) to detect cooling water leaks. Not always concentration of total iron in water in the overhead accumulator can be an indicator of corrosion because ferrous ions (Fe^{2+}) react with sulfides (S^{2-}) forming iron sulfide (FeS) deposits on carbon steel surface.

Main column at the FCCU. Corrosion problems can occur after the main column at the FCCU. Overhead stream contains cracked naphtha, light hydrocarbons, hydrogen, and contaminants (water vapor, H_2S, NH_3, cyanides, chlorides, and phenols). All contaminants are polar substances and soluble in water. Concentration of H_2S can reach 2 wt% and cause general and pitting corrosion, and hydrogen damages. Appearance of corrosive species in the overhead of the main column can be crucial, especially if the concentration of sulfur in the feed is greater than 1 wt% or if the nitrogen concentration is greater than 0.1 wt%.

Water condensate with dissolved corrosive species after main column can become highly corrosive. Therefore, using of wash water in the overhead of the main column at the FCCU can dilute or even remove all these corrosive contaminants. It is possible to add ammonium polysulfide in conjunction with a water wash to convert cyanides (CN^-) to thiocyanates (CNS^-) and thus to reduce corrosion (see Sect. 8.2.2). When wash water is injected, the pH can be controlled in the range 8–9. These pH values are explained by the fact that ammonia is more soluble in water than hydrogen sulfide and hydrogen cyanide and usually is present in higher concentrations. If hydrocarbons in the feed contain a lot of nitrogen and little sulfur, the pH of aqueous phase will be higher than 9. In this case, the pH can be lowered by increasing the wash water rate. On the contrary, if the feed contains low nitrogen, the pH of the condensed water will be low and can be increased by decreasing the wash water rate. When high concentrations of acids H_2S and HCN are formed, pH diminishes below 8, and the wash water rate can be decreased to throw out more acids.

Hydrogen sulfide can be oxidized by dissolved oxygen containing in wash water with formation of elemental sulfur and sulfurous acid (see Appendix D). If concentration of elemental sulfur in gasoline greater than 1 ppm, this gasoline is deteriorated ('corrosive') and will not pass copper strip corrosion test [2]. The reaction between rust Fe_2O_3 and H_2S can explain why after several days during shutdown and opening tanks containing gasoline it becomes 'corrosive' (see Appendix D, Equation D.11).

In addition to diminishing corrosion and fouling, wash water injection at the FCCU can solve the problems of hydrogen blistering, the accumulation of heat stable salts in the amine treating units, and deterioration of gasoline by sulfur compounds. Hydrogen sulfide in this water can cause severe general and pitting corrosion of carbon steel tubes in the overhead heat exchangers (in spite of the presence of ammonia which is the corrosion inhibitor of carbon steel). Carbon steel tubes can be completely corroded after 4 years of use. Both hydrogen sulfide and ammonia are corrosive to copper, nickel and zinc alloys. If tube bundles are made from copper-nickel (70Cu–30Ni or 90Cu–10Ni) or Admiralty brass (70Cu–30Zn), they can corrode to holes during 4 years. Titanium alloys are resistant to this environment and are recommended for use as a material for tube bundles in the overhead condensers at the FCCU. In one refinery, Ti Gr.2 tubes serve not less than 25 years with success (see Sect. 8.1.1).

The source of wash water is similar to that as in the overhead at crude distillation units, even with more strict requirements to its purity, as the purpose of the wash water in the overhead of main column at FCCU is to remove corrosive contaminants (Table 8.6). Therefore if sour water is used as a source of wash water, at least 50 % must be pure water (boiler feed water or condensed steam).

Table 8.6 The requirements to the contaminants in sour wash water

Contaminant or parameter	Chemical formula	Concentration
pH		8–9
Ammonium bisulfide	$(NH)_4HS$	<0.8 wt%
Cyanides	CN^-	<30 ppm
Phenols	ArOH	50–500 ppm
Hydrocarbons	C_nH_m	<0.2 wt%
Iron	$Fe^{2+} + Fe^{3+}$	<1 ppm

Hydrotreater, Hydrocracker, HDS, and Amine Treating Units. Wash water is injected also in the air coolers and heat exchangers in hydrotreater and hydrocracker reactor effluent system, HDS, catalytic reforming, and amine treating units to remove ammonium chloride and ammonium bisulfide deposits and to dilute aqueous condensates containing these salts (see Sect. 3.1.1.1). Use of high S- and N-containing feedstocks can cause higher concentrations of corrosives and necessity of wash water injection. Condensation, evaporation and impingement are the most common causes of corrosion in injection points. Evaporation and condensation can concentrate corrosives. Thermal fatigue and caustic SCC (in the point of caustic injection) can occur. Fatigue of injection point components can arise from thermal-, pressure-, or vibration-induced stresses (see Sect. 6.3).

Catalytic Reforming Unit Water Washes. There are differences in operation between semi-regenerative, fully regenerative, and continuous catalytic reformers, namely, in amounts of organochlorine compounds injected to maintain catalyst activity and continuous or intermittent their injection. Water wash is injected continuously or intermittently to remove ammonium chloride salt deposits in the pretreater section of the reformer.

General practices for water wash, corrosion and other problems and solutions are described in [3, 23].

8.2.2 *Use of Corrosion Inhibitors*

Wide spectrum of corrosion inhibitors is used at refinery units. They can be divided on corrosion inhibitors injected into processing streams, water cooling systems, and boiler feed water. In processing streams (hydrocarbons), high molecular weight amines and their derivatives, naphthenic acid corrosion inhibitors (see Sect. 4.2.4), inhibitors for diminishing and prevention hydrogen absorption by metals (see Sect. 4.3.1.2), and polysulfides are used. Special package of chemicals for corrosion control in cooling water systems usually includes corrosion inhibitors (zinc-phosphate-phosphonate are common), biocides, and anti-scaling agents. Oxygen scavengers are used in boiler feed water. In small cooling water systems (5–30 m^3) for the cooling pumps and compressors, inorganic corrosion inhibitors (sodium nitrate, sodium silicate or sodium molybdate) are used [1]. Corrosion inhibitors in cooling water systems and two-phase fuel-water mixtures are described in [1, 2, 24–26]. We will analyze the use of corrosion inhibitors in the overhead of crude distillation system and main column at FCCU.

8.2.2.1 Corrosion Inhibitors in the Overhead of Crude Distillation System

Corrosion at the overhead of crude oil distillation units has been the main problem at oil refineries since their existence. Acid dew point corrosion, under deposit corrosion and flow enhanced corrosion are the most important localized corrosion phenomena occurring at the overhead. Usually five actions are used for corrosion control at the overhead: neutralization of a crude oil after desalter by means of caustic soda injection; neutralization at the overhead by means of ammonia/amines injection; addition of corrosion inhibitors; water washing of corrosive deposits at the overhead; and use of alloys compatible with corrosive media. Blending of corrosive crude oils with non-corrosive ones and desalting could be added to the premium tools of anti-corrosion protection of process equipment. Carbon steel is the main material of pipes, air coolers, heat exchangers, and collectors in the overhead of the crude oil distillation unit. Use of carbon steel requires injecting neutralizers and inhibitors in the overhead.

Primary amines, diamines, quarternary ammonium salts, alkyl pyridines, amides, aminoamides, imidazolines, and products of synthesis of amines with organic acids are used as inhibitors in the overhead. Corrosion inhibitors are usually long-chain (C_{12}–C_{24}) amines and polyamines. They are solids at ambient temperatures and are dissolved in hydrocarbons (heavy aromatic naphtha, kerosene, cycle oil). After dissolving the mixtures are viscous liquids. As a rule they should be diluted before injection.

Corrosion protection mechanism by amine inhibitors is following. Amine molecule ($R\text{-}NH_2$) is a polar owing to $-NH_2$ group. This group is attached to the carbon steel surface, while the hydrocarbon tail is inside of flow and does not allow water and other corrosive molecules to reach the metal surface. In other words, amines form one or two molecule protective film on a metal surface and repel corrosive substances. Therefore, they are called 'filming amines' or 'filmers'. There is a dynamic equilibrium between corrosion inhibitor in the volume of stream and on the metal surface (8.3).

$$R\text{-}NH_{2(\text{in volume})} \leftrightarrow R\text{-}NH_{2(\text{adsorbed})} \tag{8.3}$$

Barrier film can be removed by stream. Consequently, constant injection of corrosion inhibitor is needed for continuous filming renew. Iron sulfides as corrosion products can be tightly adhered if they form tenacious thin film (5–10 μm in thickness) and be integrated in adsorbed filming amine, and thus together they can improve the effectiveness of the barrier protective film. Some amine corrosion inhibitors posses by detergency properties (if contain surfactants) and as a result loose corrosion products (iron sulfides) can be removed from the metal surface. Sometimes excessive amounts of sulfides and chloride ions in the stream can interfere with barrier film formation.

Amine inhibitors are sometimes confused with neutralizing amines because both types of substances diminish corrosion but by difference mechanism and their function in corrosion suppression in the overhead is different. Both chemicals are related to amines or their derivatives. The following differences between amine inhibitors and neutralizing amines exist (Table 8.7).

Corrosion inhibitors must be applied in the correct concentrations and at the correct location. They must be injected in the middle of the pipe and not close to its wall. Some amine corrosion inhibitors in concentrated form (insufficiently diluted) can cause corrosion of carbon steel pipes (especially at elevated temperatures), or form amine salts on the surface and then can increase under deposit corrosion. Therefore, corrosion inhibitors should not be injected into any place (e.g., valve or tee) which is convenient. Such use can result in bad distribution of the inhibitor and as a result can cause corrosion. Only special injection device (quill or nozzle) should be used.

Amine corrosion inhibitors are dissolved in hydrocarbon (usually naphtha) or in water. A very fine dispersion of the inhibitor is required to assure uniform distribution in the pipe. Bad dispersion that is large droplets of injected inhibitor can impinge metal surface. Atomizing nozzles provide very fine droplets but because their orifices are generally small can be plugged by tramp material found in the delivery system (tanks and piping) unless fine filters are used. The choice of material of the injector is important because it must be suitable and usually used of SS 316. An injector in an overhead line acts as a 'cold finger' (a small condenser) and can cause condensation of corrosive substances on the outside of the injector. Therefore, it should have thermal insulation.

Table 8.7 Properties of amine inhibitors and neutralizing amines used in the overhead at crude distillation systems

Amine inhibitors	Neutralizing amines
The most of them are long-chain (C_{12}–C_{24}) amines or their derivatives, solids at ambient conditions, and should be dissolved in suitable solvent for injection	The majority of them are short-chain (C_2–C_6) amines, liquids at ambient conditions
They form adsorbed protective film on the metal surface and isolate it from corrosive species	They neutralize acid media by increasing pH into the range of 5.5–6.5
The most of them are hydrocarbon soluble and not water soluble. They can be formulated as a water-base (amine salts) or water-dispersible (with addition of surfactants) product for injection into predominantly water condensate or wash water	The most of them are water soluble. They can be formulated as a naphtha-base or naphtha-dispersible product for injection into predominantly naphtha streams

The criterion of selecting corrosion inhibitors in the twenty-first century changed significantly. Before the 1960s it was efficiency, then until the 1980s it was economy, now the main criterion is ecology [1, 2, 24, 27]. Nowadays there is the tendency to use 'green' corrosion inhibitors [24, 27]. There is no clear and accepted definition of 'environmentally friendly', or 'green' corrosion inhibitors. They must be assessed from the health, safety, and environmental point of view. Corrosion inhibitors must be low toxic (better non-toxic), biodegradable, with low bioaccumulation, and should not contain harmful elements and substances. In other words, corrosion inhibitors used nowadays should meet requirements of safety, risk, reliability assessment and quality assurance in all equipment.

8.2.2.2 Use of Polysulfides as Corrosion Inhibitor at the FCCU

Polysulfides are compounds that contain several sulfur atoms connected to each other in the middle of the molecule: M–S(–S)$_n$–S–M (where M is usually ammonium or sodium).

Polysulfides are injected at FCCU in order to prevent hydrogen damages. Polysulfides protect by two mechanisms. First, polysulfides change and stabilize the nature of iron sulfide film on the metal surface. Polysulfides transform the iron sulfides into iron polysulfides which might be more resistant to further corrosion than the original iron sulfide film. Secondly, polysulfides can react with cyanide ions (CN^-) to form thiocyanate (SCN^-) (8.4).

$$(NH_4)_2S_{x(aq)} + 2HCN_{(aq)} \rightarrow 2NH_4SCN_{(aq)} + S_{(x-1)(s)} \quad (8.4)$$

To be certain that there is no unreacted cyanide present, there must be an excess of polysulfide at all times. The excess polysulfide can be detected by the yellow color of the accumulator water. Here are following disadvantages of ammonium polysulfide use [3].

(a) The pH of the system should always be higher than 8. At pH < 8 and at T ≈ 120 °C the polysulfide decomposes into NH_3, H_2S and elemental sulfur (8.5).

$$(NH_4)_2S_{x(aq)} \rightarrow 2NH_{3(g)} + H_2S_{(aq)} + yS_{8(s)} \quad (8.5)$$

As a result, the H_2S will increase corrosion and the elemental sulfur will deteriorate gasoline and can cause equipment plugging.

(b) It is very important to inject correct polysulfide concentration. An excess of 10–15 ppm polysulfides over the stoichiometric amount of cyanide is needed. On the other hand, if the concentration polysulfide is above 30 ppm, the excess polysulfide can decompose into NH_3, H_2S and S_8 which can result in microorganisms' kill and equipment plugging in biotreating units.

(c) Polysulfide must be in the solution. Otherwise, it can plug equipment. Therefore, requirements to wash water injection are higher to assure that the polysulfide stays in solution.

If wash water acts well polysulfide injection is not necessary in the overhead main column FCCU.

8.3 Corrosion Monitoring

Corrosion monitoring of efficiency of water wash, neutralization, and inhibitors' injection includes analytical and corrosion measurements. Analytical methods include quantitative determination of concentrations of iron, chlorides, and pH of water condensate; hydrogen activity; ammonium bisulfide and residual ammonium polysulfide in sour water. Corrosion monitoring includes measurements of corrosion rates by means of weight loss (WL) coupons and ER-probes [1, 28].

8.3.1 Corrosion Monitoring at the Overhead of the Crude Distillation System

The example of corrosion monitoring at the overhead of the atmospheric and vacuum crude distillation columns is shown in Figs. 8.7, 8.8 and 8.9. Reliable data about corrosion situation are obtained when we use several corrosion monitoring methods and as much as possible points (locations) where corrosion rates are measured. Each valve in air cooler and heat exchangers in the overhead can be used for installation WL corrosion coupons and ER-probes. We should remember that corrosion is random process because both alloy and environment are heterogeneous and the latter can vary. Therefore, corrosion rates can change from point to point and with time. We reveal that when monitor the corrosion situation in different locations in the air cooler during long period. Even in one air cooler, corrosion rates may change by an order (0.03 and 0.25 mm/year) in two points located at a distance about 2 m. It is important to emphasize good correlation between WL and ER data. The latter allow to obtain information on-line in real time. Reliable ER data can be obtained if we choose correct sensing element, and this selection can be made on the basis of WL coupons preliminary corrosion rates. WL coupons give important information about chemical composition of deposits and under deposit corrosion including measurements of pitting corrosion rate (see Fig. 8.9). Sometimes general corrosion rate is under allowable value 0.11 mm/year but pitting corrosion can reach 0.55 mm/year (see Fig. 8.9e). Therefore, it is important to carry out analytical

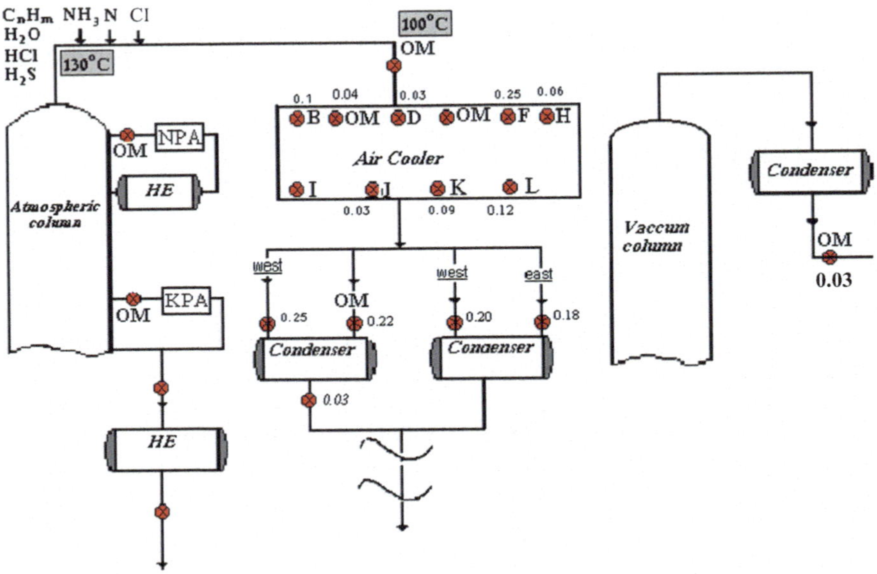

Fig. 8.7 Corrosion monitoring at the overhead of the crude distillation column. Red points—locations of WL coupons (B, D, F, H, I, J, K, L) and ER-probes (OM—on-line monitoring). NPA naphtha pump-around. KPA kerosene pump-around. HE heat exchanger. N neutralizer. CI corrosion inhibitor. Corrosion rates (mm/year) are shown near read points of the coupons' and probes' locations. Period—30 days (for this case)

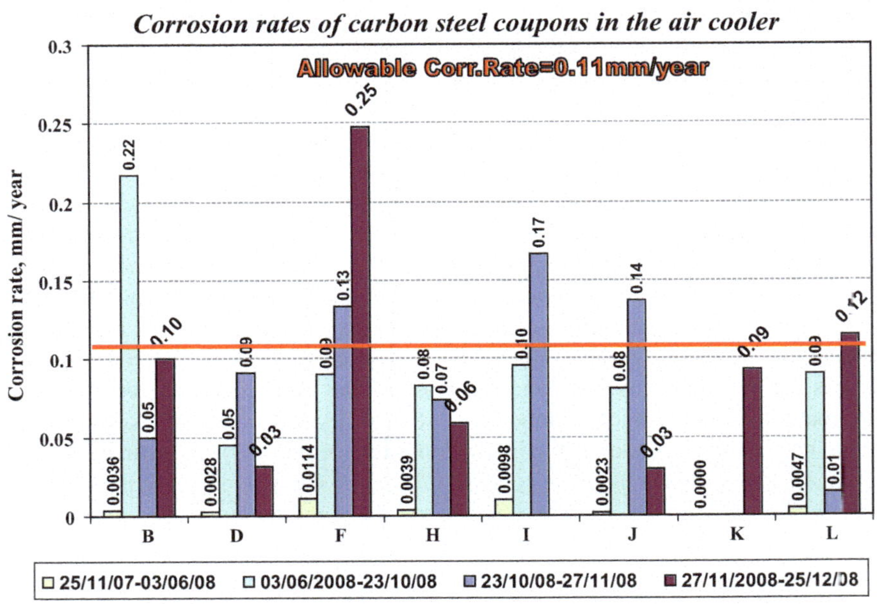

Fig. 8.8 Corrosion rates of carbon steel coupons in different locations of the air cooler (see Fig. 8.7) for different periods during 13 months

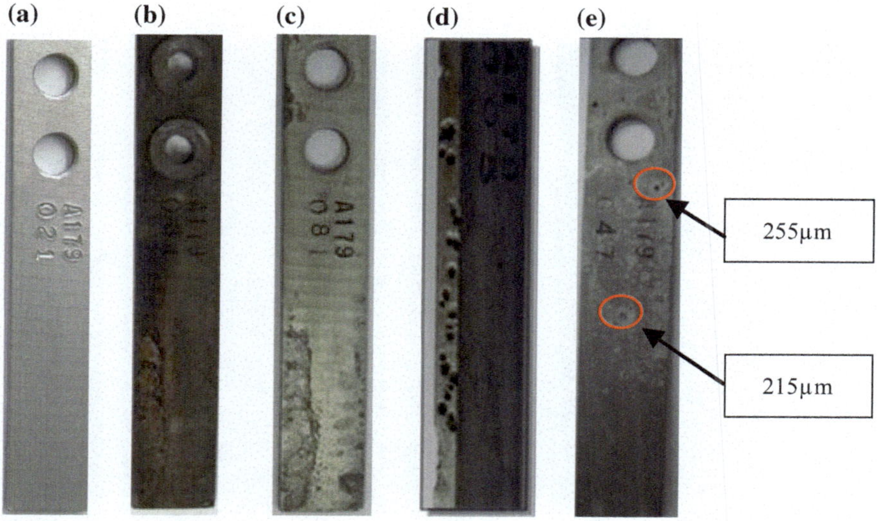

Fig. 8.9 Corrosion WL coupons. **a** Original coupon. **b** Coupon after 30 days of testing in the air cooler before cleaning. **c** The same coupon after cleaning (under deposit corrosion). **d**, **e** Pits under deposits (iron sulfide and chlorides up to 25 wt%). **e** Pits formed during 6 months (the values are the deepness of pits measured by means of optical microscope)

measurements of chemical composition of deposits on coupons and water condensate in accumulator.

These data can establish the reasons of corrosion and efficiency of anti-corrosion treatment. For instance, deposits formed on coupons contained mixture of iron sulfides (~ 75 wt%) and ammonium chlorides (~ 25 wt%) in concrete example (see Fig. 8.9).

Water condensate contained chlorides (2–290 ppm, when recommended <20 ppm), sulfates (0.5–155 ppm); pH values changed in 5.4–7.7 (when recommended 5.5–6.5). The presence of sulfates points out that oxygen was in the system and oxidized hydrogen sulfide (see Appendix D). Exposition of coupons can vary from 30 days till 6 months. This period depends on corrosion rate that we expect. The greater corrosion rate, the less duration period of coupons' exposition. After each period, report about corrosion situation, analysis, and recommendations should be send to all departments at the oil refinery.

8.3.2 Corrosion Monitoring in Petrochemical Plant

The example of corrosion monitoring at two units (BTX and TPX) in petrochemical plant is shown in Figs. 8.10 and 8.11. As a rule corrosion rates are less than allowable value 0.11 mm/year. However, sometimes corrosion rate exceeds this

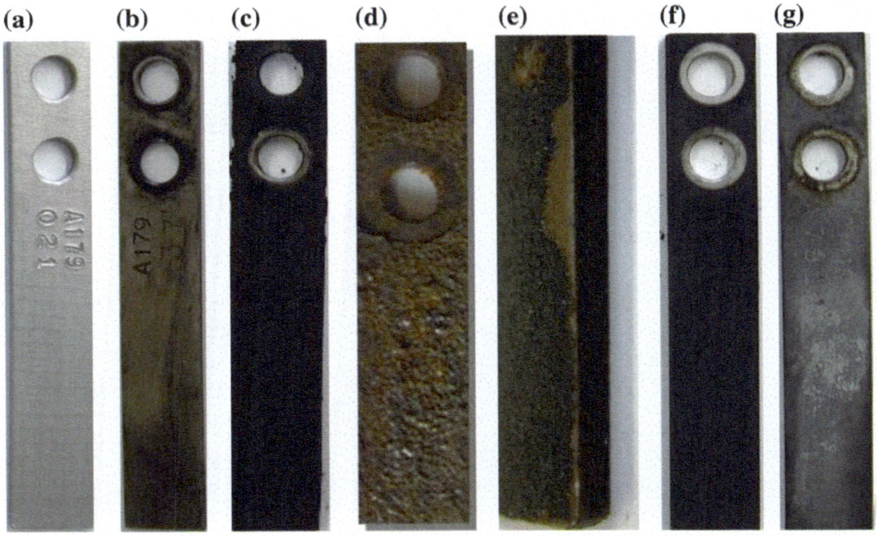

Fig. 8.10 Corrosion WL coupons after 90 days of examination in petrochemical plant. **a** Original coupon. **b–e** BTX unit. **b** Lean solvent (lower controller), 0.095 mm/year. **c** Rich solvent (extractor), 0.002 mm/year. **d**, **e** Overhead reflux, 0.113 mm/year. **f**, **g** TPX unit. **f** Tank receiver, 0.008 mm/year. **g** Aromatic tower (reflux), 0.003 mm/year

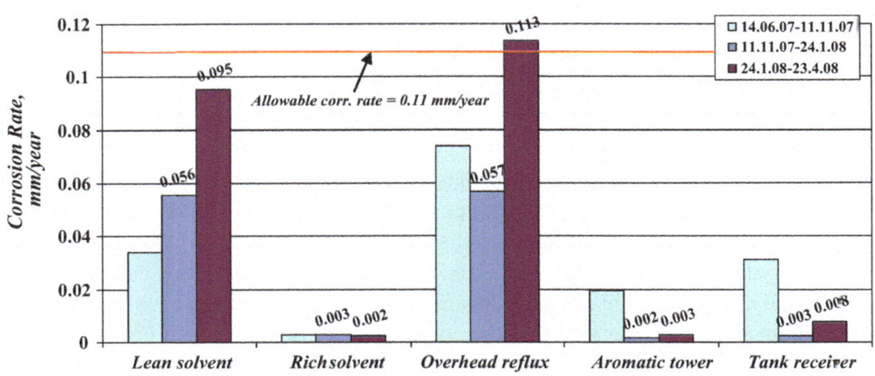

Fig. 8.11 Corrosion rates of carbon steel coupons in BTX and TPX units at petrochemical plant for different periods during 10 months

value. In these cases, corrosion products contain iron sulfides (70–80 wt%) and chlorides (20–30 wt%). Reformate feed from the CCR unit can contain chlorides. The thickness of corrosion product layer can reach 60–75 μm. This situation points out the solution how to diminish corrosion rates: purification of reformate from chloride by installation of chloride scavenger at the CCR unit.

Corrosion monitoring should be integrated with the other inspection and plant programs. These actions are an integral part of corrosion management which will be discussed in the next chapter.

References

1. Groysman A (2010) Corrosion for everybody. Springer, Dordrecht
2. Groysman A (2014) Corrosion in systems for transportation and storage of petroleum products and biofuels. Springer, Dordrecht
3. Gutzeit J (2006) Crude unit corrosion guide. PCC Process Corrosion Consultants, USA
4. Gutzeit J, Merrick RD, Scharfstein LR (1987) Corrosion in petroleum refining and petrochemical operations. Corrosion, vol 13. ASM Handbook, ASM International, USA, pp 1262–1287
5. White RA, Ehmke EF (1991) Materials selection for refineries and associated facilities. NACE, USA
6. Schutz RW (1987) Corrosion of titanium and titanium alloys. Metals Handbook, Corrosion, vol 13. ASM International, USA, pp 669–706
7. Been J, Grauman JS (2006) Titanium and titanium alloys. In: Revie RW (ed) Uhlig's Corrosion Handbook, 2nd edn. Wiley, Hoboken, New Jersey, USA, pp 863–885
8. Cotton JB, Hanson BH (1994) Titanium and zirconium. In: Shreir LL, Jarman RA, Burshtein GT (eds) Corrosion, vol 1. Butterworth Heinemann, UK, pp 5:36–5:61
9. Boyer R, Welsh G, Collins EW (eds) (1994) Materials Properties Handbook: Titanium Alloys. ASM International, USA
10. Donachie MJ (2000) Titanium: a technical guide, 2nd edn. ASM International, USA
11. Leyens C, Peters M (eds) (2003) Titanium and titanium alloys: fundamentals and applications. Wiley-VCH Verlag GmbH & Co. KGaA, Weinheim
12. Fan D, Fort WC, Horvath RJ (2000) Experience with titanium heat exchangers in refinery services. Paper 00691 presented at the CORROSION 2000, NACE International, Houston, TX, USA
13. Fort WC, Fan D (1999) Titanium experience in refining and chemical processes. http://twca.com/businesses/atispecialtyalloysandcomponents/Documents/Tech-Serv-Library/1999-Conf-Proceedings/1999002.pdf
14. Mendoza-Canales J, Marín-Cruz J (2008) Corrosion behavior of titanium and nickel-based alloys in HCl and HCl + H_2S environments. Int J Electrochem Sci 3:346–355
15. Klink DR, Schutz RW (1992) Engineering incentives for utilizing Ti-3Al-8V-6Cr-4Zr-4Mo alloy tubulars and highly aggressive deep sour wells. Paper 63 presented at the Corrosion 92, NACE, Houston, TX, USA
16. Blauvelt P, Rowan D, Willey B, McMaster M (2000) Titanium—proven performance in hydrocarbon processing. Paper 00684 presented at the CORROSION 2000, NACE international, Houston, TX, USA
17. Neill WJ (1980) Experience with titanium tubing in oil refinery heat exchangers. Mater Perform 19(9):57–63
18. Bird DB (1976) Titanium exchanger tubing works well in refinery service. Hydrocarbon Process 55(5):105–107
19. Palraj S, Venkatachari G (2008) Effect of biofouling on corrosion behavior of grade 2 titanium in Mandapam seawaters. Desalination (Elsevier) 230(1–3):92–99
20. Sigalovskaya TM et al (1976) Corrosion-electrochemical behavior of titanium and its alloys in alkaline solutions of hydrogen peroxide. Zashchita Metallov (Protection of Metals) 12 (4):363–367

21. ASTM B338-14 (2014) Standard specification for seamless and welded titanium and titanium alloy tubes for condensers and heat exchangers. ASTM International, West Conshohocken, PA, USA
22. Petersen PR (1996) The use of corrosion inhibitors in refining industry. Paper 594 presented at the NACE Corrosion 96 conference, Houston, TX, USA
23. Refinery injection and process mixing point (2001) NACE International Publication 43101. Houston, TX, USA
24. Sastri VS (2011) Green corrosion inhibitors: theory and practice. Wiley, Hoboken, New Jersey, USA
25. Papavinasam S (2006) Corrosion inhibitors. In: Revie RW (ed) Uhlig's corrosion handbook, 2nd edn. Wiley, Hoboken, New Jersey, USA, pp 1089–1105
26. Boffardi BP (1987) Control of environmental variables in water-recirculating systems. Corrosion, vol 13, 9th edn. ASM Handbook, USA, pp 487–497
27. Groysman A et al (2012) Examination of a "green" corrosion inhibitor and corrosion resistant alloys for the overhead of the crude oil distillation unit. Paper 2134 presented at the NACE 2013 conference, NACE International, Houston, TX, USA
28. Groysman A (2009) Corrosion monitoring. Corr Rev 27(4–5):205–343

Chapter 9
Corrosion Management

No one should have to sacrifice their life for their livelihood, because a nation built on the dignity of work must provide safe working conditions for its people.
Thomas Edward Perez (1961-), an American politician.

Abstract The aim of this chapter is to analyze the role of corrosion management in prevention of corrosion accidents related primarily to safety of personnel and the environment. Corrosion management is achieved by use of anti-corrosion control measures, corrosion monitoring, regular inspection, study of each accident, implementation of meetings, publications of minutes, education, and knowledge transfer. The reasons of inevitability of corrosion failures and ways for their decreasing are analyzed. The human factor is a key issue in corrosion management and reducing corrosion accidents. The reasons of humans' mistakes are the lack of awareness, education, knowledge, and training, incorrect design, insufficient control and supervision, lack of incentives to reduce corrosion risk, and wrong operation. Risk-based inspection is described as well.

The oil refining and petrochemical industry is high hazard, dealing with considerable amounts of flammable, explosive and toxic substances, which are dangerous both to people and the environment. Corrosion accidents worsen image of this industry and technological achievements. Corrosion (if not managed, namely, controlled or prevented) can result in catastrophes. Therefore, oil refineries and petrochemical plants should manage their units safely. For this, they have to make efforts to control technological processes and organize corrosion management strategies in order to diminish corrosion accident risks to a minimum.

We unite two wide phenomena-concepts: *corrosion* and *management*. *Corrosion* is an interaction between a metal and its environment that results in deterioration of both the metal, and the environment [1]. *Management* in organization is the function that coordinates the efforts of people to accomplish goals and objectives using available resources efficiently and effectively. To manage is to plan, to organize, to coordinate, to command, to forecast, and to control [2].

A. Groysman, *Corrosion Problems and Solutions in Oil Refining and Petrochemical Industry*, Topics in Safety, Risk, Reliability and Quality 32,
DOI 10.1007/978-3-319-45256-2_9

Corrosion management includes planning actions for determination of corrosion risk, performance requirements of standards, recommended practices and specifications for correct selection of materials, corrosion protection and monitoring methods. This should be carried out at the stage of design and then their realization in practice under inspection, control, and monitoring. Study of each accident with failure analysis, coordination, education, and knowledge transfer are also necessary components of *corrosion management methodology*.

Thus, corrosion management is a wide spectrum of measures and procedures of interaction between different services and specialties, from a worker to a head, from a chemical engineer to a purchaser and a lawyer.

Corrosion management program does not necessarily require establishing a "zero corrosion rate" system, but rather establishing operating limits, monitoring (including corrosion monitoring) and inspection plans to assure acceptable corrosion rates which allow the system to function reliably and safely.

Corrosion management appeared immediately after the estimation of the cost of corrosion in different industries and countries beginning since 1949. It was shown that without all the above mentioned actions corrosion will worsen the economy of countries [3–5]. In fact, some of the functions of corrosion management appeared much earlier. For instance, corrosion control actions were the first in long list: coatings have been used since ancient times, cathodic protection since 1820s and intensively in 1920s, corrosion inhibitors in 1930s, the first text-book on corrosion was published in 1924 [1]. *Corrosion management* entered and occupied such an important place in the policy of many companies that this word combination is under the title on websites of some companies [6]. Many publications about corrosion management appeared during the last 10 years [7–25]. In spite of numerous publications in this field, not all specialists dealing with the subject of corrosion in oil refining and petrochemical industry consider corrosion management as a niche (probably even discipline) combining multi-faceted and interdisciplinary topics. These topics include chemistry, metallurgy, instrumentation, anti-corrosion measures, corrosion monitoring, education, knowledge transfer, and psychology.

It was established that humans were responsible in 65–85 % of corrosion cases in oil refining industry [16–19]. In this chapter, the role of corrosion management in prevention of corrosion failures will be discussed. The reasons of inevitability of corrosion failures and ways of their decreasing are analyzed. The human factor is a key issue in corrosion management and diminishing corrosion accidents. The causes of humans' mistakes are the lack of awareness, education, knowledge and training, incorrect design, insufficient control and supervision, lack of motivation and incentives to decrease the corrosion risk, wrong operation, and element of change.

Improving corrosion management we reduce the corrosion risk in oil refining and petrochemical industry and its negative influence on the environment and society. It is suggested to establish legislation about corrosion management on the state and federal levels (like legislation about pollutions and environmental protection), to improve education and knowledge transfer. Penalties of management can force them to put more attention on corrosion problems and make it priority issue at oil refineries and petrochemical plants.

9.1 Corrosion Failures

Success is the result of learning from failure.
Colin Powell (1937-), an American statesman

Corrosion is an inevitable process in accordance with the 2nd law of thermodynamics. How to resist unavoidable natural route? Our task is to understand these paths, their mechanism, influence of different inner and outer factors, in order to maintain service functions of metallic constructions during their economic life period, usually 10–25 years [26]. Unfortunately, equipment in some cases is not used the time period nominated for them. The reason is *corrosion failure*. *Failure* is a sudden cessation of functioning, or a lack or deficiency of a desirable quality. The synonyms of *failure* are malfunction, catastrophe, disaster, or collapse. Certainly, the level of corrosion failure and its consequences may be different: the leak of crude oil, petroleum products, fuels, water, liquid and gaseous hazardous chemicals, fire, explosion, damages and deterioration of the environment, injuries, and even death of people. The paradox is that in most cases, corrosion failures occur as a result of corrosion phenomena well known and registered several times [27–36]. Corrosion can result in diminishing metal thickness, pits, and cracks, which in their turn can give rise to loss of functionality of equipment or construction, or in other word to *failure*.

The aging of oil refinery equipment, wide variation in crude oil type, high corrosiveness of media and harsh technological parameters result in high corrosion risks. Important analysis of 99 corrosion-related accidents occurring in refineries in 1962–2012 was carried out by European Commission in Institute for the Protection and Security of the Citizen in Italy [37]. This analysis showed that corrosion failure was responsible for one of each five major refinery accidents that have occurred in European countries since 2000. Thus, refinery equipment is at particular risk of corrosion failures. Half of the accidents had very serious consequences on the environment, economics of the refinery, and surrounding community. Significant corrosion failures occur either because the corrosion hazard and risk was not identified or ignored.

In spite of the achievements in corrosion science and corrosion technology, many questions still have no answer. Here are some of them:

(a) Can we predict the date of corrosion failure of an equipment or a construction?
(b) What is the minimum flow velocity of sea water which will not cause pitting corrosion of stainless steels?
(c) How to calculate the corrosion rate of MIC in storage tanks containing crude oil, petroleum products, and fuels?

Why do corrosion failures occur? Corrosion phenomena in most cases and corrosion failures in all cases are stochastic processes, namely, random processes, representing the evolution of a system metal-environment of random values over time. This means that in a stochastic process there is some indeterminacy: even if the initial conditions are known, there are several directions in which the corrosion process may evolve. The past of corrosion system in most cases cannot help forecast its behavior in future and possible failure. The corrosion phenomena depend on many inner (properties of metals) and outer (properties and conditions of the environment) factors. Inner factors are physicochemical, metallurgical and mechanical properties of metals. Heterogeneity of alloys and the environment (including the changes of the latter) can give arise to unpredicted corrosion rate and as a result to unexpected failures. Our knowledge about the behavior of the metal-environment dynamic system is not perfect. Any variation in properties of both can lead to unknown and unforeseen behavior of the metal.

Existing anti-corrosion protective measures do not guarantee complete protection and prevention from failure. For instance, use of corrosion inhibitors usually provides protection efficiency lower than 100 %. Protective properties and as a result durability of coating systems depend on so many factors (including correct surface preparation and application) that in rare cases needed conditions for providing protection without failure can be achieved. Most technological regimes and properties of media at refinery units are not keeping constant. Small changes in these conditions can drastically influence efficiency of use of corrosion inhibitors, cathodic protection, or coating system.

Existing corrosion monitoring methods do not reflect real situation and the behavior of the metal-environment system. Any corrosion sensor shows how this sensor reacts with the environment and not real equipment behavior. Even when we place coupon, ER-probe or LPR (Linear Polarization Resistance)-sensor into the media, their surface and "history" are different from the surface and "history" of real pipe, tank, reactor or equipment wall.

Corrosion management including many topics and sub-issues is not efficiently and effectively used in oil refining and petrochemical industry. Here are main subjects of insufficient corrosion management:

(a) Specialists are not always familiar with existing standards, codes of practice, specifications, recommended practices, reports, in short, all literature related to corrosion problems. As a result, these documents are not used for selection of correct materials, corrosion protection and monitoring at the design stage, and then are not applied for realization of corrosion mitigation, prevention and monitoring methods in practice during service life of equipment and constructions.
(b) Inspection and control measures are not always carried out on a regular basis and in time.
(c) About 30 % of corrosion accidents are not registered in most plants [38], and as a result failure analyses are not carried out. Thus, mistakes and corrosion failures are repeated.
(d) The *human factor* plays a vital role in *corrosion management* and as a result in prevention corrosion failures. We should pay significant attention on education, dissemination of information, knowledge transfer, collaboration, communication, documentation, writing down reports, implementation of

meetings, publications of minutes, and their distribution. Due to purely biological and psychological properties of human nature, we are unlikely to achieve perfection in humans' behavior and fully exclude mistakes.

Thus, we come to the conclusion that corrosion failures are inevitable as most phenomena in nature. For instance, earthquakes, tsunamis, and typhoons. Learning deeper and understanding mechanism and origins of these processes we will be able to predict and probably to control their occurring. Similar situation occurs with corrosion failures, even more complicated, because in addition to understanding of the principles of corrosion we should take into consideration *corrosion management* and *human factor* as its essential component.

We will describe and explain deeper the reasons of corrosion failures inevitability and then we will give the ways of their diminishing. First, physicochemical nature of a system metal-environment gives arise to thermodynamic possibility of corrosion.

9.1.1 *Thermodynamic Possibility and Inevitability of Corrosion*

If to put a piece of pure iron into pure water contained dissolved oxygen, rust is formed (reaction 9.1).

$$4Fe_{(s)} + 6H_2O_{(l)} + 3O_{2(g)} \leftrightarrow 4Fe(OH)_{3(s)} \quad \Delta G^\circ_{298} = -1397.2 \text{ kJ/mol} \qquad (9.1)$$

Change of Gibbs energy ΔG°_{298} is negative for (9.1), and thermodynamics shows that corrosion is possible. Put a piece of pure iron into pure water which does not contain dissolved oxygen (reaction 9.2).

$$Fe_{(s)} + 2H_2O_{(l)} \leftrightarrow Fe(OH)_{2(s)} + H_{2(g)} \quad \Delta G^\circ_{298} = -15.7 \text{ kJ/mol} \qquad (9.2)$$

The change of Gibbs energy for (9.2) is also negative. Thus, corrosion is also possible in the absence of dissolved oxygen in water. Certainly it is possible to prevent the occurrence of corrosion reactions (9.1) and (9.2), if to change conditions: concentrations of reactants, temperature or pressure. Removing dissolved oxygen in (9.1) we can significantly decrease its occurrence and even stop it. Increasing hydrogen pressure more than 570 bar in (9.2) we can stop this corrosion process [1]. Thus, we should apply energy to prevent corrosion.

When the surface of iron is coated or corrosion inhibitor is injected, 'metastable' film can protect iron from corrosion. The term 'metastable' means 'quasi-stable', not absolutely stable, namely, stable at its present condition and can exit this state at any moment as a result of changes of outer conditions or inner properties of the metal-environment system (Fig. 9.1).

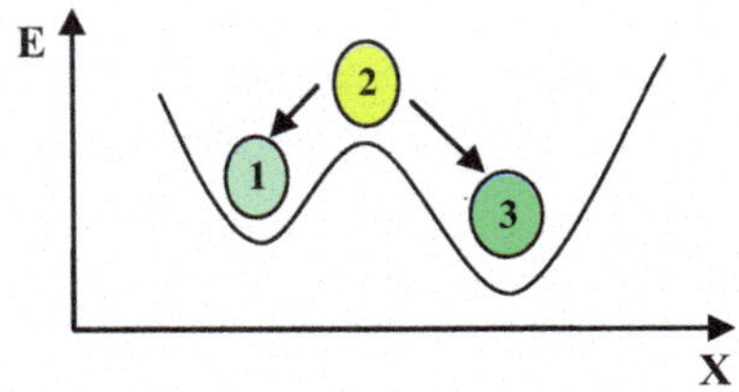

Fig. 9.1 Difference between stable (*1* and *3*) and 'metastable' (*2*) state. *E*—the energy of a system. *X*—the direction of the reaction

Any change, for instance, temperature, flow rate, agitation, appearing and concentration of aggressive components in the environment, type and concentration of inhibitor, and appearing of stresses in metal can force the system to leave 'metastable' state and turn into active corrosion process. Really, metals in 'metastable' state thermodynamically can corrode, moreover, suddenly (peculiar to failure), and this is a matter of time. Pitting, crevice and intergranular corrosion, SCC, erosion and cavitation, corrosion fatigue are developed from 'metastable' state.

In spite of thermodynamic possibility of corrosion of some metals (Fe, Al, Mg, Ti, Cr, Ni) in some media under certain conditions, kinetic restrictions do not allow realization of corrosion because of passive film formation. This means that kinetic restrictions prevail over thermodynamic possibility.

9.1.2 *Kinetics of Corrosion*

Time does not exist in thermodynamics. Thermodynamics does not consider corrosion kinetics—the study of rates and mechanisms of corrosion reactions—the main parameter and issue that we are interested in order to predict corrosion failure. Only in some particular cases, namely, general corrosion of amphotheric metals in acids and in alkalis can be predicted, when corrosion losses are subject to a strict linear dependence on time. There were suggested empirical kinetic equations for uniform and pitting corrosion (9.3) and influence of concentrations and temperature on corrosion rate (9.4) [39]:

$$\mathrm{CR} = \mathrm{AC}^{\mathrm{n}} \tag{9.3}$$

$$\log \mathrm{CR} = \mathrm{K} \pm n \cdot \log \mathrm{C} - \mathrm{A}/\mathrm{T} \tag{9.4}$$

where CR is a corrosion rate; A, n, and K are experimental constants; C is a concentration of an aggressive reactant; T is an absolute temperature.

Corrosion rates decrease with time in most media at oil refineries (aqueous solutions, two-phase mixtures hydrocarbons-water, atmosphere, and soil, see Fig. 4.1) [26]:

$$CR = 0.5 \cdot t^{-0.5} \tag{9.5}$$

where CR is corrosion rate, g·m^{-2}·day^{-1}; t is time, hours. An exception is erosion.

Diverse equations were proposed for estimation corrosion rate of different alloys as a function of influence of concentrations of sulfur compounds containing in crude oil (high-temperature sulfidic corrosion) [40]:

$$CR = a \cdot b \cdot c \cdot d \cdot C_{TRS} \tag{9.6}$$

where a, b, c, d are coefficients taking into consideration type of steel, nature of the hydrocarbon stream (vapor, liquid or two-phase flow), velocity, flow regime (degree of turbulence), and temperature; C_{TRS} is concentration of total reactive sulfur (wt%). It is clear that all these coefficients are not easy to define.

Corrosion rates defining by weight loss, ER, LPR, or any other method are very approximate, possess certain limitations and are changed from one location to another in the unit, even in the same equipment (see Sect. 8.2.3) All corrosion charts and guides contain squares with 'questionable resistance' of alloys [41–44]. 'Iso-corrosion rates' figures can be used for apparent estimation of corrosion rates. New media and new materials appear all time, for instance, new crude oils, oil sands, different biofuels; alloys, corrosion inhibitors, and paints. The slightest impurities, the lack of control of appearing of corrosive species, aeration, galvanic couplings, tensile stresses, surface conditions, disturbances of seals, and changes of technological parameters, constructions and schemes (e.g., adding new pumps) can significantly change kinetics of corrosion. Most chemical compounds when their concentration is increased (e.g., H_2S, salts, dissolved oxygen, peroxides) non-linearly influence corrosion rate. Sometimes presence of two or more substances can accelerate or inhibit corrosion (e.g., H_2S can inhibit NAC). All corrosion reactions are heterogeneous (take place on the phase boundary), complex, and contain many steps. Kinetic parameters of any corrosion reaction can be determined only providing experiment which is nearly impossible to repeat under identical conditions. All metals and alloys have inclusions and defects which cannot be defined on the micro (atomic) levels when metals are used as whole constructions. For instance, if we produce an iron of high purity (99.99999 wt% Fe), this means that there are 10^{15} impure atoms in one gram of iron [1]. This electrochemical heterogeneity influences significantly on corrosion kinetics and as a result on a corrosion failure time. It is nearly impossible to predict existence of defects in alloys and thus practically difficult to foresee their behavior.

Metal-environment systems belong to dynamical ones which can be described by the chaos theory. The latter is the field of mathematics and studies the behavior of dynamical systems which are highly sensitive to initial conditions: small differences in initial conditions yield widely diverging outcomes, rendering long-term prediction impossible in general [45]. Chaotic behavior exists in many natural systems and phenomena, such as weather, climate, corrosion, and even … politics and history [46]. There are two types of chaotic systems. In the first type, chaos does not react to predictions about it. The weather belongs to such systems. In spite of many-many factors influencing weather, it is possible to create mathematical models that take more and more of variables into consideration, which allow produce better and better weather forecasts and for longer period. Mathematical modeling is also used for the prediction corrosion rates and the life estimation of metallic constructions in different systems [47–52]. The more data (about metal and environment) will be included in corrosion model, the better prediction of its behavior will be reached. This is a mutual problem of corrosion and computer science. However, there are no yet satisfactory kinetic theories and models for most corrosion phenomena, and as a result it is impossible to predict yet occurrence of corrosion with enough high probability, and eventually corrosion failures are little predictable.

Corrosion systems are complex, non-stable, nonlinear ones. Corrosion phenomena in most cases relate to the first type of chaotic systems (like weather) that does not react to prediction. Like with forecast of weather, we sometimes can predict the behavior and state of corrosion system on short-term period. However, the prediction on long-term period is made with low probability. It is very difficult to take into consideration all the initial conditions of a complex system in sufficient details. Therefore, we cannot hope to predict the final fate of corrosion system yet for a long period. Even slight errors in measuring the state of a non-stable corrosion system or lack of data make difficult long-term behavior prediction.

In the second type of chaotic systems, chaos reacts to predictions about them, and therefore can never be predicted accurately. Markets, politics, history relate to such systems.

The conclusion is that both thermodynamics and kinetics may only partly and roughly answer some of our questions (see Sect. 9.1).

Unfortunately, we cannot exclude human mistakes in corrosion accidents, even when accurate mathematical models will be created. The role of knowledge management and human factor as main components in corrosion management and occurrence of corrosion failures are discussed in the next section.

9.2 Knowledge Management and Human Factor in Oil Refining Industry

Cadres decide everything.

The motto of communist leaders in the former USSR

Humanity faces severe global problems of health and environment. New problem of a shortage of skilled knowledgeable specialists in every sphere have appeared. Educated corrosion experts will be in shortage in the near future too. One can see the aging of the workforce in any country. The number of elderly experts after the age of 60 is increasing while the influx of young people (at the age of 25–30) who would have to become experts is decreasing [38, 53]. In order to work out this problem, we should improve knowledge management which includes education and knowledge transfer from older specialists to the younger generation. Knowledge management is an important component of corrosion management. The term knowledge management began be popular in the late 1990s and early 2000s, becoming one of the main buzzwords of the time.

Information and human knowledge in modern world like snowball' drastically increases according to exponential growth and doubles every 1.5 years [54]. There are two kinds of knowledge: explicit and tacit. Explicit knowledge contains in books, encyclopedias, and journals. Tacit knowledge contains in memory of specialists, is gained and learned with experience, usually not documented, is hardest to find, easiest to lose and, often, most valuable. Tacit knowledge is not easy to get, as many experts are reluctant to share it.

9.2.1 Education and Knowledge Transfer

Corrosion education for the oil refining and petrochemical industry is of paramount significance. Usually students studying chemistry, chemical engineering, refining and petroleum engineering, oil and gas engineering, mechanical engineering, materials science and engineering, metallurgy, learn the subject of corrosion briefly or do not study this subject at all. The most refineries do not have position of corrosion engineer. Usually an engineer becomes the corrosion engineer during his work at a plant. It takes a long time until a specialist will be familiar with all the processes and technologies, standards, codes of practice, specifications, reports, recommended practices and other documentation in the plant and with all problems in the oil refining and petrochemical industry.

The cost of corrosion education was estimated less than 1 % of the total cost of corrosion in the USA [55]. Survey among corrosion specialists in the USA showed that three-quarters of the respondents were between 41 and 65 years old, with 41 % between the ages of 51 and 65 [56]. "Ageing" of corrosion specialists occurs in other countries too. Therefore, corrosion world of specialists needs in young generation. This is common paradox. On the one hand, we want to catch the attention of new generations of students and engineers to be involved in corrosion science and engineering. On the other hand, there is real shortage of specialists in corrosion teaching and education, and we can mark only some publications which deal with this problem [38, 53, 57–62].

Knowledge transfer includes identification and evaluation, validation and documentation, publishing and sharing, transfer and application, learning, capturing and storage the knowledge. Knowledge transfer methods include formal education and training, seminars, verbal communication, discussions, regular meetings, conferences, seminars, mentoring, apprenticeships, interviews, memos, storytelling, research papers, reports, video, publishing on websites, intranets, newsletters, simulations and games, excursions, instant messages, peer assists, job transfer, wikis, and blogs. Knowledge is complex and diverse, is needed in constant replenishment of new information and its treatment.

Most oil refineries do not have a plan to manage the transfer of knowledge. We should take into account that usually conflict between generations exists: about 50 % of employees have no respect and good relationships among generations [63]. It was described how to capture and transfer knowledge across generations and create useful network to absorb and keep knowledge in oil refining and petrochemical industry [38, 53]. The more employees know about corrosion problems and their solutions, the better will be anti-corrosion situation and lower corrosion risk.

Four target audiences for corrosion teaching and education in oil refining and petrochemical industry are considered: (a) practiced engineers and technical personal, (b) educators, (c) managers, and (d) students. Our experience shows that using different curricula for these four groups of people it is possible to improve awareness, knowledge and thus decrease corrosion risk and failures. We emphasize that educators should be instructed how efficiently and effectively teach different groups of listeners.

9.2.2 Human Factor

The reasons of humans' mistakes are:

a. Lack of awareness, knowledge, education, and training.
b. Incorrect design.
c. Wrong operation.
d. Insufficient control and supervision.
e. Lack of motivation and incentives to reduce corrosion risk.
f. Element of change.

Eventually people decide what to do in each situation when corrosion can happen or occurs, namely, corrosion risk exists. Humans are responsible up to 85 % of corrosion accidents in oil refining and petrochemical industry [16, 17]. About 75 % of all corrosion failures had happened because of insufficient information, communication, interaction, and knowledge. Staff usually does not report about 30 % of corrosion accidents [38].

Penalties and financial incentives can drastically change situation in decrease of corrosion failures. More attention should be paid to human factor in order to decrease corrosion failures. This can be made by several ways: first, set up of legislation in the field of corrosion management; second, improve education and knowledge transfer.

Corrosionists concentrated on awareness about corrosion for public and governmental organizations, showed economical losses in different industrial spheres and countries. It is necessary to go forward to legislation on the state and federal levels, namely, to establish penalties for the lack of anti-corrosive preventive actions and corrosion monitoring (similar to penalties for pollutions of the environment). Penalties and fear of courts can force management to put attention on corrosion problems and make them priority issue because corrosion failures are related to safety of personnel and the environment, and this cost prevails significantly more than the cost loss of equipment and production because of corrosion.

9.2.2.1 Documentation

Refineries order and obtain equipment made from different materials. Unfortunately not always refineries require and insist on accompanying technical documents about chemical, physical and mechanical properties of materials. Sometimes equipment (mostly aboveground storage tanks containing different petroleum products and fuels) are not inspected and examined during long period (10–20 and even more years), and thus are exploited with "closed" eyes … till failure [30].

Use of standards, codes of practice, recommended practices, and specifications. Corrosion management includes also constant pursuit of standards, codes of practice, recommended practices and specifications, and updating them in the field of corrosion: NACE International, ASTM International, API, ANSI, EN, ISO, SSPC, GOST (Russia), BS (UK), DIN (Germany), and AFNOR (France). Some of them give general recommendations and approach to the problem, for instance, use of coatings for tanks containing crude oil and fuels, or in atmospheric conditions. In such cases, refineries should create their own specifications applied to concrete climate and industrial conditions [30].

9.2.2.2 Element of Change

One of the problems of human factor as a part of corrosion management is the *element of change*. Examples of such element of change in oil refineries are a change in types of crude oil, demulsifiers, neutralizers, corrosion inhibitors and their dosage, wash water place injection and rate, addition new pumps, valves or heat exchangers, changes in their function, changes in flow regime and rate, appearance of dead legs, use of water with unknown chemical and microbiological composition. Any changes of equipment design, operations, processes, treatment, conditions, materials, and chemicals, additions or removing of equipment, in short, any interference in the operation and process, should be examined by corrosionist expert for identification of a new corrosion risk. Occasionally changes carried out during shutdown and repair are not recorded. Each equipment should have passport like a person: the date of born, chemical, physical and mechanical data about equipment materials and all changes which are made during its service.

9.2.2.3 Examples of Human Errors

Who is Guilty?

Herzen A.I. (1812–1870), a Russian writer and thinker

Despite of numerous cases described in literature [1, 26–36], information about new corrosion failures is dispersed in reports and papers, some of them are not reported at all and exists as tacit knowledge. Several corrosion cases are presented in this section as examples of human mistakes (Figs. 9.2, 9.3, 9.4). The conclusion from these lessons is that using corrosion management as complex actions with an emphasis on the human factor and abiding proper relationship to corrosion

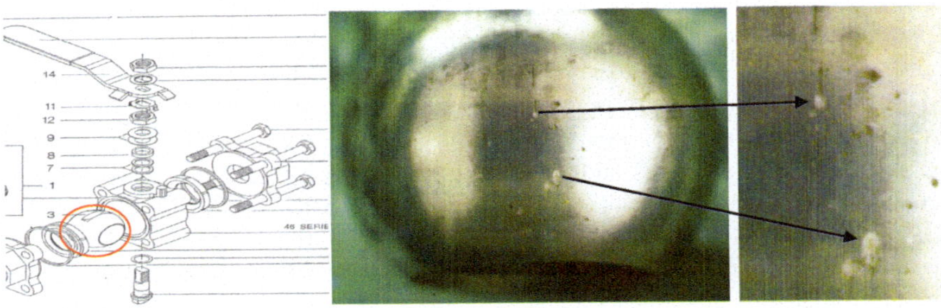

Fig. 9.2 Pitting corrosion of SS316 valve ball by chlorides after 3 months of contact with sea water (stagnation; fire-fighting system; ambient temperature). It was possible to select ball material from Naval Brass UNS C46400 resistant to pitting corrosion in these conditions. Reason: incorrect design, lack of awareness and knowledge. Extremely very high risk t < 1 year

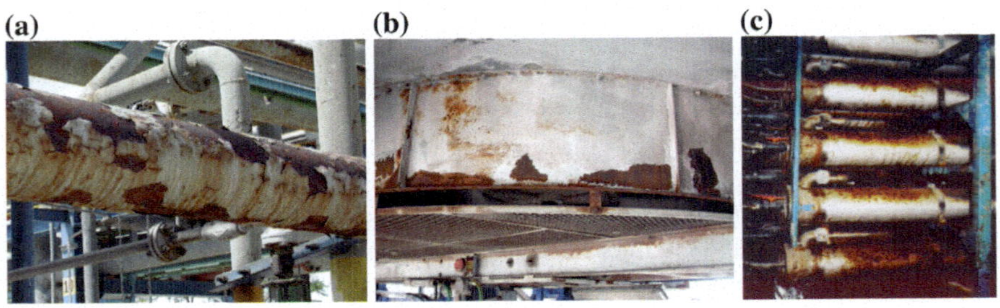

Fig. 9.3 **a**, **b** Failure of coating systems after 2 years of service. **a** Pipeline with flowing water from the desalter (90 °C) in the atmosphere of the oil refinery. **b** Vent under the air cooler (80 °C). Reason: incorrect design—coating systems (not resistant to high temperatures) were selected wrongly. **c** Failure of SS304 pipes after 2 years of service. Storage tank situated in 50 m from the SS pipes was filling with HCl acid twice a week. Volatile HCl acid gas from the tank was condensed together with water vapor on the SS pipes surface and attacked it. Reason: lack of awareness and knowledge; wrong design and operation; insufficient control. **a**, **b** and **c** very high risk 1 < t < 3 years

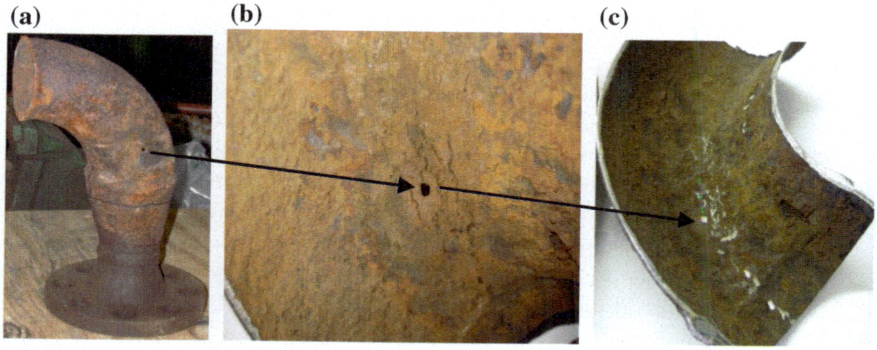

Fig. 9.4 Oxygen corrosion of reducer (pipe for naphtha inlet to the furnace at 110 °C and 4–10 bar) after 16 years of service. Reason: lack of awareness and knowledge, insufficient control. Low risk t > 10 years

problems many corrosion failures can be prevented and as a result corrosion risk will be diminished. Based on the time (t) estimation to the failure (e.g., crack, thinning of the thickness of equipment less than allowable value, loss of protective properties of coatings, leak) [20], and our experience, it is possible to suggest following qualitative categories of corrosion risk (without consequences): extremely very high risk t < 1 year; very high risk 1 < t < 3 years; high risk 3 < t < 5 years; medium risk 6 < t < 10 years; and low risk t > 10 years [64].

9.3 Corrosion as a Hazard

Analysis of corrosion cases at Fig. 9.2 (extremely very high risk, t < 1 year before the failure) and at Figs. 9.4 and 4.16 (low risk, t > 10 years before the failure) lead to the paradox and wrong decision about inspection and monitoring for corrosion situation. The problem is that consequences were not taken into consideration. Pitting corrosion of the valve ball after 3 months of service (see Fig. 9.2) lead to the leak of water in the fire-fighting system and if this leak was not detected in time, the fire-fighting system could remain without water which in its turn in the case of fire would lead to large destructions and disaster. The hole in the reducer (after 16 years) caused the leak of naphtha and fire. Hydrogen embrittlement of carbon steel pipe (after 16 years, see Fig. 4.16c) where hydrogen gas flowed out at ~300 °C and 46 bar lead to explosion and fire. The consequences were several millions $. Thus, if this corrosion accident happened after 16 years of service this does not mean that we can talk about low risk (t > 10 years). In order to overcome such contradictions *risk-based inspection* (RBI) was developed in 1993 and introduced into practice in the end 1990s–beginning 2000s [65–67]. Certainly the use of RBI helps in diminishing and prevention corrosion failures.

9.3.1 Risk-Based Inspection

In order to explain the principles of RBI, we will give definitions of some terms which were used previously. A 'hazard' is a substance, an object or a situation with a potential for an accident or damage. Corrosion can be considered as one of the biggest 'hazards' in oil refining and petrochemical industry, because it related to substances (including environment), objects (equipment and constructions made from alloys and other materials) and situation leading under suitable conditions to interaction between them (equipment and environment) and deterioration of both. A 'risk' is the likelihood that the accident or damage will occur [68]. Risk is the result of the probability of an event occurring during a period of interest times the negative consequences of the event (9.7):

$$\text{Risk} = \text{Probability} \times \text{Consequence} \tag{9.7}$$

In contrast to 'hazard' which is a real essence, a 'risk' is a mathematical construction which is not simple to evaluate. Risk analysis (named also hazard analysis, risk assessment, probabilistic risk assessment, quantitative or quantified risk assessment) came from the nuclear and chemical industries [68], was standardized [65, 66, 69–71] and found wide application in corrosion management [20–25, 72]. Similar to hazard identification and assessment in chemical industry [68], it was suggested to use *corrosion risk management* including identifying, analysis, assessment and management of corrosion hazards [20, 65, 66]. RBI is a methodology that uses risk analysis to prioritize and manage the inspection program for plant equipment [65–67]. The aim of RBI is to develop a cost-effective inspection program that limits the risk associated with potential failures of equipment at oil refinery. Inspection and maintenance program should ensure reliable assessment of corrosion risk and solving corrosion problems in order to carry out processes safely. Two factors are taken into account: the probability of failure and its consequences. There are four levels of risk-based assessment: lower risk, medium risk, medium-high risk, and high risk. They can be evaluated as qualitative, semi-quantitative, and quantitative [67]. The qualitative estimation is used to determine risk associated with whole or large portions of process units. The semi-quantitative and quantitative evaluations are used to determine risk associated with individual equipment pieces. Thus, identifying the higher-risk equipment we can concentrate inspection and maintenance on these items in order to diminish the risk. Two standards [65, 66] give detailed methodology of definition of probability of failure and its consequences.

The probability of corrosion failure depends on the rate of corrosion process and the effectiveness of the inspection. The type of inspection method and its frequency affect the efficiency of the inspection and thus probability of failure.

Consequence of corrosion failure is the second important component of RBI (see Eq. 9.7) and reflects an impact of release of hazardous media might have. An impact can be explosion, fire, destruction, and exposure to toxic substances. Three factors play leading role in evaluation of consequence: the type of medium and its hazard that may be released, its stock available for release, and the rate of release [67]. The latter is defined by the dimensions of the breaches (holes and cracks) in equipment. The 'consequence' is not easy to evaluate if many breaches are formed simultaneously.

The standards [65, 66] suggest three methods of realization RBI: optimizing the monitoring and inspection plan, changes of materials construction, and control of key process parameters.

The monitoring and inspection plan includes definition of the most vulnerable locations where corrosion can occur and intervals for examination (see Sect. 8.2.3). This concerns on-line real-time and periodical measurements of corrosion rates by means of ER-probes, LPR-sensors, WL coupons, ultrasonic, eddy current, acoustic emission and radiography measurements, physicochemical analytical determination of composition of media (including corrosive contaminants, microorganisms, etc.) and alloys, metallurgical examinations [1, 73].

The change of materials construction. Suitable materials are chosen on the stage of design (see Sect. 8.1). This selection is usually made by chemical and mechanical engineers. Unfortunately corrosion and materials engineers do not always take part in this step, although it is necessary that each project should be signed by them. This issue relates to monitoring and inspection plan, because corrosion and materials specialists consider also susceptible locations to corrosion for materials of constructions and define where, by which method and how often materials must be examined from corrosion point of view. Correct dimensions of impingement plates, absence of locations vulnerable to erosion by changes of flow direction, to name of few. Economist takes part in the change of materials construction because we should select among alternative materials, and RBI program provides means of assessment between initial cost and corrosion risk. For instance, gradual replace of heat exchanger tubes material in the overhead of main column at the FCCU (corrosion by H_2S, NH_3, and H_2O) carbon steel with copper-nickel alloy and then with Ti Gr.2. In spite of increasing of materials cost in this case, corrosion risk was diminished to zero and titanium heat exchangers serve at the FCCU more than 20 years without failure. Certainly economists together with corrosion and materials specialists justify the added cost of a material upgrade.

Key process parameters. Many corrosion failures occur during start-up, shut-downs and other changers of process technological parameters, e.g. temperature fluctuations: dew point acid corrosion, hydrogen damages, SCC by polythionic acids, to name of few. The key process parameters include conditions of proceeding of technological processes: temperature, pressure, flow rate and flow regime (one-, two- or three-phase media), composition of media (pH, corrosive contaminants, presence of hard particles). Keeping these parameters in the recommended limits is very important for diminishing corrosion risk and relates to monitoring strategy and corrosion control. RBI methodology allows to identify the key process parameters which can influence corrosion, its intensity and rate.

RBI has been successfully used for quantifying, prioritizing and managing corrosion risk in oil refineries and petrochemical plants [67, 74–76], in oil and gas systems [77], and FRP tanks [78].

However, it was marked that standardized methodology [65, 66] of quantifying of corrosion risk is based on several flawed assumptions, and main of them that "*risk is an elusive thing, in particular in probability part*" [72, 79]. Certainly such criticism can help in improving RBI methodology and wider implementation into practice at the oil refining and petrochemical industry for prevention corrosion accidents.

What is important that for effective employing of RBI, a multidiscipline group should act together: corrosion and materials engineers, inspectors, maintenance, process, operation, equipment, and environmental (ecological) departments, economist, and of course managers.

It is important to emphasize that RBI is not really 'inspection' and it is better to name risk-based management (RBM) which is a part of corrosion management.

9.4 Conclusion

What is to be done?
Chernyshevsky N. (1828–1889), a Russian philosopher

The main question of any refinery is HOW TO PREVENT CORROSION FAILURES or at least to diminish their occurrence and corrosion risk? Searching for the answer on this question we analyzed reasons and came to following conclusions and recommendations.

A. The main reason of corrosion failures is improper corrosion management which does not define possible corrosion risk. We should improve and sustain all efforts in corrosion management culture.

B. If we know the reasons of occurring of failures, we can outline the activities to decrease them:

- Searching for the ways how chaotic corrosion processes result in failure. This is the field of computer science. We should remember that even if we come to mathematical models allowing to define time of corrosion failure with 100 % probability, humans will be responsible for introducing of all the data in models and even in this case the definition of corrosion failure time will depend on human factor.
- Forcing managers to establish penalties and incentives for the staff in oil refining and petrochemical industry. At a higher stage, legislation in the field of corrosion management must be created on the state or federal levels. This legislation will oblige managers to take actions for prevention of corrosion failures.
- We should change relationship of society to corrosion like relationship to ecology was changed; show the strict influence of corrosion failures on safety, health, pollutions, and well-being.

C. It is needed to work out corrosion management programs with planned investments at each stage of design, fabrication, implementation of anti-corrosion measures, operation, monitoring, inspection, maintenance, education, knowledge transfer, training, and research.

References

1. Groysman A (2010) Corrosion for everybody. Springer, Dordrecht
2. Prasad L, Gulshan SS (2011) Management principles and practices. Excel Books, New Delhi
3. Uhlig HH (1949) The cost of corrosion in the United States. Chem Eng News 27:2764
4. Corrosion Costs and Preventive Strategies in the United States (2002) Report FHWA-RD-01-156 (Federal Highway Administration), USA
5. Sastri VS (2011) Green corrosion inhibitors. Wiley, Hoboken, New Jersey, USA
6. http://www.cosasco.com/index.php
7. Morshed A (2013) The evolution of the corrosion management concept. Mater Perform 52(8):66–69
8. Morshed A (2011) Case history: results of corrosion management applications. Mater Perform 50(5):66–69
9. Morshed A (2008) Case history: enhancing asset value through corrosion management. Mater Perform 47(12):66–71
10. Morshed A (2011) An introduction to asset corrosion management in the oil and gas industry. NACE International, Houston, TX, USA
11. Morshed A (2015) Corrosion management and the significance of regular reporting. Mater Perform 54(1):68–71
12. Morshed A, Mousavi MS (2013) Corrosion management applications for seawater injection systems. Mater Perform 52(12):70–75
13. Morshed A (2011) Asset corrosion management contracts: issues and improvements. Mater Perform 50(1):56–59
14. Morshed A (2010) Creating an asset corrosion management strategy document. Mater Perform 49(6):56–60
15. Morshed A (2009) Asset corrosion management and the existing training gap. Mater Perform 48(4):56–58
16. Groysman A, Brodsky N (2006) Corrosion and quality, accreditation and quality assurance. J Qual Comp Reliab Chem Meas 10(10):537–542
17. Groysman A (2005) Anti-Corrosion Management and Environment at the Oil Refining Industry. In: Proceedings of international conference on corrosion CORCON2005, 28–30 November, Chennai, India
18. Groysman A, Brodsky N (2006) Corrosion and quality in the oil refining industry. In: Proceedings of EUROCORR 2006, 25–28 September, Maastricht, the Netherlands
19. Groysman A (2004) Corrosion problems and their solutions in the refining industry. In: Proceedings of EUROCORR 2004, 12–16 September, Nice, France
20. Singh B, Britton J, Poblete B, Smith G (2005) The 3 R's—risk rust reliability. Paper 05553 presented at the CORROSION 2005, NACE International, Houston, TX, USA
21. Wilson P, Susic O (2010) Corrosion management in a safety case regulatory regime. Paper 10190 presented at the CORROSION 2010, NACE International, Houston, TX, USA
22. Kermani B, Chevrot T (2014) Pipeline corrosion management; a compendium. Paper 3723 presented at the CORROSION 2014, NACE International, Houston, TX, USA
23. Morshed A (2014) Corrosion management and the widening competency gap. Paper 3740 presented at the CORROSION 2014, NACE International, Houston, TX, USA

24. Ogosi E, McKenny S (2014) Indexing model for pipeline risk assessment and corrosion management. Paper 3781 presented at the CORROSION 2014, NACE International, Houston, TX, USA
25. Whittaker J, Liu Q (2014) Corrosion management and mechanism study on SAGD Brackish water system. Paper 3809 presented at the CORROSION 2014, NACE International, Houston, TX, USA
26. White RA, Ehmke EF (1991) Materials selection for refineries and associated facilities. NACE, Houston, TX, USA
27. Corrosion (1987) ASM handbook, vol 13, 9th edn. ASM International, USA
28. Linda Garverick (ed) (1994) Corrosion in the petrochemical industry. ASM International, USA
29. API RP 571 (2011) Damage mechanisms affecting fixed equipment in the refining and petrochemical industries, 2nd edn. American Petroleum Institute, Washington, D.C., USA
30. Groysman A (2014) Corrosion in systems for transportation and storage of petroleum products. Springer, Dordrecht
31. Heidersbach R (2011) Metallurgy and corrosion control on oil and gas production. Wiley, Hoboken, New Jersey, USA
32. Revie RW (ed) (2006) Uhlig's corrosion handbook, 2nd edn. Wiley, Hoboken, New Jersey, USA
33. Fontana MG (1986) Corrosion engineering, 3rd edn. McGraw-Hill, New York
34. Mattsson E (1996) Basic corrosion technology for scientists and engineers, 2nd edn. The Institute of Materials, London
35. Shreir LL, Jarman RA, Burstein GT (eds) (1994) Corrosion, vol 1 and 2, 2nd edn. Butterworth-Heinemann Ltd, Oxford
36. Esaklul KA (ed) (1992) Handbook of case histories in failure analysis, vol 1 and 2. ASM International, USA
37. Wood MH, Vetere Arellano AL, Van Wijk L (2013) Corrosion-related accidents in petroleum refineries. JRC Scientific and Policy Reports, European Commission EUR 26331 EN, Institute for the Protection and Security of the Citizen, Italy
38. Groysman A (2014) Corrosion education: present and future. Paper presented at the 19th international corrosion congress, Jeju, Korea, 2–6 November, 2014
39. Boden PJ (1994) Chemicals. In: Shreir LL, Jarman RA, Burstein GT (eds) Corrosion, vol 1, 3rd edn. Butterworth-Heinemann Ltd, Oxford, pp 2.99–2.113
40. Gutzeit J (2006) Crude unit corrosion guide. PCC Process Corrosion Consultants, USA
41. Rabald E (2012) Corrosion guide. Elsevier, London
42. Berg FF (1965) Corrosion diagrams. V.D.I.-Verlag, G.m.b.H., Düsseldorf
43. DECHEMA Corrosion Handbook. Wiley VCH, 1988–2007, vol 1–11
44. Nelson GA (1960) Corrosion data survey. Shell Development Co, Emeryville, California, USA
45. Kellert SH (1993) In the wake of chaos: unpredictable order in dynamical systems. University of Chicago Press, USA
46. Harari YN (2011) Sapiens. A brief history of humankind. Vintage Books, London
47. STP 1154 (1992) Computer modeling in corrosion. Munn RS (ed) ASTM International, USA
48. Addis K, Singer M, Nešić (2014) A corrosion model for oil and gas mild steel production tubing. Corrosion 70(12):1175–1176
49. Nouri M, Mesbah J, Shaibani S (2008) A verified computer model for corrosion rate prediction in a di-ethanol amine system using field data. Paper 08417 presented at the CORROSION 2008, NACE International, Houston, TX, USA
50. Pidaparti RM (2009) Corrosion modeling, simulations and assessment through intelligent computations. US Army Corrosion Summit 2009, February 3–5, 2009, Clearwater Beach Resort, FL, USA. file:///C:/Users/Alec/Downloads/ADA508789.pdf
51. Garber JD et al (2004) Internal corrosion rate prediction in pipelines and flowlines using a computer model. Paper 04155 presented at the CORROSION 2004, NACE International, Houston, TX, USA

52. Rafiee A (2012) Computer modeling and investigation on the steel corrosion in cracked ultra high performance concrete. Kassel University Press, Germany
53. Groysman A (2014) Corrosion education in oil and gas industry: experience and insight into the future. Corr Mater Prot, Moscow, Russia 4:1–13
54. Tritton TT (2011) From the 18th century until today: great names, great steps. Launch of the International Year of Chemistry 2011, 27–28 January, UNESCO and IUPAC, Paris, France
55. http://www.corrosioncost.com/pdf/main.pdf
56. Assessment of Corrosion Education, Committee on Assessing Corrosion Education (2008) National Research Council. The National Academies Press, Washington, D.C., USA
57. Landolt D (2004) The role of corrosion science in engineering education. In: Trends in electrochemistry and corrosion at the beginning of the 21st century, Dedicated to Professor Dr. Josep M. Costa on the occasion of his 70th birthday, Universitat de Barcelona, Barcelona, Spain, pp 1195–1201
58. Roberge PR, Ghali E (2000) A corrosion engineering training model. Paper 00478 presented at the CORROSION 2000, NACE International, Houston, TX, USA
59. Cottis RA, McAllister AM, Wendy D (2004) Online delivery and tutor support of an introductory corrosion course. Paper 04069 presented at the CORROSION 2004, NACE International, Houston, TX, USA
60. Klassen RD, Roberge PR (2004) Early experiences with a web-based distance course on corrosion. Paper 04070 presented at the CORROSION 2004, NACE International, Houston, TX, USA
61. Faidi SE, Fay M, Cottis RA (1996) A computer aided learning package for teaching corrosion to engineering students. Paper 96377 presented at the CORROSION 1996, NACE International, Houston, TX, USA
62. Groysman A (2010) Interdisciplinary relationships in corrosion teaching and education at school, college and university. In: Proceedings of EUROCORR 2010, Moscow, Russia, 13–17 September, 2010
63. Piktialis D, Greenes KA (2008) Bridging the gaps. How to transfer knowledge in today's multigenerational workplace? Research Report R-1428-08-RR. The Conference Board, Inc, USA
64. Groysman A (2016) The role of corrosion management in prevention of corrosion failures. Paper 7252 presented at the NACE international 2016 conference, 6–10 March 2016, Vancouver, Canada
65. API RP 580 (2002) Risk-based inspection. American Petroleum Institute, Washington, D.C., USA
66. API RP 581 (2008) Risk-based inspection technology. American Petroleum Institute, Washington, D.C., USA
67. Horvath RJ (1998) The role of the corrosion engineer in the development and application of risk-based inspection for plant equipment. Mater Perform 39(7):70–75
68. Kletz TA (1992) Hazop and hazan: identifying and assessing process industry, 3rd edn. Institution of Chemical Engineers, UK
69. ISO 31000 (2009) Risk management—principles and guidelines
70. ISO/IEC 31010 (2009) Risk management—risk assessment technique
71. ISO Guide 73 (2009) Risk management—vocabulary
72. Helle HPE (2012) Integrity management beyond RBI. Paper presented at the ICPVT-13 (The 13th international conference on pressure vessel technology), 20–23 May 2012, London, UK
73. Groysman A (2009) Corrosion monitoring. Corr Rev 27(4–5):205–343
74. Krynicki JW, Bagnoli KE, McLaughlin JE (2006) Probabilistic risk based approach for performing on onstream high temperature hydrogen attack inspection. Paper 06580 presented at the Corrosion 2006, NACE International, Houston, TX, USA
75. McLaughlin JE (2007) A qualitative risk-based assessment procedure for high temperature hydrogen attack of C-1/2Mo steel. In: Proceedings of ASME Pressure Vessels and Piping Conference (PVP2007), and the Eighth International Conference on Creep and Fatigue at Elevated Temperatures (CREEP8), PVP2007-26129, 22–26 July 2007

76. Canale G, De Marco M, Pinca S (2009) The role of welded joint vulnerability and various damage mechanisms active in process and petrochemical plants in reliability analysis with risk based inspection (RBI) approach, Eurojoin 7. In: Proceedings of 7th European congress on joining technology; incorporating GNS 5, 5th edition of Italian welding days, technical session 2: reliability of welded components and structures
77. Hodges S et al (2010) High level corrosion risk assessment methodology for oil and gas systems. Paper 10367 presented at the NACE international 2010 conference, Houston, TX, USA
78. Chant E, Sims JR (2010) Risk-based inspection for polymers. Paper 10040 presented at the NACE International 2010 Conference, Houston, TX, USA
79. Helle HPE (2012) Five fatal flaws in API RP 581. Paper presented at the 14th Middle East corrosion conference and exhibition, February 12–15, 2012, Manama, Kingdom of Bahrain

Appendices

Reading makes a person knowledgeable, conversation - resourceful, and the habit of recording – accurate.

Francis Bacon (1561–1626), an English philosopher.

Twelve appendices contain diverse information about physicochemical properties of crude oils and petroleum products; physicochemical properties of sulfur compounds, acids, alkalis, and hydrogen used in refinery units; chemical composition of alloys; metallographic replication; fouling mechanism; chemical cleaning from fouling; recommended procedure for passivation of cooling water systems; and boil-out procedure (chemical cleaning and passivation of inner surfaces of boiler and steam pipelines).

A. Groysman, *Corrosion Problems and Solutions in Oil Refining and Petrochemical Industry*, Topics in Safety, Risk, Reliability and Quality 32,
DOI 10.1007/978-3-319-45256-2

Appendix A
Schematic of a Typical Oil Refinery

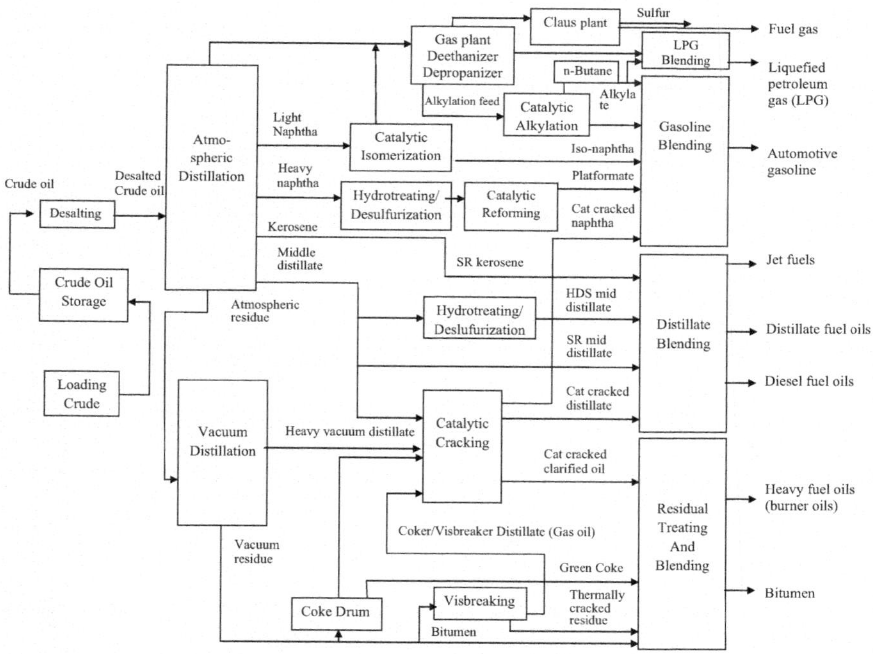

Schematic of a Typical Oil Refinery

A. Groysman, *Corrosion Problems and Solutions in Oil Refining and Petrochemical Industry*, Topics in Safety, Risk, Reliability and Quality 32,
DOI 10.1007/978-3-319-45256-2

Appendix B
Physicochemical Properties of Crude Oils and Petroleum Products

Crude oil is a mixture of numerous distinct chemical compounds, mainly hydrocarbons.

Crude oil classification. Crude oil is described by density (°API) and distillation curves (boiling point ranges). Classification of crude oils is usually made according to density (gravity), crude composition, sulfur content, and geographical origin.

API gravity: Crude oil gravity may range from less than 10 °API to over 50 °API but most crudes are in the 20 to over 45 °API [1]. Crude oils are classified as light (>30 °API; density <870 kg/m^3); intermediate or medium (20 < °API < 30; 870 < density < 920 kg/m^3); and heavy crudes (<20 °API; 920 < density < 1000 kg/m^3). Crude oils with API gravity less than 10 °API are referred to as extra heavy oil or bitumen. For instance, bitumen derived from the oil sands deposits in the Alberta (Canada) has an API gravity of ~8 °API.

Crude composition: paraffinic, naphthenic, and aromatic (see Table 2.1).

Sulfur content: sweet and sour crudes. The term 'sour' crude initially related to crudes containing only dissolved hydrogen sulfide independent of total sulfur content. There is no sharp dividing line between sweet and sour crudes. A crude is considered 'sour' if total sulfur content is ranged in 0.5–1.0 wt% S and more. A crude is considered 'sweet' if total sulfur content is less than 0.5 wt%.

Geographical origin: There are above 500 different crude oil types from different geographical regions. For instance, West Texas (USA), Maya (Mexico), Ras-Garib (Egypt), Ural (Russia), Azeri (Azerbaijan), etc. If refineries use different types of crudes and blends, this can lead to greater deterioration of equipment than when one type of crude oil is used.

Conventional oil includes crude oil (from wells) and condensates (of natural gas sources).

Opportunity crude oil refers to crude oils with a relatively high sulfur, salts and metals content, a high total acid number (TAN > 0.5 mg KOH/g), and density. These crudes give rise to problems related to corrosion, fouling, fuel quality, poison

A. Groysman, *Corrosion Problems and Solutions in Oil Refining and Petrochemical Industry*, Topics in Safety, Risk, Reliability and Quality 32,
DOI 10.1007/978-3-319-45256-2

of catalyst, and environmental protection. Therefore, *opportunity crudes* are considered as 'low quality' corrosive crudes with the price of about 80 % of the price of conventional crude oil [2]. The cost of crude accounts for about 90–95 % of the total running costs of refineries, so it is very attractive for refineries to process opportunity crudes, especially high TAN crudes. Opportunity crudes use is expected to grow up to 20 % to 2025 [3].

Unconventional oils include oil sands, synthetic crudes, heavy and extra-heavy oil, coal-to-liquids, biomass-based liquid supplies, gas to liquid (hydrocarbon liquids arising from chemical processing of natural gas). Unconventional oil is crude oil produced or extracted using techniques other than the conventional (oil well) method. Oil industries and governments across the globe are investing in unconventional oil sources due to the increasing scarcity of conventional oil reserves. Conventional oil is easier and cheaper to produce than unconventional oil. However, the categories "conventional" and "unconventional" do not remain fixed, and over time, as economic and technological conditions evolve, resources hitherto considered unconventional can migrate into the conventional category.

Oil sands (named also *bituminous sands* or *heavy oil sands*) consist of natural bitumen or extra-heavy crude oil trapped in unconsolidated sandstone. These hydrocarbons are forms of crude oil that are extremely dense and viscous, with a consistency ranging from that of molasses for some extra-heavy oil to as solid as peanut butter for some bitumen at ambient temperature, making extraction difficult. As a result of their high viscosity, they cannot be produced by conventional methods, transported without heating or dilution with lighter hydrocarbons, or refined by older oil refineries without major modifications. Such heavy crude oils often contain high concentrations of naphthenic acids, sulfur and heavy metals (particularly nickel and vanadium), which interfere with refining processes, although lighter crude oils can also suffer from these contaminants.

Synthetic crude (*syncrude* or *upgraded crude*) is an intermediate product produced when unconventional oil (usually oil sands) is upgraded into a transportable form. Synthetic crude may also refer to *shale oil* (an output from an oil shale pyrolysis). The properties of the synthetic crude depend on the processes used in the upgrading. Usually, it is low in sulfur and has ~30 °API gravity. Synthetic crude is transported to oil refineries where it is treated into finished products. Synthetic crude may also be mixed, as a diluent, with heavy oil to create *synbit*. *Synbit* is more viscous than synthetic crude, but can be a less expensive alternative for transporting heavy oil to a conventional oil refinery. *Synthetic crude* should not be confused with *synthetic oil*, *synthetic fuel*, or *biocrude*.

Tight oil is crude oil contained in crude oil-bearing formations of low permeability, often shale or tight sandstone. It should not be confused with *oil shale* which is an organic-rich fine-grained sedimentary rock rich in *kerogen* (a solid mixture of organic compounds). The kerogen in oil shale can be converted to shale oil through the chemical processes of pyrolysis, hydrogenation, or thermal dissolution.

Table B.1 Major kinds of hydrocarbons in one reference crude oil [4]

Type	Percent in crude oil
Alkanes:	
Normal	14
Iso (Branched)	18
Cycloparaffins (Naphthenes):	
Alky Cyclopentanes	10
Alky Cyclohexanes	6
Bicycloparaffins	5
Aromatics:	
Alkylbenzenes	18
Aromatic-cycloparaffins	5
Fluorenes	3
Binuclear Aromatics	17
Polynuclear Aromatics	4

Table B.2 Petroleum products obtaining from crude oil in atmospheric distillation column [5]

Petroleum product	Hydrocarbons[a]	Boiling temperature range (°C)
Gas[b]	C_1 to C_4	
Naphtha[c]	C_5 to C_{12}	20–210
Gasoline	C_5 to C_{12}	20–210
Kerosene[d]	C_9 to C_{16}	150–290
Light gas oil[e]	C_{12} to C_{20}	180–320
Heavy gas oil[e]	C_{12} to C_{24}	190–370
Fuel oil	$>C_{20}$	>340
Bitumen (Asphalt)	$>C_{40}$	>540

[a]A shorthand method of listing hydrocarbons is used for characterizing compounds by number of carbon atoms in the molecule, e.g., C_1 and C_4 are related to methane (CH_4) and butane (C_4H_{10})
[b]Liquefied Petroleum Gas (LPG, propane-butane) is obtained from this gas mixture
[c]Full-range naphtha. The naphtha obtained directly from the atmospheric crude distillation column is named straight-run naphtha. Naphtha is also produced during processing of heavier parts of the crude oil: catalytic cracker naphtha, visbreaker naphtha, cocker naphtha. As opposed to the straight-run naphtha, these naphthas contain olefinic hydrocarbons. *Light naphtha* is the fraction boiling from 30 to 90 °C, containing the C_5 and C_6 hydrocarbons. *Heavy naphtha* is the fraction boiling from 90 to 200 °C, containing the C_7 to C_{12} hydrocarbons. *Medium naphtha* is the fraction of the heavy naphtha that boils below 150 °C and contains mostly C_7 to C_9 hydrocarbons
[d]Jet fuel and aviation turbine fuel are obtained then from this kerosene
[e]Diesel fuel is obtained from these fractions

Hydrocarbons. Chemical compounds in crude oil and petroleum products are given in Tables B.1 and B.2 [4, 5]. Hydrocarbons (HC) constitute of three main types: alkanes (aliphatic HC or paraffins), the saturated ring (naphthenic ring), and the aromatic ring. Almost all of the compounds in crude oils are composed of these three kinds, alone or in combination.

Table B.3 Number of possible alkane isomers [6]

Number of carbon atoms	Number of isomers
1 or 2 or 3	1
4	2
5	3
6	5
7	9
8	18
9	35
12	355
15	4347
18	60,523
25	36,797,588
40	62×10^{12}

Hydrocarbons have huge number of isomers. The more number of carbon atoms in hydrocarbons the greater amount of isomers (Table B.3). This enormous number of possible isomers explains why crude oil contains hundreds of thousands of individual compounds.

Between an oilfield and a refinery there is a long way. Usually crude oils from wellheads at oilfields are processing (named also conditioning or handling) before sale and transport by pipelines, tankers or trucks. Oilfield operations include separation, dewatering, desalting, sweetening, and stabilization [7]. *Separation* is separating the vapor, oil, and aqueous phases of a produced wellhead stream. *Dewatering* is removing water or together with sediments from crude oil. *Desalting* is removing of salts from crude oil by diluting the entrained/emulsified water and dewatering. *Sweetening* is removing of H_2S and other sulfur compounds. *Stabilization* is removing the most volatile components of a crude oil. Thus, the scope of oilfield processing of crude oil starts at the wellhead and ends with the pipeline, storage tank, or tanker. In other words, field processing of crude oil for storage, transportation and sale involves three process objectives:

(a) *Separation* of the crude oil from free and emulsified water or brine and entrained solids (mainly sand and soil).
(b) *Stabilization* of the crude oil: removal of dissolved natural gas in order to provide crude that can be safely stored and transported.
(c) *Removal* of impurities from the crude oil to meet transport specifications according to environmental, safety and health regulations.

After field treatment, crude oil cannot be used directly. It must be processed further in refineries to produce petroleum products that are themselves useful, or that become useful when chemically processed and blend [8–10]. Some of these products are the sources for petrochemical industry producing numerous chemicals, such as polyethylene, polypropylene, organic solvents, and ethers, to name a few.

Appendix C
Hydrochloric Acid in Refinery Units

Hydrochloric acid (HCl_{aq}) is a hydrogen chloride gas (HCl_g) dissolved in water. Hydrogen chloride gas is dissolved well in water (up to 40 g in 60 g H_2O at 25 °C and 1 bar) forming strong inorganic hydrochloric acid (C.1) which dissociates (C.2) easily and its strength (pH) depends on HCl concentration in water.

$$HCl_{(g)} + H_2O_{(1)} \rightarrow HCl \cdot H_2O_{(aq)} \quad (C.1)$$

$$HCl \cdot H_2O_{(aq)} \rightarrow H_3O^+{}_{(aq)} + Cl^-{}_{(aq)} \quad (C.2)$$

It has very sharp unpleasant smell and is contained in our stomach. This acid plays positive role in digestion, descaling (dissolution of scale, i.e. $CaCO_3$) and pickling (acid cleaning from rust). Its negative role is manifested in corrosion of metals.

Hydrochloric acid is formed in refinery units as result of hydrolysis of metal chlorides (C.3–C.5),

$$CaCl_{2(aq)} + H_2O_{(l)} \rightarrow \mathbf{HCl}_{(aq)} + Ca(OH)Cl_{(aq)} \quad (C.3)$$

$$MgCl_{2(aq)} + H_2O_{(l)} \rightarrow \mathbf{HCl}_{(aq)} + Mg(OH)Cl_{(aq)} \quad (C.4)$$

$$AlCl_{3(s)} + H_2O_{(l)} \rightarrow \mathbf{HCl}_{(aq)} + Al(OH)Cl_{2(aq)} \quad (C.5)$$

hydrolysis organochlorine compounds (C.6)

$$RCl_{(aq)} + H_2O_{(1)} \rightarrow \mathbf{HCl}_{(aq)} + ROH_{(aq)} \quad (C.6)$$

and hydrolysis of solid ammonium chloride (NH_4Cl) and amine chloride $[RNH_3]^+Cl^-$ salts (C.7 and C.8).

$$NH_4Cl_{(s)} + H_2O_{(l)} \rightarrow \mathbf{HCl}_{(aq)} + NH_4OH_{(aq)} \quad (C.7)$$

A. Groysman, *Corrosion Problems and Solutions in Oil Refining and Petrochemical Industry*, Topics in Safety, Risk, Reliability and Quality 32,
DOI 10.1007/978-3-319-45256-2

$$(RNH_3)Cl_{(s)} + H_2O_{(l)} \rightarrow \mathbf{HCl}_{(aq)} + RNH_2OH_{(aq)} \quad (C.8)$$

Hydrochloric acid reacts with iron (see reaction 3.5) to form ferrous chloride salt. Then this compound can react with ammonium sulfide to form iron sulfide and ammonium chloride (C.9):

$$FeCl_{2(aq)} + (NH_4)_2S_{(aq)} \rightarrow FeS_{(s)} + 2NH_4Cl_{(aq)} \quad (C.9)$$

A vicious circle exists in such systems. On the one hand, corrosion takes place under deposits. On the other hand, the ammonium chloride salt regenerates in (C.9) and then is hydrolized in (C.7) to form HCl, and thus corrosion continues.

Appendix D
Physicochemical Properties of Sulfur Compounds

Sulfur (S_8), hydrogen sulfide (H_2S) and sulfuric acid (H_2SO_4) are so important in oil refining industry, especially in corrosion problems, that we will describe their physicochemical and hazard properties. Others, ammonium bisulfide and polythionic acids, are formed and cause corrosion problems too.

Sulfur

Elemental sulfur exists at ambient conditions in the form of ring containing eight atoms S_8. Two crystal forms of sulfur exist: orthorhombic (T < 95.4 °C) and monoclinic (95.4 °C < T < 119.3 °C). When heated above 119.3 °C at 1 bar, sulfur becomes liquid and boils at 444.7 °C.

Pure sulfur is not poisonous. Even sulfur powder is contained in some ointments, intended for skin care and treatment of skin diseases.

Sulfur is inert at low temperatures, but by heating it becomes active and reacts with oxygen, hydrogen and metals (D.1, D.2):

$$8Fe_{(s)} + S_{8(s)} \rightarrow 8FeS_{(s)} \quad (D.1)$$

$$8H_{2(g)} + S_{8(s)} \leftrightarrow 8H_2S_{(g)} \quad (D.2)$$

Equilibrium of the reaction (D.2) is shifted to the right direction at T < 350 °C, and to the left at T > 350 °C.

Hydrogen sulfide begins thermally dissociate at T > 350 °C and completely dissociates at 1700 °C.

A. Groysman, *Corrosion Problems and Solutions in Oil Refining and Petrochemical Industry*, Topics in Safety, Risk, Reliability and Quality 32,
DOI 10.1007/978-3-319-45256-2

Hydrogen Sulfide

Hydrogen sulfide (H_2S) is also named hydrosulfuric acid or stink damp or sewer gas. Physicochemical properties of H_2S: $T_{melting} = -85.49$ °C; $T_{boiling} = -60.33$ °C [11]. Hydrogen sulfide in air ignites at ~300 °C. Its mixtures with air are explosive at 4.3–46 vol.% H_2S [11]. Hydrogen sulfide is colorless and very poisonous gas. The presence of one part H_2S in 100,000 parts of air is detected in his characteristic odor of rotten eggs. The recognition threshold concentration is 4.7 ppb at which 50 % of humans can detect the characteristic odor of H_2S [12]. The value 10 ppm H_2S is permissible exposure limit (at 8 h time-weighted average) and 20 ppm H_2S is the acceptable ceiling concentrations established by Occupational Safety and Health Administration (OSHA) [11]. Many personal safety gas detectors used by personal at oil refineries and petrochemical plants are set to alarm at 5–10 ppm. At 100–150 ppm H_2S the olfactory nerve is paralyzed after a few inhalations, and the sense of smell disappears, often together with awareness of danger, so victims may be unaware of its presence until it is too late [13, 14]. Concentration 320–530 ppm H_2S can lead to pulmonary edema with possible death [15]. Concentration 530–1 000 ppm H_2S causes strong stimulation of the central nervous system and rapid breathing, leading to loss of breathing. Concentration 800 ppm H_2S is the lethal for 50 % of humans for 5 min exposure (LC50). Concentrations over 1000 ppm H_2S cause immediate collapse with loss of breathing, even after inhalation of a single breath.

The toxicity of H_2S is sometimes underestimated and works with this gas occasionally carry out without adequate precautions. Hydrogen sulfide is heavier than air, therefore accumulates in pitches and other low ground places at the bottoms of poorly ventilated spaces. Hydrogen sulfide forms a complex bond with Fe^{2+} in the enzymes responsible for respiration. At some threshold concentration of ~300 ppm the oxidative enzymes become overwhelmed. Here are only several sad examples. In 1975, a hydrogen sulfide release from an oil drilling operation in Texas (USA), killed nine people. In 2010, two workers dead from contact with gas containing H_2S in oil refinery. In 2014, a substantial amount of H_2S shrouded some regions of Moscow (Russia), and blame had been placed on a Moscow oil refinery.

Hydrogen sulfide is corrosive substance. It is interesting to note that corrosion of coins containing copper and silver in the pockets of the victim helps to diagnose extreme poisoning by H_2S. Copper and silver in contact with H_2S corrode with formation black sulfides (D.3, D.4):

$$Cu_{(s)} + H_2S_{(g)} + 1/2O_{2(g)} \rightarrow CuS_{(s)} + H_2O_{(l)} \quad (D.3)$$

$$2Ag_{(s)} + H_2S_{(g)} + 1/2O_{2(g)} \rightarrow Ag_2S_{(s)} + H_2O_{(l)} \quad (D.4)$$

This is an example of uniform corrosion. Therefore, it is recommended to take off silver jewelry when taking water baths containing H_2S. Under high temperature, gaseous H_2S reacts with metals in the absence of oxygen to form black corrosion products of sulfides (D.5, D.6):

$$Fe_{(s)} + H_2S_{(g)} \rightarrow FeS_{(s)} + H_{2(g)} \tag{D.5}$$

$$Cu_{(s)} + H_2S_{(g)} \rightarrow CuS_{(s)} + H_{2(g)} \tag{D.6}$$

Hydrogen sulfide is produced by many bacteria around us [16]. This is the reason why silver jewelry blacken with time and we should clean them periodically [5].

Hydrogen sulfide burns and this is the basics of obtaining sulfur in the Claus process at SRU (D.7, D.8):

$$2H_2S_{(g)} + 3O_{2(g)} \leftrightarrow 2H_2O_{(g)} + 2SO_{2(g)} \quad \text{(with an excess of oxygen)} \tag{D.7}$$

$$2H_2S_{(g)} + O_{2(g)} \leftrightarrow 2H_2O_{(g)} + 1/4S_{8(s)} \quad \text{(with a lack of oxygen)} \tag{D.8}$$

Not only in the Claus process H_2S is oxidized but H_2S dissolved in aqueous solution is easily oxidized by dissolved oxygen according to reactions (D.7) and (D.8). For instance, even when standing in air, the solution gradually becomes cloudy as elemental sulfur is released according to (D.8). Thus, H_2S is a strong reducing agent. Sulfur dioxide (SO_2) forming in aqueous solution (see D.7) can react with water to form sulfurous acid H_2SO_3 (D.9):

$$SO_{2(g)} + H_2O_{(l)} \leftrightarrow H_2SO_{3(aq)} \tag{D.9}$$

Elemental sulfur (S_8) can be formed as a result of reaction between H_2SO_3 and H_2S (D.10):

$$16H_2S_{(g)} + 8H_2SO_{3(aq)} \rightarrow 3S_{8(s)} + 24H_2O_{(l)} \tag{D.10}$$

Moreover, S_8 can be formed as a result of reaction between H_2S and Fe_2O_3 on metal surface (D.11).

$$8Fe_2O_{3(s)} + 24H_2S_{(g)} \rightarrow 16FeS_{(s)} + S_{8(s)} + 24H_2O_{(l)} \tag{D.11}$$

One volume of water dissolves three volumes H_2S under standard conditions, thereby forming 0.1 M solution with pH = 4.3 [17]. The solubility of H_2S in water is ~3000 ppm at 30 °C and 0.1 MPa [18]. Temperature increase causes diminishing of H_2S solubility in water and at 80 °C the solubility nearly three times lower than at 20 °C. Solubility of H_2S in liquid hydrocarbons and other organic liquids is higher than in water [5]. For instance, the solubility of H_2S in liquid alkanes and aromatics is 25–35 times greater than in water. Moreover, the greater is the molecular weight of alkane the greater is H_2S solubility in it.

Aqueous solution of H_2S is a weak acid, a little bit weaker than carbonic acid H_2CO_3. Hydrogen sulfide forms two types of salts with anion S^{2-} (sulfides) and with anion HS^- (hydrosulfides). In spite of colorless ions S^{2-} and HS^- the most their salts have characteristic colors. The majority of sulfides are non-dissolved in water. Contrary the most hydrosulfides are well dissolved in water. Ions S^{2-} exist only in strongly alkaline aqueous solutions.

Hydrogen sulfide reacts with metal ions to form sulfides. They often have a dark color. Some sulfides (Na_2S, K_2S, BaS) are dissolved in water. Some sulfides (FeS, ZnS, MnS) are not dissolved in water but dissolved well in diluted HCl (5–15 wt%). Others (CuS, PbS, HgS) are neither dissolved in water nor in diluted acids.

When sulfides of metals react with acids, H_2S releases (D.12):

$$FeS_{(s)} + 2HCl_{(aq)} \rightarrow H_2S_{(g)} + FeCl_{2(aq)} \quad (D.12)$$

Lead(II) acetate paper is used to detect H_2S because grey-black lead(II) sulfide (PbS) is obtained.

Iron sulfides can be oxidized to form ferric and ferrous sulfate acid salts (D.13):

$$8FeS_{(s)} + 17O_{2(g)} \rightarrow 2FeSO_{4(s)} + 2Fe_2(SO_4)_{3(s)} + 2FeO_{(s)} \quad (D.13)$$

Like acids, cyanides ($\sim$4 ppm, even at concentrations ≤ 1 ppm) dissolve protective iron sulfide layer or render it porous, non-adherent and, as a result, non-protective (D.14):

$$FeS_{(s)} + 6NH_4CN_{(aq)} \rightarrow (NH_4)_4Fe(CN)_{6(aq)} + (NH_4)_2S_{(aq)} \quad (D.14)$$

Sulfur-containing compounds (H_2S, mercaptans, thiols, di- and polysulfides) can react with oxygen to form inorganic and organic acids containing sulfur. These acids, especially sulfuric (H_2SO_4) and sulfurous (H_2SO_3) acids can cause severe corrosion (like HCl) of refinery equipment.

The iron sulfide deposits posses some protection against corrosion at pH > 5.5. At lower pH the dissolved hydrogen cyanide (HCN) and ammonium cyanide (NH_4CN) accelerates corrosion by destroying the protective iron sulfide film, converting it into soluble ferrocyanide complexes (D.15, D.16):

$$FeS_{(s)} + 6\,HCN_{(aq)} \rightarrow H_4\left[Fe(CN)_6\right]_{(aq)} + H_2S_{(aq)} \quad (D.15)$$

$$FeS_{(s)} + 6NH_4{}^+{}_{(aq)} + 6CN^-{}_{(aq)} \rightarrow (NH_4)_4\left[Fe(CN)_6\right]_{(aq)} + (NH_4)_2S_{(aq)} \quad (D.16)$$

Appearing of sulfide ions in reactions (D.15 and D.16) causes further corrosion of iron, and these processes are autocatalytic. The presence of ferrocyanide [$Fe(CN)_6^{4-}$] ions can be recognized by its blue color (known as iron blues or Prussian blue pigment) after drying of the water sample. Both cyanide and hydrogen sulfide contribute to hydrogen absorption by steel and blistering. During maintenance, it is important to look for a bluish color when the unit is first opened. This is a good indication of areas where corrosion and hydrogen blistering may be found. In order to prevent harmful action of cyanide ammonium polysulfides are injected (see Sect. 8.2.2).

Sulfuric acid and Some Other Acids

Concentrated sulfuric acid is a strong oxidizer and many metals (e.g., copper, excluding iron) corrode in it (D.17):

$$Cu_{(s)} + 2H_2SO_{4(aq)} \rightarrow CuSO_{4(aq)} + SO_{2(g)} + 2H_2O_{(l)} \quad (D.17)$$

Diluted H_2SO_4 (<75 wt%) dissolve iron (D.18):

$$Fe_{(s)} + 2H_2SO_{4(aq)} \rightarrow FeSO_{4(aq)} + SO_{2(g)} + 2H_2O_{(l)} \quad (D.18)$$

However, some metals (Au, Pt) are resistant to H_2SO_4. Not only pH value but also anion influences significantly corrosion of metals. Iron corrodes in diluted acids HCl, H_2SO_4, H_2SO_3, HNO_3, HCOOH, CH_3COOH thereby forming ferrous (Fe^{2+}) water soluble salts. In the presence of oxygen, oxidation proceeds to ferric (Fe^{3+}). Iron surface is remained clean and 'active' for further corrosion. Opposite, when iron is in contact with aqueous solutions of H_2S, H_3PO_4 and H_2CO_3, insoluble (or having limited solubility) iron salts are formed on the surface of iron. These insoluble salts can form metastable protective layers capable deter corrosion if they are even thin without any defects. 'Metastable' means that these films are not really stable, and any changes of conditions (temperature, flow velocity, concentration of acids or other components) can destroy tenacity of these films and cause localized corrosion.

Ammonium Bisulfide

Ammonium bisulfide (ammonium hydrosulfide)—$(NH_4)SH$. Ammonia (NH_3) or its hydroxide (NH_4OH) is often injected in the overhead vapor line of distillation column to neutralize acid gases (HCl and H_2S) containing in light hydrocarbons. Ammonia can also appear in catalytic cracking, hydrotreating, and hydrocracking units after destruction of N-containing organic compounds. Ammonium hydroxide NH_4OH can react with H_2S forming two salts $(NH_4)SH$ and $(NH_4)_2S$. Aqueous solution of these salts are slightly alkaline with pH ~9–9.3 [17].

Two gases react to form ammonium bisulfide (D.19):

$$NH_{3(g)} + H_2S_{(g)} \leftrightarrow (NH_4)SH_{(s)} \quad (D.19)$$

Both NH_3 and H_2S gases have a strongly unpleasant smell. Ammonium hydrosulfide decomposes when slightly heated and by diluted hydrochloric acid (D.20):

$$(NH_4)SH_{(aq)} + HCl_{(aq)} \rightarrow NH_4Cl_{(aq)} + H_2S_{(g)} \tag{D.20}$$

This is the basis of the 'stink bomb', a device for creation of an unpleasant smell. Opposite reaction to D.19 takes place in this 'bomb'.

The bisulfide ions (HS^-) are produced by the dissociation of H_2S and their quantity depends on the pH, temperature and H_2S partial pressure (D.21, D.22, D.23):

$$H_2S_{(aq)} \leftrightarrow H^+{}_{(aq)} + HS^-{}_{(aq)} \tag{D.21}$$

$$HS^-{}_{(aq)} \leftrightarrow H^+{}_{(aq)} + HS^-{}_{(aq)} \tag{D.22}$$

$$H_2S_{(aq)} + NH_{3(aq)} \leftrightarrow (NH_4)HS_{(aq)} \leftrightarrow NH_4{}^+{}_{(aq)} + HS^-{}_{(aq)} \tag{D.23}$$

Carbon steel corrodes in contact with bisulfide ion (D.24):

$$Fe_{(s)} + 2HS^-{}_{(aq)} \rightarrow FeS_{(s)} + S^{2-}{}_{(aq)} + 2H_{ads} \tag{D.24}$$

Ammonium bisulfide is oxidized by air to form S_8, $(NH_4)_2S_n$ and $(NH_4)_2SO_3S$. Therefore, oxygen content is a critical variable of the wash water depending on the process water pH. Oxygen is a strong oxidizer and reacts with the bisulfide ion to form elemental sulfur S_8 (D.25):

$$9(NH_4)SH_{(aq)} + 4O_{2(g)} \rightarrow S_{8(s)} + 9NH_{3(g)} + H_2S_{(g)} + 8H_2O_{(l)} \tag{D.25}$$

At certain lower pH and temperature conditions, the elemental sulfur is stable and will cause fouling and corrosion in this wet sour system. However, at pH > 8, the elemental sulfur will react with $(NH_4)SH$ in the excess of ammonia in the sour water to form ammonium polysulfide $(NH_4)_2S_n$ (D.26), which can act as a corrosion inhibitor further downstream in this wet sour system (see Sect. 8.2.2).

$$(NH_4)SH_{(aq)} + NH_{3(g)} + (n-1)S_{8(s)} \rightarrow (NH_4)_2S_{n(aq)} \tag{D.26}$$

Polythionic Acids

Polythionic acids are oxoacids $H_2S_xO_6$ ($x \geq 2$) which have a straight chain of sulfur atoms (–S–S–) attached to an end SO_3H-groups (Table D.1). Polythionic acids are unstable and possess strong reducing properties. The bond S–S is stronger than that of O–O in peroxides. Polythionic acids with x = 3 − 6 are the most stable.

Table D.1 Polythionic acids

The formula of polythionic acid	The name of polythionic acid[a]
$H_2S_2O_6$	Dithionic acid
$H_2S_3O_6$	Trithionic acid
$H_2S_4O_6$	Tetrathionic acid
$H_2S_5O_6$	Pentathionic acid
$H_2S_6O_6$	Sexathionic acid

[a]The name of polythionic acids is determined by the number of sulfur atoms in a chain

Polythionic acids exist only in solution and in course of time they decompose with yielding sulfuric acid (H_2SO_4), sulfur dioxide (SO_2) and elementary sulfur (S_8) (D.27) [17, 19]:

$$H_2S_3O_{6(aq)} \rightarrow H_2SO_{4(aq)} + SO_{2(g)} + 1/8S_{8(s)} \quad (D.27)$$

Polythionic acids can form in refinery process units during shutdowns when equipment (inner and outer surfaces of tubes and others) with sulfide deposits (mostly FeS, but also Cr_2S_3 and NiS formed as a result of high temperature sulfidic corrosion) are opened, cooled and exposed to moisture air (D.28). Water and oxygen can appear also from steam and wash water used to release tubes or other equipment from hydrocarbons before shutdown and inspection.

$$2FeS_{(s)} + 4O_{2(g)} + H_2O_{(l)} \rightarrow H_2S_2O_{6(aq)} + Fe_2O_{3(s)} \quad (D.28)$$

It is interesting to note that the reaction (D.28) was used in the Middle Ages for obtaining sulfuric acid through intermediate salt (D.29) [17]:

$$4FeS_{(s)} + 9O_{2(g)} + 2H_2O_{(l)} \rightarrow 4Fe(OH)SO_{4(aq)} \quad (D.29)$$

and further heating (D.30):

$$2Fe(OH)SO_{4(aq)} \rightarrow \mathbf{H_2SO_4}_{(aq)} + Fe_2O_{3(s)} + SO_{3(g)} \quad (D.30)$$

Thus, most likely that sulfuric acid accompanies polythionic acids.

Polythionic acids also can be formed by the reaction of water and oxygen with oxidizable sulfur species (S_8 or H_2S), namely, as a result of combustion of H_2S in refinery flares. Hydrogen sulfide can react also with sulfurous acid to form polythionic acids. Usually flare tips are made of austenitic stainless steels or high-nickel alloys and are prone to attack by polythionic acids. These acids are easily cathodically reduced and thus function as cathodic depolarizers which promote dissolution of the metal at the grain boundaries depleted by chromium

We will explain the state of alloys named by *sensitization* [20]. When austenitic stainless steels are heated in the range 400–815 °C (production, fabrication, welding, heat treating, slow cooling through this temperature range, and process

Table D.2 Sensitized temperature range for some austenitic alloys[a] [21]

Alloy[a]	General name of austenitic alloys	Sensitization temperature range (°C)
SS 304L	Low-carbon stainless steels (<0.03 wt% C)	400–815
SS 316L		
SS 317L		
SS 304	Standard-carbon stainless steels (0.08 wt% max C)	370–815
SS 316		
SS 317		
SS 304H	High-carbon stainless steels (0.1 wt% max C)	
SS 316H		
SS 321	Chemically stabilized stainless steels	400–815
SS 347		
Incoloy 800	Nickel alloys	370–815
Incoloy 825		650–760
Inconel 625		

[a]Chemical composition of alloys is given in Appendix F

operations at refinery units), chromium and carbon react to form carbides $Cr_{23}C_5$ and Cr_7C_3. These carbides precipitate at the grain boundaries where nucleation is favored. As a result, grains become depleted by chromium, namely grains turn into areas without this element which provided formerly the corrosion resistance to the alloy. Grains become unprotected against corrosive components (i.e. polythionic acids) in the environment, and alloys are susceptible to intergranular corrosion and SCC. This state is named *sensitization*. The higher the temperature (in the range given in the Table D.2), the shorter the time of exposure required for sensitization. The maximum rate of carbide precipitation is 675 °C. Polythionic acid SCC is a cracking of a sensitized stainless steel or other sensitized austenitic alloys under combined action of tensile stress and corrosion in the presence of polythionic acids which are usually formed from sulfide deposits, water and oxygen.

Appendix E
Physicochemical Properties of Alkalis

Sodium hydroxide. Similar to iron, magnetite (Fe_3O_4) has amphoteric properties that means it can be dissolved by both acid and concentrated alkali NaOH solution (E.1) [22]:

$$Fe_3O_{4(s)} + 4NaOH_{(aq)} \rightarrow 2NaFeO_{2(aq)} + Na_2FeO_{2(aq)} + 2H_2O_{(l)} \quad (E.1)$$

When magnetite is removed, NaOH may react directly with the iron (E.2):

$$Fe_{(s)} + 2NaOH_{(aq)} \rightarrow Na_2FeO_{2(aq)} + H_{2(g)} \quad (E.2)$$

Sodium hydroxide reacts with aluminum alloys causing severe and quick corrosion and producing flammable hydrogen gas (E.3):

$$Al_{(s)} + NaOH_{(aq)} + H_2O_{(l)} \rightarrow NaAlO_{2(aq)} + 3/2H_{2(g)} \quad (E.3)$$

Sodium carbonate (Na_2CO_3, soda ash, washing soda) dissolved in water forms strong alkaline aqueous solutions up to pH ~ 13 (E.4):

$$Na_2CO_{3(s)} + H_2O_{(l)} \rightarrow \mathbf{NaOH}_{(aq)} + NaHCO_{3(aq)} \quad (E.4)$$

It was produced in olden times from the ashes of many plants growing in sodium-rich soils, therefore the salt was named soda ash.

Ammonia and ammonium hydroxide. Ammonia (NH_3) is a colorless gas at normal conditions with a characteristic pungent suffocating smell. Ammonia has alkali properties and hazardous in its concentrated form: causes severe skin burns, eye damage and very toxic to aquatic life. Ammonia is dissolved well in water to form ammonium hydroxide NH_4OH (30 wt% maximum at 20 °C) (E.5):

$$NH_{3(g)} + H_2O_{(l)} \leftrightarrow NH_3 \cdot H_2O_{(aq)} \leftrightarrow NH_4OH_{(aq)} \leftrightarrow NH_4{}^{+}{}_{(aq)} + OH^{-}{}_{(aq)} \quad (E.5)$$

A. Groysman, *Corrosion Problems and Solutions in Oil Refining and Petrochemical Industry*, Topics in Safety, Risk, Reliability and Quality 32,
DOI 10.1007/978-3-319-45256-2

Amines. Amines are organic substances, derivatives of ammonia (NH_3), wherein one or more hydrogen atoms are replaced by alkyl (R, e.g., CH_3) or aryl (Ar, e.g., C_6H_5) group. Amines can be differentiated into three categories.

A. Primary amines—one of three hydrogen atoms in ammonia is replaced by alkyl or aryl. Examples: methylamine (CH_3–NH_2), phenylamine (C_6H_5–NH_2, named also aniline).
B. Secondary amines—two of three hydrogen atoms in ammonia are replaced by alkyl or aryl. Examples: dimethylamine $(CH_3)_2NH$, diphenylamine $(C_6H_5)_2NH$.
C. Tertiary amines—all three hydrogen atoms in ammonia are replaced by alkyl or aryl. Examples: trimethylamine $(CH_3)_3N$, dimethylethanol amine $(CH_3)_2NC_2H_4OH$.

Like ammonia, amines are alkalis as they produce OH^- anions in water (E.6–E.8).

$$\begin{aligned} CH_3NH_{2(g)} + H_2O_{(l)} &\leftrightarrow CH_3NH_2 \cdot H_2O_{(aq)} \leftrightarrow CH_3NH_3OH_{(aq)} \\ &\leftrightarrow CH_3NH_3{}^+{}_{(aq)} + \mathbf{OH}^-{}_{(aq)} \end{aligned} \tag{E.6}$$

$$(CH_3)_2NH + H_2O_{(l)} \leftrightarrow (CH_3)_2NH_2OH_{(aq)} \leftrightarrow (CH_3)_2NH_2{}^+{}_{(aq)} + \mathbf{OH}^-{}_{(aq)} \tag{E.7}$$

$$\begin{aligned} (CH_3)_2NC_2H_4OH + H_2O_{(l)} &\leftrightarrow [(CH_3)_2NHC_2H_4OH]OH_{(aq)} \\ &\leftrightarrow [(CH_3)_2NHC_2H_4OH]^+{}_{(aq)} + \mathbf{OH}^-{}_{(aq)} \end{aligned} \tag{E.8}$$

Their strength depends on the dimensions and structure of a radical. Alkyl amines are stronger alkalis than ammonia. Aryl amines are weaker alkalis than ammonia. Amines can be dissolved both in water and in hydrocarbons. The more molar weight of radical, the better solubility of amine in hydrocarbons. Amines are used as neutralizers of acids (see reaction E.9), as corrosion inhibitors, and as absorbers-neutralizers of acid gases (H_2S and CO_2) at gas-treating, sulfur recovery, and amine treating units (see reaction E.10) (Tables E.1 and E.2).

$$C_2H_5NH_{2(g)} + HCl_{(g)} \rightarrow C_2H_5NH_3Cl_{(s)} \tag{E.9}$$

$$\begin{aligned} NH_2C_2H_4OH_{(aq)} + H_2S_{(g)} &\leftrightarrow (NH_3C_2H_4OH)HS_{(aq)} \\ &\leftrightarrow (NH_3C_2H_4OH)^+{}_{(aq)} + HS^-{}_{(aq)} \end{aligned} \tag{E.10}$$

Then H_2S is desorbed from the amine solution in the regenerator (E.11):

$$(HS)^-{}_{(aq)} + (NH_3C_2H_4OH)^+{}_{(aq)} \rightarrow H_2S_{(g)} + NH_2C_2H_4OH_{(aq)} \tag{E 11}$$

Table E.1 Amine neutralizers

Name	Formula
Ammonia	NH_3
Ethylamine	$CH_3CH_2NH_2$
Diethylamine	$CH_3CH_2NHCH_2CH_3$
Tripropylamine	$(CH_3CH_2CH_2)_3N$
Monoethanolamine (MEA)	$NH_2CH_2CH_2OH$
Dimethylethanolamine (DMEA)	$(CH_3)_2NCH_2CH_2OH$
Ethylenediamine (EDA)	$C_2H_4(NH_2)_2$
Methoxypropylamine (MOPA)	$CH_3OCH_2CH_2CH_2NH_2$
Morpholine	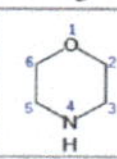
Diethylaminoethanol (DEAE)	$(C_2H_5)_2NCH_2CH_2OH$
Dimethylisopropanolamine (DMIA)	$(CH_3)_2NCH(CH_3)CH_2OH$
N-methylmorpholine	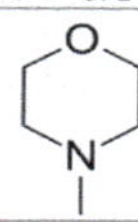
Cyclohexylamine	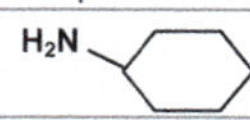
Benzylamine	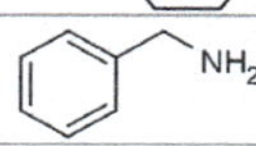

Table E.2 Amines for absorption H_2S and CO_2

Name	Formula
Monoethanolamine (MEA)	$H_2NC_2H_4OH$
Diethanolamine (DEA)	$HN(C_2H_4OH)_2$
Methyl diethanolamine (MDEA)	$CH_3N(C_2H_4OH)_2$ 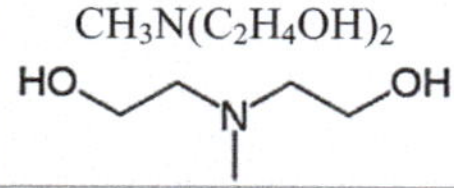
Diisopropanolamine (DIPA)	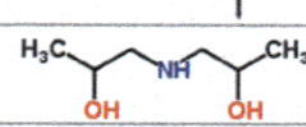

In addition to H_2S, different contaminants (formic acid, acetic acid, etc.) can react with amine (E.12):

$$RCOOH_{(aq)} + NH_2C_2H_4OH_{(aq)} \rightarrow (RCOO)^-{}_{(aq)} + (NH_3C_2H_4OH)^+{}_{(aq)} \quad (E.12)$$

producing heat stable amine salts (HSAS) (see Sect. 4.4).

Appendix F
Chemical Composition of Alloys

Chemical composition of alloys using for equipment, piping and accessories in oil refining and petrochemical industry is shown (Tables F.1-F.13). These alloys are ferrous ones, nickel alloys and non-ferrous alloys (based on aluminum, copper and titanium). Influence of oxygen and sulfur on steel properties, detailed explanation of killed steel, elucidation regarding niobium and columbium in stainless steels are given. Anti-abrasive materials (Table F.9) and materials resistant to galling (Table F.10) are shown as well.

Table F.1 Chemical composition of carbon steels (wt%)

Alloy			C	Mn	P	S	Si	Cu	Fe
UNS	AISI/ ASTM/ ASME	EN[a]							
G10100	C1010		0.10	0.3–0.5	Max 0.04	Max 0.05	0.1	–	Rem.[b]
K02801	A285 Gr.C		Max 0.28	Max 0.90	Max 0.035	Max 0.035	–	–	Rem.[b]
K03101	A515 Gr.70		Max 0.35	Max 0.90	Max 0.035	Max 0.040	0.15–0.30	–	Rem.[b]
K02700	A516 Gr.70		0.27–0.31	0.79–1.30	Max 0.035	Max 0.035	0.13–0.45	–	Rem.[b]
	A516 Gr.70 HIC[c]		0.27–0.31	0.79–1.30	Max 0.001	Max 0.001	0.13–0.45	–	Rem.[b]
K01200	A 179	(DIN 1629; St 35.4)	0.06–018	0.27–0.63	Max 0.035	Max 0.035	Max 0.25	–	Rem.[b]
K03504	A 105[d]		Max 0.35	0.60–1.05	Max 0.035	Max 0.040	0.10–0.35	Max 0.040	Rem.[b]
K03006	A 106[e] Gr. B		Max 0.30	0.27–0.93	Max 0.035	Max 0.035	Min 0.10	Max 0.040	Rem.[b]

(continued)

A. Groysman, *Corrosion Problems and Solutions in Oil Refining and Petrochemical Industry*, Topics in Safety, Risk, Reliability and Quality 32,
DOI 10.1007/978-3-319-45256-2

Table F.1 (continued)

Alloy			C	Mn	P	S	Si	Cu	Fe
UNS	AISI/ ASTM/ ASME	EN[a]							
K01807	SA 214		Max 0.18	0.27–0.63	Max 0.035	Max 0.035	Max 0.5		

[a]EN—European Standard
[b]Rem. = Remainder
[c]HIC—Hydrogen Induced Cracking resistant (for sour service). HIC resistant steel is manufactured via an electric arc furnace with desulfurisation, dephosphorisation, ladle refining and vacuum degassing to provide ultra clean and homogeneous steel. They are carbon steels characterized with ultra low sulfur concentration (below 0.001 wt% instead 0.035 wt% maximum in usual steels), vacuum degassing, normalizing heat treatment and calcium addition to cause sulfide shape control, and fully killed (oxygen concentration below 0.002 wt%); maximum carbon concentration 0.18 %; maximum carbon equivalent 0.43; and maximum hardness of 22 HRC (Hardness Rockwell Scale)
Carbon Equivalent: $C_{eq} = [C + Mn/6 + (Cr + Mo + V)/5 + (Ni + Cu)/15]$ wt%
HIC Tested NACE TM-0284 (2003)/NACE MR-0175/ISO 15156-2; SSCC Tested (Sulfide Stress Corrosion Cracking) NACE TM-0177 (2005)/NACE MR-0103 (2007), SOHIC (Stress Oriented Hydrogen Induced Cracking)
[d]Ni (0.40 max); Cr (0.30 max); Mo (0.12 max); V (0.08 max). These four elements and copper combined shall not exceed 1 wt%
[e]Ni (0.40 max); Cr (0.40 max); Mo (0.15 max); V (0.08 max). These four elements and copper combined shall not exceed 1 wt%

Influence of Elements on Steel Properties

Oxygen. Oxygen may become dissolved in liquid metal during the steel making process. This dissolved oxygen can combine with carbon (which is added to the steel as an alloying element) to form carbon oxides' bubbles and with iron to precipitate as iron oxides. They are trapped in the casting forming blowholes and can act as initiation points for failure. In other words, these oxide inclusions lower mechanical properties of steel, and toughness among them. Formation of the carbon oxides' bubbles and other deleterious oxides can be eliminated through the addition of deoxidizing elements (Si, Ca, Al, Mn)—removing of oxygen from the melt during solidification. This steel is named killed steel.

Killed steel is a steel that has been completely deoxidized by the addition of silicon, calcium, aluminum, ferrosilicon (an alloy of iron and silicon), or manganese, before casting and forging. Why this steel is called killed? When we compare a fresh non-killed slab and killed one, the first has bubbles coming out (as if alive) and the other is dead, hence killed. Killed steels usually quietly solidify in the mould, with no gas bubbling out. As a result, killed steel is characterized by dense in structure, uniform in composition (a high degree of chemical homogeneity), free from gas porosity (blowholes) and not so segregative as other types of steel. If a steel has silicon content of more than 0.10 wt% (usually 0.15–0.30 wt%) it is considered to be killed. Usually silicon-killed steels contain 0.15 wt% Si and more than 0.24 wt% carbon. Killed steel is also called deoxidized or silicon-killed carbon steel.

Table F.2 Chemical composition of carbon steels for piping (wt%)

Alloy	Fe	C[a]	Si	Mn[a]	P[a]	S[a]	Nb + V[a]	Ti[a]	Cr[a]	Mo[a]	Ni[a]
API 5L X65	Rem.[b]	0.16	0.45	1.65	0.02	0.01	0.05	0.04			
API 5L Gr. B	Rem.[b]	0.26	0.45	1.20	0.03	0.03	0.04	0.04			
ASTM A234 WPB[c]	Rem.[b]	0.30	0.10[d]	0.29–1.06	0.05	0.058	0.10		0.4	0.15	0.4
ASTM A216 Gr. WCB[e]	Rem.[b]	0.16	0.10[d]	1.0	0.04	0.045	0.03		0.5	0.20	0.5
ASTM A209-03[f]	Rem.[b]	0.10–0.20		0.30–0.80	0.025	0.10–0.50	0.01			0.44–0.65	

[a]Maximum content
[b]Rem. = Remainder
[c]ASTM A234 WPB; contains also Cu (0.4 max) and V (0.08 max) [24]
[d]Minimum
[e]UNS J03002, contains also Cu (0.30 max) and V (0.03 max)
[f]0.5 Mo seamless carbon-molybdenum alloy-steel boiler and superheater tubes [25]

Table F.3 Chemical composition of carbon steel stud bolts[a]

ASTM	Fe	C	Mo	Cr	Si	Mn	P_{max}	S_{max}	Hardness, HRC
A 193 Grade B7	Rem.	0.38–0.48	0.15–0.25	0.8–1.1	0.2–0.35	0.75–1	0.04	0.04	35
A 193 Grade B7M	Rem.	0.38–0.48	0.15–0.25	0.8–1.1	0.2–0.35	0.75–1	0.04	0.04	21.7

[a]Stud bolts (AISI 4140/4142)—low carbon alloy steel grades followed by quenching and tempering (a heat-treated chromium-molybdenum steel widely used for medium high-temperature service)

Sulfur. Sulfur affects the toughness of steels due to the presence of manganese sulfide (MnS) stringers and other sulfide inclusions. Modern steels (after 1980s) contain significantly less sulfur (0.010 wt% vs. 0.025 wt%). This is one reason why modern pipeline and vessel steels are considered to be less susceptible to brittle behavior of all types (Table F.2) [23].

Table F.4 Chemical composition of low-alloy Cr–Mo steels (wt%)

ASTM	UNS	Alloy	Cr	Mo	C_{max}	Si_{max}	Mn	P_{max}	S_{max}
A213 Grade T22	K21590	2.25Cr–1Mo	1.9–2.6	0.87–1.13	0.05–0.15	0.5	0.3–0.6	0.025	0.025
A199 Grade T5	K41545	5Cr–0.5Mo	4.0–6.0	0.45–0.65	0.15	0.5	0.3–0.6	0.025	0.025
A213 Grade T5	K41545	5Cr–0.5Mo	4.0–6.0	0.45–0.65	0.15	0.5	0.3–0.6	0.035	0.030
A213 Grade T9	K90941	9Cr–1Mo	8.0–10.0	0.90–1.10	0.15	0.25–1.00	0.3–0.6	0.025	0.025

Niobium and Columbium in Stainless Steel

We should elucidate regarding the use of two names of the same element: niobium (Nb) and columbium (Cb). In chemistry it is widespread that one substance or element has many names (synonyms). For instance, tungsten and wolfram (W), hydrochloric acid and muriatic acid (HCl), ammonia and hydrogen nitride (or tri-hydrogen nitride) (NH_3), water and dihydrogen monoxide (H_2O).

The English chemist Charles Hatchett in 1801 found new element in a mineral sample which was sent from the United States and named it columbium (Cb) (after Columbia, the poetical name for the United States) [27]. In 1846, the German chemist Heinrich Rose rediscovered new element in the columbite ore and named it niobium (Nb) (a daughter of Tantalus in Greek mythology). Thus 100 years the name columbium was used in America and niobium in Europe. To end this confusion, the name niobium was adopted in 1950 by IUPAC [28]. Some metallurgists, metal societies and commercial producers in the USA still use the name columbium instead niobium … but you know that these two elements are the same (Tables F.5, F.6, and F.8).

Anti-abrasive Materials

Following anti-abrasive (wear-resistant) coating materials can be used: fluorinated coatings (e.g., Teflon); cermet coatings like tungsten carbide/cobalt; chromium carbide/nickel chromium; oxide ceramics like chromium oxide and alumina; fused self fluxing alloys (NiCrSiB); various hard alloys of iron, nickel, chromium or cobalt (Table F.9). Hardness of these materials is proportional to wear resistance and must be >50 HRC.

Table F.5 Chemical composition of stainless steels (wt%)

Alloy		Cr	Ni	C[a]	Mo	Mn[a]	P[a]	S[a]	Si[a]	N[a]	Fe
UNS	AISI										
S30400[b]	304	18–20	8–12	0.08		2.0	0.045	0.03	0.75	0.1	Rem.[c]
S30403[d]	304L	18–20	8–12	0.03		2.0	0.045	0.03	0.75	0.1	Rem.[c]
S30409[e]	304H	18–20	8–11	0.04–0.10		2.0	0.04	0.03	1.0		Rem.[c]
S30900	309	22–24	12–15	0.20		2.0	0.045	0.03	1.0		Rem.[c]
S31600[b]	316	16–18	10–14	0.08	2–3	2.0	0.045	0.03	0.75	0.1	Rem.[c]
S31603[d]	316L	16–18	10–14	0.03	2–3	2.0	0.045	0.03	0.75	0.1	Rem.[c]
S31609[e]	316H	16.5–18.5	11–14	0.04–0.10	2.0–2.5	2.0	0.05	0.02	0.75		Rem.[c]
S31640[f]	316Cb	16–18	10–14	0.08	2–3	2.0	0.045	0.015	1.0		Rem.[c]
S31700[b]	317	18–20	11–15	0.08	3–4	2.0	0.045	0.03	0.75	0.1	Rem.[c]
S31703[d]	317L	18–20	11–15	0.03	3–4	2.0	0.045	0.03	0.75	0.1	Rem.[c]
S32100[g]	321	17–19	9–12	0.08		2.0	0.045	0.03	0.75		Rem.[c]
S34700[h]	347	17–19	9–12	0.08		2.0	0.045	0.03	1.0		Rem.[c]
S44400[i]	444	17.5–19.5	1 max	0.025	1.75–2.5	1.0	0.04	0.03	1.0	0.035	Rem.[c]
S41000[j]	410	11.5–13.5		0.015		1.0	0.04	0.03	1.0		Rem.[c]
S42000[k]	420	12.0–14.0		0.015		1.0	0.04	0.03	1.0		Rem.[c]
	UN 316 ELC[l]	18	11.5	0.025	2.8	0.8	0.04	0.03	0.[illegible]	0.035	Rem.[c]
	A217 Gr. CA15[m]	11.5–14.0	1	0.015	0.5	1.0	0.04	0.04	1.[illegible]		Rem.[c]

[a]Maximum content
[b]Standard-carbon stainless steels (0.08 wt% max C)
[c]Rem. = Remainder
[d]Low-carbon stainless steels (<0.03 % C)
[e]High-carbon stainless steels (0.1 wt% max C)
[f]Cb (Nb) 10 × C
[g]Chemically stabilized stainless steel: Ti is present in content 5 × C (0.70 max)
[h]Chemically stabilized stainless steel: Nb + Ta are present in content 10 × C (0.80 max)
[i]Ferritic stainless steel. It contains also titanium + columbium [0.20 + 4 × (C + N) min.] − 0.8 wt% max
[j]Martensitic SS 410 (ASTM A268 TP410)
[k]Martensitic SS 420 (13 Cr)
[l]Electrodes austenitic SS with rutile coating containing very low carbon amount for welding austenitic SS (316, 316L, 317, 317L): UN 316 ELC (Z-316L) (EN 1600; BS 2926; DIN 8556; AWS A5.4)
[m]Martensitic SS [26]

Table F.6 Chemical composition of proprietary austenitic stainless steels (wt%)

Alloy		Cr	Ni	C[a]	Mo	Mn[a]	P[a]	S[a]	Si[a]	N[a]	Fe
Proprietary	UNS										
AL-6XN	N08367	20–22	23.5–25.5	0.03	6–7	2.0	0.04	0.03	1.0	0.18–0.25	Rem.[d]
20Mo-6[b]	N08026	22–26	33–37	0.06	5–6.7	1.0	0.03	0.03	0.5	0.17–0.40	Rem.[d]
20Cb-3[c]	N08020	19–21	32.5–35.0	0.06	2–3	2.0	0.035	0.035	1.0	0.18–0.25	Rem.[d]

[a]Maximum content
[b]It contains also 1–3 Cu wt%
[c]Another name is Alloy 20; it contains also (wt%): 3–4 Cu, 8 × C_{min} (1 max) Nb + Ta. Proprietary alloy produced by Carpenter company
[d]Rem. = Remainder

Table F.7 Chemical composition of duplex stainless steels (wt%)

Alloy		Cr	Ni	C[a]	Mo	Mn[a]	P[a]	S[a]	Si[a]	N	Cu	Fe
UNS	Proprietary designation											
S31803	SAF 2205	21–23	4.5–6.5	0.3	2.5–3.5	2.0	0.03	0.02	1	0.08–0.2		Rem.[b]
S32205	SAF 2205	22–23	4.5–6.5	0.3	3–3.5	2.0	0.03	0.02	1	0.14–0.2		Rem.[b]
S32003	2003	19.5–22.5	3–4	0.03	1.5–2.0	2.0	0.03	0.02	1	0.14–0.2		Rem.[b]
S32101	2101	21–22	1.35–1.7	0.04	0.1–0.8	4–6			1	0.2–0.25	0.1–0.8	Rem.[b]
S32304	2304	23	4	0.02	0.2			0.001		0.1		Rem.[b]
S32550	Ferralium alloy 255	24–27	4.5–6.5	0.04	2.9–3.9	1.5			1			Rem.[b]
S32520	255+	24–26	5.5–8	0.03	3–4	1.5	0.035	0.02	0.8	0.02–0.35	0.5–2	Rem.[b]
S32750	SAF 2507	24–26	6–8	0.03		1.2	0.035	0.02	0.8	0.24–0.32	0.5	Rem.[b]
S32760[c]	Zeron 100[c]	24–26	6–8	0.03	3–4	1.0	0.03	0.02	1	0.2–0.3	0.5–1	Rem.[b]
S32950	7Mo-PLUS	26–29	3.5–5.2	0.03	1–2.5	2.0	0.035	0.01	0.6	0.15–0.35		Rem.[b]

[a]Maximum content
[b]Rem. = Remainder
[c]Contains tungsten W (0.5–1 wt%)

Table F.8 Chemical composition of nickel alloys (wt%)

Alloy		Ni	Fe	Cr	Mo	W	Co	Cu	Ti	Al	C	Mn	S	Si
Proprietary name	UNS						max	max			max	max	max	max
Incoloy 800H (Alloy 800H)	N08810[a]	30–35	39.5 min	19–23				0.75	0.15–0.60	0.15–0.60	0.05–0.10	1.5	0.015	1.0
Incoloy 800 (Alloy 800)	N08800	30–35	39.5 min	19–23				0.75	0.15–0.60	0.15–0.60	0.10	1.5	0.015	1.0
Inconel 600 (Alloy 600)	N06600	72	8	15.5				0.50			0.15	1.0	0.015	0.5
Inconel 625 (Alloy 625)	N06625[b]	Rem.[c]	2.5	21.5	9.0						0.10			
Incoloy 825 (Alloy 825)	N08825	Rem.[c]	30.0	21.5	3.0			2.2			0.03			
Hastelloy[d] C 276	N10276[e]	Rem.[c]	4–7	14.5–16.5	15–17	3.0–4.5	2.5				0.01	1.5	0.03	0.08
Hastelloy B-2 (Alloy B-2)	N10665	Rem.[c]	2.0 max	1.0 max	26–30		1.0				0.02	1.0	0.03	0.1

[a]ASTM B408-06 (2011)—Ni–Fe–Cr alloy
[b]Contains also Nb (3.6 wt%)
[c]Rem.= Remainder
[d]Hastelloy 'C-type'
[e]It contains also V (0.35 max) and P (0.04 max)

Table F.9 Abrasive (wear) resistant alloys (wt%)

Alloy	Ni	Co	Cr	W	Fe	Mo	Si	B	C	Hardness, HRC
Colmonoy[a] 6	Rem.		14		4.5		4.2	3	0.7	56–62
Colmonoy[a] 88	Rem.		15	16.5	3.5		4.0	3	0.8	59–64
Stellite[b] 190	<3	Rem.	26	14	<3				3.2	55–62
Stellite[b] 720	<3	Rem.	33		<3	18			2.4	65
HC-250[c]	0.92		30.5		Rem.		0.97			

[a]Colmonoy is registered trademark of Wall Colmonoy
[b]Stellite is registered trademark of Deloro Stellite
[c]Abrasion resistant cast iron (UNS F45009)

Table F.10 Chemical composition of alloys resistant to galling[a] (wt%)

Alloy		Fe	Cr	Mn	Ni	Si	N	Mo	C
Proprietary name	UNS							max	max
Nitronic 60[b] (Alloy 218)	S21800	[d]Rem.	16–18	7–9	8–9	3.5–4.5	0.018–0.08	0.75	0.10
Gall-Tough Stainless[c]	S20161	[d]Rem.	15–18	4–6	4–6	3.0–4.0	0.08–0.20	0.75	0.15
Gall-Tough Plus Stainless[c]	S21800/ S20162	[d]Rem.	16.5–21	4–8	3.0–4.0	2.5–4.5	0.05–0.25	0.5–2.5	0.15

[a]The additions of Mn and Si gave this alloy a matrix to resist galling, wear, and fretting
[b]Nitronic is a trademark of AK Steel
[c]Manufactured by Carpenter company [29]
[d]Rem. = Remainder

Non-ferrous Alloys

Chemical composition of non-ferrous alloys based on aluminum, copper and titanium are given in Tables F.11, F.12 and F.13 respectively.

Table F.11 Chemical composition of aluminum alloys (wt%)

Alloy		Mg	Cu	Mn	Si	Fe	Cr	Other metals	Al
ASTM	UNS								
Al 1050[a]	A91050	0.05 max	0.05 max	0.05 max	0.25 max	0.40 max	0.05 max	Ti < 0.03 Zn < 0.05 V < 0.05	Rem.[b]
Al 1100	A91100		0.06						Rem.[b]
AlSi9Mg[c]		0.25–0.45	0.05 max	0.1 max	9–10	0.19 max		Ti < 0.15 Zn < 0.07	Rem.[b]

[a]Aluminum EN AW-1050A (Al 99.5)—commercially pure aluminum
[b]Rem.= Remainder
[c]Aluminum EN AC-43300

Table F.12 Chemical composition of copper alloys (wt%)

Alloy		Cu	Zn	Fe	Ni	Pb	Sn	As
Name	UNS							
Admiralty brass[a]	C44300	70–73	27–30	0.06 max		0.07	0.9–1.2	0.02–0.06
Naval brass[b]	C46400	59.5–63.5	39.2				0.5–1.5	
Al-Bronze[c]	C95800							
90-10 Copper-Nickel[d]	C70600	Rem.	1 max	1–1.8	9–11	0.05		
70-30 Copper-Nickel	C71500	Rem.	1 max	0.04–1	29–33	0.05		
Monel[e]	N04400	28–34		2.5 max	63 min			

[a]CDA 443(70Cu-30Zn) [30]
[b]CDA 464(CuZn38Sn1) [30]
[c]Al-Bronze [31]
[d]CDA 706; 1 wt% Mg max [30]
[e]Nickel-copper Alloy 400. In addition to Ni, Cu, and Fe, the following elements are present (wt%): Mn (2 max) Si (0.5 max) S (0.024 max) C (0.3 max)

Table F.13 Chemical composition of titanium alloys (wt%) [32, 33]

Alloy		C_{max}	N_{max}	O_{max}	H_{max}	Fe_{max}	Mo	Ni	Pd	Ti
Grade	UNS									
1	R50250	0.1	0.03	0.18	0.015	0.2				Rem.[a]
2	R50400	0.080	0.03	0.25	0.015	0.3				Rem.[a]
4	R50700	0.1	0.05	0.4	0.015	0.5				Rem.[a]
5[b]	R56400			0.2		0.25				90
7	R52400								0.12–0.25	Rem.[a]
12	R53400	0.080	0.03	0.25	0.015	0.3	0.2–0.4	0.6–0.9		Rem.[a]
16	R52402								0.04–0.08	Rem.[a]

[a]Rem. = Remainder
[b]Gr.5—Ti–6Al–4V (6 % Al; 4 % V)

Appendix G
Hydrogen

In order to understand origin and occurring of hydrogen attack and damages of metals, we will describe physicochemical properties of hydrogen and mechanism of its appearing on metal surface and some characteristics of its behavior.

Hydrogen (H) is the only element that can exist without neutrons. Under ordinary conditions on the Earth, hydrogen exists as the diatomic gas H_2. Hydrogen is the lightest, the smallest and the most abundant element in the universe. However, hydrogen gas is very rare in the Earth's atmosphere (0.000055 wt%, or 0.55 ppmv) because of its light weight, which enables it to escape from the Earth more easily than heavier gases. In spite of this, hydrogen is the third most abundant element on the Earth's surface, mostly in the form of hydrocarbons and water.

Probably the first report that hydrogen had a deleterious effect on mechanical properties (decrease of toughness and strength) of iron and steels immersed in acids was made by Johnson W.H. in 1873–1875 [34, 35]. Since that time advances have been achieved in understanding of mechanism of hydrogen damages, preventive actions, and monitoring measures. New alloys possessing increased resistance to hydrogen attack (having outstanding combinations of high tensile strength and high fracture toughness) and standards of their use in hydrogen service were created [36–40].

We will describe how hydrogen atoms are formed on metal surface in aqueous solutions of electrolytes (T < 100 °C) and in hydrocarbon non-electrolyte phase containing gaseous hydrogen H_2 at high temperature (T > 230 °C) and high pressure.

In liquid phase, hydrogen is formed on a metal surface as a result of an electrochemical reaction in acids

$$H_3O^+{}_{(aq)} + e^- \rightarrow H_{ads} + H_2O_{(l)} \qquad \text{(G.1)}$$

A. Groysman, *Corrosion Problems and Solutions in Oil Refining and Petrochemical Industry*, Topics in Safety, Risk, Reliability and Quality 32,
DOI 10.1007/978-3-319-45256-2

or in neutral and alkaline solutions:

$$H_2O_{(l)} + e^- \rightarrow H_{ads} + OH^-{}_{(aq)} \quad (G.2)$$

During cathodic polarization of metal surface (i.e. applying of impressed current in cathodic protection) adsorbed H atoms are formed, which partly are removed because of recombination and partly intrude into steel. The rate of hydrogen entering into steel depends on the surface concentration of H atoms, energy of absorption and energy of hydrogen desorption.

Most part of hydrogen atoms are recombined on metal surface and removed as H_2 gas molecules in two following processes:

(a) Recombination on the surface of the metal:

$$H_{ads} + H_{ads} \rightarrow H_{2(g)} \quad (G.3)$$

(b) Electrochemical desorption:

$$H_3O^+{}_{(aq)} + H_{ads} + e^- \rightarrow H_{2(g)} + H_2O_{(l)} \quad (G.4)$$

$$H_2O_{(l)} + H_{ads} + e^- \rightarrow H_{2(g)} + OH^-{}_{(aq)} \quad (G.5)$$

The mechanism of dissolution of hydrogen in metals from a gas phase ($T > 230\ °C$) includes adsorption of H_2 molecules (or other gas containing hydrogen, for instance, H_2S), its dissociation into adsorbed atoms H_{ads} on the metal surface and the absorption (diffusion, penetration) into the metal lattice. At first, hydrogen gas can be broken down at particular conditions to form atomic hydrogen radicals (H·) on metal surface:

$$H_{2(g)} \leftrightarrow H\cdot + H\cdot \quad (G.6)$$

In chemical equations, 'free radicals' are denoted by a dot placed immediately to the right of the atomic symbol. A 'radical' in chemistry is an atom that has unpaired valence electrons. This situation makes 'free radicals' highly chemically reactive. The reaction (G.6) is reversible and the opposite reaction is a recombination (see G.3).

When hydrogen atoms are inside of the volume of metals they also can recombine:

$$H_{abs} + H_{abs} \rightarrow H_{2(g)} \quad (G.7)$$

Some important physicochemical parameters of hydrogen and iron, solubility and diffusion coefficients of hydrogen in iron are shown in Tables G.1, G.2 and G.3.

Solubility of hydrogen in iron is 3×10^{-8} mole H/mole Fe at 25 °C and P_{H2} = 0.0981 MPa [18]. Solubility increases with increase of temperature and partial pressure of hydrogen in the environment (Table G.2).

Table G.1 Dimensions of hydrogen species and iron atoms [41]

Species		Diameter (D) ($\times 10^{12}$ m)	Ratio $D_{species}/D_{H+}$	Bond length ($\times 10^{12}$ m)
Name	Symbol			
Hydron[a]	H^+	0.0016	1	–
Atom (Radical)	H^o (H·)	13	8125	–
Hydride	H^-	50	31,250	–
Hydrogen	H_2	120	77,500	74[b]
Iron	Fe	248	155,000	248[c]

[a]The term 'hydron' is recommended by IUPAC in 1988 [42, 43]. It is used instead 'proton' without distinguishing between the hydrogen isotopes protium, deuterium and tritium
[b]The bond length H–H
[c]The distance Fe–Fe in the iron crystal

Table G.2 Solubility of hydrogen in iron at different temperatures [17]

T (°C)	500	700	900	1100	1200	1350	1450	1550
Solubility, volume of H_2 in one volume of Fe	0.05	0.14	0.37	0.55	0.65	0.80	0.87	2.05

Note Significant increase of solubility of hydrogen between 1450 and 1550 °C occurs because of changing aggregation state of iron ($T_{melting}$ = 1536 °C). The molten iron contains hydrogen in atomic state (H)

Table G.3 Diffusion coefficient of hydrogen in iron and steels

Iron or Steel	Diffusion coefficient of hydrogen (cm^2/sec)	Reference
Pure iron	1500×10^{-8}	[44]
12 % Cr steel (ferritic steel)	100×10^{-8}	[45]
Austenitic steel	2.3×10^{-8}	[46]

Appendix H
Metallographic Replication of Heat Exchanger Surface (Case Study 4.3.4.3)

Metallographic (microscopic) examination was carried out according to the standard ASTM E 1351 [47]. Replication is a nondestructive sampling procedure that records and preserves the topography of a metallographic prepared surface as a negative relief on a plastic film (replica). The replica permits the examination and analysis of the metallographic prepared surface with a light optical microscope or SEM. Surface nondestructive examinations (liquid-penetrate testing, wet fluorescent magnetic-particle testing, and ultrasonic testing) are recommended for detection hydrogen blistering and HIC [48]. What is important, that on-site metallurgical analysis (metallographic replication) unlike traditional nondestructive examination methods (for instance, acoustic emission, penetrant examination and ultrasonic flaw detection), can predict such dangerous failures as SCC and hydrogen embrittlement (namely, depth or width of decarburization, fissures, and cracks). These methods are especially vital for metallic constructions after a fire accident. On-site metallurgical testing can reveal which metallic components are actually damaged and must be replaced, and which components can be put back into operation again without risk. This form of nondestructive testing has been in use for several decades, and is standardized [47]. In-situ surface replication helps to determine remaining life of metallic constructions, structures and equipment [49].

Cellulose acetate replicas and light optical microscope Olympus BX60M were used. In order to examine if similar damage (decarburization) occurred with equipment worked under analogous hydrogen service, on-site nondestructive metallurgical analysis in the form of 16 replicas was carried out in different areas of welds in heat exchanger which has been in service for 16 years at the HDS unit (Fig. H.1). The results of replicas' examination are summarized in Table H.1.

<u>Results of replicas examination of the surface of the heat exchanger in HDS unit.</u>

(a) Cracks were found on the surface and after grinding 3 mm from the surface of the periphery weld no. 1 (inside) and the weld between the flange and the shell (Figs. H.2 and H.3).

A. Groysman, *Corrosion Problems and Solutions in Oil Refining and Petrochemical Industry*, Topics in Safety, Risk, Reliability and Quality 32,
DOI 10.1007/978-3-319-45256-2

Fig. H.1 General view of the heat exchanger and the locations of replicas examination: *1*, *2*—periphery welds; *3*—fasten; *4*—shell surface; *5*—flange; *6*—longitudinal weld

Table H.1 Results of replica's examination in heat exchanger [50]

Replica's number	Location of the replica's examination (see Fig. H.1)	Presence of cracks	Figure's number
1	The periphery weld no. 2, outside 1	No crack	–
2	The periphery weld no. 2, outside 2	No crack	–
3	The periphery weld no. 1, cross-section with fasten, inside	Crack	H.2a
4	The periphery weld no. 1, cross-section with fasten, inside, after polishing 3 mm of surface	Crack	H.2b
5	The periphery weld no. 1, cross-section with fasten, inside, after polishing 8 mm of surface	No crack	H.2c
6	The periphery weld no. 1, cross-section with shell, inside	Crack	–
7	The periphery weld no. 1, cross-section with shell, inside, after polishing 3 mm of surface	Crack	–
8	The periphery weld no. 1, cross-section with shell, inside, after polishing 8 mm of surface	No crack	–
9	The weld between flange and shell, inside	Crack	H.3a
10	The weld between flange and shell, inside, after polishing 3 mm of surface	Crack	H.3b
11	The longitudinal weld close to the periphery weld no. 1, inside	No crack	–
12	The longitudinal weld close to the periphery weld no. 2, inside	No crack	–
13	The periphery weld no. 2, inside 1	No crack	–
14	The periphery weld no. 2, inside 2	No crack	–
15	The cross-section between longitudinal weld and periphery weld no. 1, inside, after polishing 8 mm of surface	No crack	–
16		No crack	–

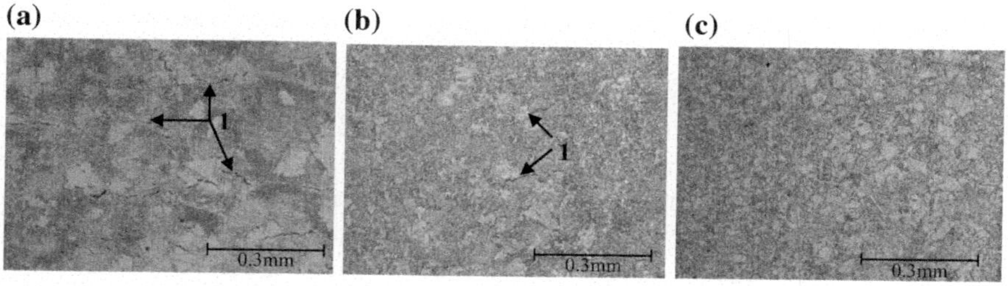

Fig. H.2 Replica images: the periphery weld no. 1, cross-section through the fasten, inside (HAZ) no. 1, (see Fig. H.1): **a** the surface; **b** after grinding 3 mm of the surface; *1*—cracks. **c** after grinding 8 mm of the surface (no cracks)

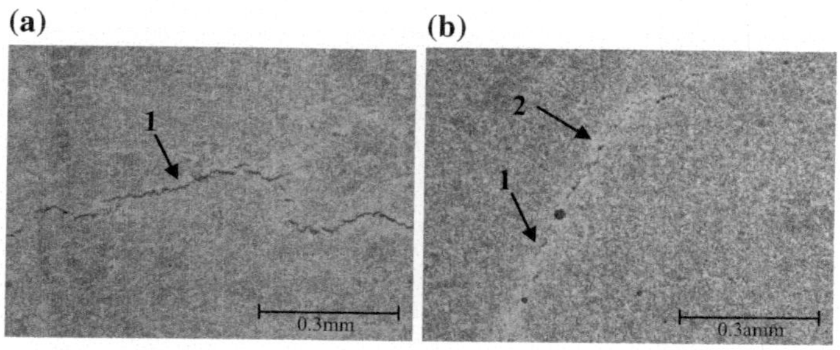

Fig. H.3 Replica images: the weld between the flange and the shell (HAZ), inside (see Fig. H.1): **a** surface; **b** after grinding 3 mm of the surface. *1*—cracks; *2*—characteristic structure of decarburization

(b) Cracks were not found on the inside and outside surfaces of the periphery weld no. 2 and on the longitudinal weld surface.
(c) After grinding 8 mm from the surface of the periphery weld no. 1 and the weld between the flange and shell, cracks were not found.

Discussion of results.

Results of replicas examination gave important information about the nature of HTHA, deepness of this attack and possibility to continue using this carbon steel heat exchanger after 16 years of service and fire accident. Figure H.3b showed decarburization in the carbon steel structure. Such a phenomenon (decarburization) was well documented as a result of HTHA [18, 39, 51–53]. Total result of the HTHA was loss of carbon in the phase of pearlite (decarburization), and as a result loss of mechanical properties (tensile strength and toughness) which in its turn resulted in fissures and cracks, as can be seen in Figs. H.2 and H.3. Thus, any hit or stress could result in a crack. This was detected when the pipes around this heat exchanger were examined. A hit with a hammer caused immediate crack of the pipes in the HAZ.

Hydrogen atoms penetrated up to the depth of 3 mm (because cracks were detected after the polishing of 3 mm of thickness—see Fig. H.2b) and maybe more, but less than 8 mm (cracks were not detected after polishing of 8 mm of thickness—see Fig. H.2c). This fact allowed making the conclusion that it was dangerous to continue using this heat exchanger and the pipes around it in the HDS unit.

Usually HTHA begins in the areas possessing (where are concentrated) high stresses, as hydrogen atoms diffuse to these areas. Therefore, HAZ, which have high stresses, are more sensitive to hydrogen attack [54].

Appendix I
Recommended Procedure for Passivation of Cooling Water Systems

This procedure is based on the proprietary standard of several oil refineries. Any renovation or connecting new piping and heat exchangers should be accomplished by proper cleaning and passivation (pretreatment) of a cooling water system to reduce corrosion and fouling. The procedure is described below.

1. Examination of inner surface of piping. Visual examination and by means of a boroscope.
2. Alkaline cleaning from oil and grease contaminants.
   - 2.1 Soda ash (1.8–2.2 wt%).
   - 2.2. Tri-sodium phosphate (0.15–0.5 wt%).
   - 2.3. Liquid detergent (0.15 wt%).
     Fresh tap water is used for the preparation of the solution. Alkaline cleaning is carried out by circulation at ambient temperature for 24 h.
   - 2.4. Drain and flush the system.
3. Passivation.
   - 3.1. Refill the system with fresh tap water and add corrosion inhibitors of iron (zinc-phosphate, 250–300 ppm) and of copper (benzotriazole, 5–10 ppm).
   - 3.2. Adjust pH to 6.8–7.2 and start the passivation.
   - 3.3. Add organic biocide and bio-dispersant (30–50 ppm) and circulate for 3 days.
   - 3.4. Additional corrosion zinc-phosphate inhibitor should be added to maintain the recommended value 250–300 ppm (19–24 ppm orto-phosphate, 23–27 ppm total phosphate).

A. Groysman, *Corrosion Problems and Solutions in Oil Refining and Petrochemical Industry*, Topics in Safety, Risk, Reliability and Quality 32,
DOI 10.1007/978-3-319-45256-2

4. Analytical program during passivation.
   4.1. Conductivity, total hardness, calcium hardness, chlorides, sulfates, total phosphate, total iron and copper concentration in circulated water should be measured daily; orto-phosphate and pH—three times a day.
5. If final start-up of cooling water system is delayed, circulate water 8 h once a week until normal operation begins. Maintain corrosion inhibitor concentration in water.
6. Before starting with normal operation the system should be emptied and refilled with fresh cooling water and injected of needed chemicals (corrosion inhibitors, anti-scaling agents, and biocides).

Appendix J
Fouling and Corrosion

Fouling formation mechanisms and an example of physicochemical composition of fouling in different equipment and units at oil refineries and petrochemical plants (Table J.1) are given in this Appendix.

Five Mechanisms of Fouling

Crystallization fouling can occur as a result of solidification and deposition. Crystallization fouling occurs due to cooling below the solidification temperature of a dissolved component (e.g., solidification of wax from crude oil). Deposition of dissolved salts occurs due to supersaturation of aqueous solutions which in its turn may be caused by following processes [55]:

(a) Water heating: $CaCO_3$, $MgCO_3$, $CaSO_4$, $MgSO_4$, $Ca_3(PO_4)_2$, $CaSiO_3$, $FeSiO_3$, $Fe_2(SiO_3)_3$, $Fe_3(PO_4)_2$, $FePO_4$.
(b) Cooling below the solubility limit.
(c) Variation of pH.
(d) Mixing of streams with different chemical composition.
(e) Evaporation of solvent.

Particulate fouling occurs as a deposition of small suspended particles by gravitational settling.

Chemical reaction fouling occurs in the volume of stream without participation of surface material and depends on crude oil composition and process conditions. Oxygen present in very small concentrations (few ppm) in petroleum products in storage tanks can form insoluble gums and deposits by the auto-oxidation mechanism including polymerization [56]. This mechanism is realized through formation of free organic radicals (R·) and peroxides (R_1–O–O–R_2) which also increase corrosiveness of petroleum products.

Polymerization of monomeric organic compounds can happen when cracked feedstock have been added to the crude. Olefines are absent in original crude oils

A. Groysman, *Corrosion Problems and Solutions in Oil Refining and Petrochemical Industry*, Topics in Safety, Risk, Reliability and Quality 32,
DOI 10.1007/978-3-319-45256-2

Table J.1 Physicochemical composition of fouling in different equipment and units at oil refineries and petrochemical plants

Equipment	Unit	Medium and conditions	Volatile substances[a] at 110 °C (wt%)	Burned substances[b] at 550 °C (wt%)	Ash[c] (wt%)	Chemical content of ash
Compressor	CDU[d]	Sour gases (H_2S)[e], 60 °C	32	19	49	FeS, $FeSO_4$
Compressor	VDU[f]	HC[g], H_2O, H_2S	54.2	34.2	11.6	FeS
Air cooler	FCCU	HC[g], H_2O, H_2S, NH_3	23	21	56	FeS (FeOOH traces)
Compressors, Heat exchangers	Isomerization	HC[g], H_2O, H_2S, 78 °C	0	0	100	FeS
Heat exchanger	VDU[f]	HVGO, 200–250 °C	0.5	87.5 (high molecular weight HC[f])	12	FeS, chlorides
Tank	VDU[f]	Residue Atm. column, 320 °C	0.4	56.8 (high molecular weight HC[g])	42.8	FeS, Ca, Si, Cl, Al
Pump	TPX[h]	HC[g] (C_7–C_9) 160 °C	26.2	70.5	3.3	FeS, Cl, Cu
Pump[i]	TPX[h]	HC[g] (C_5–C_7) 100 °C	7.3	42.4	50.4	
Pump	TPX[h]	HC[g] C_9 310 °C	3	88 (coke)	9	Rust
Tank	Tank farm	Aromatic C_{9+} 16 °C	33	6	61	Rust, chlorides
Pump[i]	TPX[h]	HC[g] C_6–C_9 20 °C	0	8	92	Rust
Heat exchanger	TPX[h]	HC C5–C9 60-90 °C	41.1	48.5	10.4	Rust
Furnace	TPX[h]	HC[g] C8–C9 310 °C	1.2	17.0	81.8	Fe, O, Si, S, Ni, Ca
Pump	BTX[j]	HC[g] C5–C6 70 °C	1.6	0.6	97.8	FeS
Pump	TPX[h]	HC[g] C7–C9 160 °C	17.1	56.9	26.0	FeS, Cl
Pump (after flash drum)	CDU[d]	HC[g] (H_2O, H_2S)	3.1	84.8	12.1	FeS
Pipeline[k]		Reformate	4	40.6	55.4	

[a]Water and light hydrocarbons
[b]Mostly organic substances and some inorganic compounds decomposing at T < 550 °C
[c]Inorganic substances
[d]CDU—Atmospheric crude distillation unit
[e]Sour gases contain also water vapor
[f]VDU—Vacuum distillation unit
[g]HC—Hydrocarbons
[h]TPX—Toluene Para-Xylene unit
[i]The filter was clogged every three months
[j]BTX—Benzene, Toluene, Xylene unit
[k]Pipeline from the CCR unit to the petrochemical plant

and are formed in cracked processes at refinery units and are prone to polymerization. As a result of polymerization, high molecular weight organic compounds are formed and deposit. The mechanism of polymerization is similar to auto-oxidation and involves radical formation (chain initiation), reaction propagation, and chain termination. Similar to auto-oxidation, dissolved oxygen, halides, sulfides, N-organic compounds, certain metals, and metallic compounds are capable to initiate polymerization. Under certain conditions, mercaptans (R–SH) and sulfides (R–S–R) can take part in corrosion reactions to produce free radicals (R·) that could initiate polymerization [56]:

$$Fe_{(s)} + \text{R-SH}_{(l)} \rightarrow FeSH_{(s)} + R\cdot \tag{J.1}$$

When the polymer becomes insoluble in its solution it precipitates out on metal surface.

Chemical compounds in the crude and petroleum products might undergo thermal decomposition (cracking) at high temperatures (T > 340 °C) to ultimately form coke. These temperatures are more likely to arise in furnaces rather than in heat exchangers. The thermal cracking mechanism involves the breaking down of higher molecular weight hydrocarbons into lighter alkanes and alkenes (olefins). The latter can take part in both auto-oxidation and polymerization.

Asphaltenes are high molecular weight polar molecules, which normally are present in heavy crude oils in a colloidal state (as an emulsion). They can precipitate when the solubility properties of the surrounding liquid changes due to blending of crudes, or changes in temperature and pressure. When heavy oil products are blended with aliphatic diluents, asphaltenes are precipitated and heavy fouling occurs. Thus, if incompatible crude oils are blended, asphaltenes may flocculate (clustering of individual dispersed droplets together) and come out of solution resulting in severe fouling. If diluents contain significant amount of aromatics (e.g., BTX), much less precipitation occurs and fouling is reduced. Colloidal instability index (C.I.I.) was proposed for defining the crude oil compatibility on asphaltene stability [57]:

$$\text{C.I.I.} = \frac{[Asphaltenes + Saturates]}{[Aromatics + \text{Resins}]} \tag{J.2}$$

When C.I.I. < 1, the amounts of aromatics and resins is sufficient to keep the asphaltenes in solution. When C.I.I. > 1, asphaltenes are expected to precipitate out. Physicochemical composition of media (liquid or gas) and contaminants play leading role in fouling formation.

<u>Oxygen</u>. Not only dissolved molecular oxygen but O-, N-, S-organic compounds and metal traces influence gum formation. Rigorous exclusion of dissolved oxygen (or air) can substantially reduce or even eliminate deposition from hydrocarbon phase. However, the extent depends on the type and concentration of sulfur compounds.

<u>Sulfur and nitrogen</u>. Free sulfur, disulfides, polysulfides and thiophenol promote sludge formation in storage tanks whereas thiophenes, aliphatic mercaptans, and

Table J.2 Recommended water flow velocities for some metals and alloys [59]

Alloy/Metal	Minimum velocity[a] (m/s)	Recommended velocity[b] (m/s)
Carbon steel	1.2	2.0–4.0
Copper	1.2	1.5–2.0
Admiralty brass	1.2	1.4–2.0
Aluminum brass	1.2	1.8–2.2
Arsenical copper	1.2	<1.5
90-10 Copper-Nickel	1.5	1.8–2.5
70-30 Copper-Nickel	1.8	2.4–3.5
Titanium	1.2	7.0–9.2
Stainless steel	1.2	7.0–9.2

[a]Minimum water velocity is associated with the value when suspended solids do not precipitate on metal surface

[b]The upper value of recommended range of water velocity is associated with the velocity when erosion cannot occur

aliphatic sulfides show little effect [58]. Thiols, sulfides, disulfides, and some condensed thiophenes which can break down to form free radicals can be problematic but diphenyl sulfide and dibenzothiophene are not. The presence of small amounts of N-organic compounds (pyridines and pyrroles) can be deleterious to hydrocarbon stability.

Salts and metals. Salts $CaCl_2$, $MgCl_2$ and $FeCl_2$ in the presence of water in crude oil can hydrolyze producing acids (see Appendix C) and promote fouling. Copper and vanadium are also deleterious.

Corrosion fouling occurs in all three streams: liquid organic, aqueous, and gaseous phases. Any fouling can cause under deposit corrosion but corrosion fouling containing metals are capable initiate and intensify chemical reaction fouling, particulate fouling, crystallization, and biofouling.

Biofouling (biological fouling) can occur both in organic streams and cooling water systems. Organic streams include crude oil (mostly in tanks and sometimes in pipelines), fuels, such as diesel fuel and kerosene, and biofuels. Microorganisms do not "like" light hydrocarbons, therefore they do not proliferate in gasoline and naphtha. Cooling water systems if not properly treated with biocides also can be fouled by diverse microorganisms. We should remember that not only crude oil, fuels, water but the ambient air may be the source of penetration of microorganisms. Temperatures between 20 and 50 °C are ideal for microbial growth and proliferation. Dozens and even hundreds of different types of microorganisms exist in friendly cohabitation. Metabolic products and excretion of some of them represent nutrients (food) for others. Milliards of them dye and together with even more live bacteria amass, form uneven very thick layer of slime (sludge) on metal surface both in organic and water phases and can cause severe MIC [5, 20]. Minimum water velocities for some metals and alloys when suspended solids (organic substances and microorganisms among them) do not precipitate on metal surface and recommended velocity values are given in Table J.2.

Appendix K
Chemical Cleaning from Fouling

Chemical cleaning procedures consist of five stages:

A. Alkaline cleaning (degreasing).
B. Rinsing.
C. Acid cleaning.
D. Neutralization and rinsing.
E. Passivation.

Alkaline Cleaning

For removing only organic fouling (crude distillation units, FCCU, visbreakers), simply the first step (degreasing) is used and following solvents are applied: naphtha, heavy aromatic naphtha, kerosene, gas oil, light cycle oil, and organic solvents based on aromatics. Organic deposits in most cases are composed from degraded organic compounds, polymerized substances, coke, asphaltenes, and tars. One part of organic fouling materials is dissolved in polar, and another part is dissolved in non-polar solvent. Therefore, organic solvent mixture should contain alkaline and aromatic components. Well-established solvents are mixtures of aromatics (10–90 vol%), kerosene (10–90 vol%) and orange terpenes (10–30 vol%) (D-Limonene and surfactants). Corrosion products (iron sulfides, iron oxides and hydroxides) and inorganic salts are often present in organic deposits. There is no solvent which can dissolve coke. It can be removed only mechanically. When coke is mixed with heavy viscous organic deposits, the latter can be dissolved and coke is loosened and removed with liquid solution.

Organic substances generally are very viscous and severely adhered to metal surface. Quiescent zones exist in heat exchangers, and often solvent cannot reach these 'dead' areas. Therefore, chemical cleaning from 'heavy' organic fouling compounds is significantly less effective than mechanical cleaning. Special safety measures should be carried out because organic solvents are flammable, some of

A. Groysman, *Corrosion Problems and Solutions in Oil Refining and Petrochemical Industry*, Topics in Safety, Risk, Reliability and Quality 32,
DOI 10.1007/978-3-319-45256-2

them are toxic, and volatile organic compounds (VOC) may be dangerous for human health and harmful to the environment.

We should remember that rubber, some other polymeric materials and organic coatings used in different equipment at oil refinery units are not resistant to aromatic solvents. Therefore, before using of organic solvents, it is necessary to check this possible contact in order to prevent destruction of polymeric materials. It is desirable to replace materials of gaskets and hoses with materials resistant to organic solvents during cleaning. Usually circulation is carried out during several hours (5–6) at ambient temperature, and organic fouling is washed away. Rinsing is not carried out in these systems.

In cooling water systems, boilers, compressors, and in preoperational cleaning, rinsing must be carried out. Otherwise the next stage 'acid cleaning' would not be effective.

Cooling water systems (heat exchangers and piping), boilers, and compressors which need all five steps of cleaning, the degreasing involves the use of alkaline solution (Appendices I and L). Degreasing (alkaline) solution is similar for treatment fouling formed on all alloys (excepting aluminum). It includes sodium carbonate, trisodium phosphate, and low-sudsing detergent. These components are added to pure water in concentrations of 1.8–2.2 wt% and circulated for 6–24 h at 20–60 °C (see Appendices I and L). It is not recommended to use NaOH because of the possibility of caustic embrittlement of carbon steel. It is important to measure increased concentration of oil in solution every half an hour. When this concentration reaches maximum non-changed value, this means that all oil is removed from the inner surface of the equipment. Then the equipment should be rinsed (flushed) well with tap water twice and then with pure water in order to remove alkaline solution containing dissolved organic components (oil, grease, fat, and others). Drainage water can be discharged into separators at the oil refinery (ecological facility).

Acid Cleaning (Pickling)

The third step (acid cleaning, named also pickling) is intended for removing inorganic deposits consisting of different salts and corrosion products. For this, inorganic (hydrochloric or sulfamic) or organic (citric) acids are used (see Appendix L). If silica or silicates are present in deposits, hydrochloric acid is used with ammonium bifluoride ($NH_4F{\cdot}HF$ or NH_4HF_2, 1 wt%) to dissolve silica and silicates (K.1):

$$SiO_{2(s)} + 4\,NH_4F \cdot HF_{(aq)} \rightarrow SiF_{4(aq)} + 4\,NH_4F_{(aq)} + 2H_2O_{(l)} \qquad (K.1)$$

Titanium should never be exposed an acid containing ammonium bifluoride. The latter dissolves titanium well. It is important to remember that ammonium bifluoride is most hazardous substance (comparing to other compounds used in cleaning) and appropriate precautions must be performed. Chemical cleaning by acids by

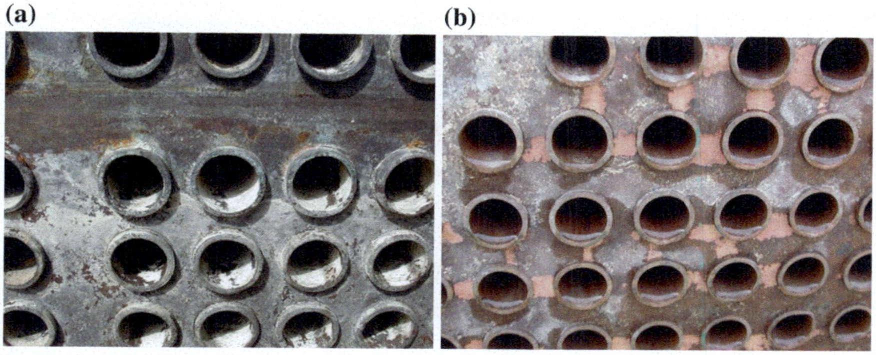

Fig. K.1 **a** Admiralty brass tube sheet and tubes with carbonate scale before acid cleaning by means of citric acid. **b** After acid cleaning and rinsing

circulation at 50 °C is applied to heat exchangers and piping in cooling water systems during 6 h (Fig. K.1).

Scale ($CaCO_3$, $MgCO_3$, $CaSO_4$) and corrosion products (FeS, FeOOH, Fe_2O_3) are removed in this procedure. The type of cleaning solution and its concentration must be chosen according to the type of the fouling (scale), its extent, and materials of construction.

Hydrochloric acid is significantly stronger scale dissolver than organic acids. Sulfamic acid is less corrosive toward metals than hydrochloric acid; it can be used for cleaning carbon steel, stainless steel, copper alloys, but not galvanized steel and aluminum. Sulfamic acid (8–10 wt%) is used at 60 °C to remove both carbonate scale and rust. It is not as effective as HCl to remove rust. An advantage of citric acid is that it can be disposed of by incineration.

Surfactants are added to both alkali and acidic cleaning solutions to improve their wetting characteristics (acting as detergents).

Hydrochloric acid is the least expensive and most widely used compound for chemical cleaning from scale and rust of carbon steel and cast iron equipment in water cooling systems. Amine corrosion inhibitors (0.2 wt%) must be added to HCl solution to protect bare carbon steel surface. It is possible to use HCl (5 wt%, not more!) for cleaning copper and titanium alloys too. In no case to use HCl solution for chemical cleaning equipment made from stainless steel, galvanized steel, aluminum, Incoloy 800, and Inconel 600.

Detailed procedure is described in Appendix L. It is very important to control temperature of cleaning solution (50 °C), concentrations of acid (5–8 wt% HCl, pH ~2.8), hardness ions (Ca^{2+}, Mg^{2+}), and total iron in cleaning solution. These parameters should be measured every half an hour. Acidic solution for maintaining its efficiency must be replenished in two cases: if concentration of acid diminished less than 5 wt% HCl or total iron content in solution increased above 1 wt%. The duration of acid cleaning depends on the extent of scaling and usually proceeds 6 h. The acid cleaning is considered complete, if concentrations of hardness ions remain constant for three consecutive readings (in half an hour each). This means that the entire scale is dissolved from the inner surface. It is desirable to install corrosion coupons in order to determine the corrosive effect of the cleaning solutions on the equipment materials.

After cleaning, the equipment is drained, rinsed, neutralized or flushed with water (till pH $\sim$7), and passivated (if needed). Most chemical cleaning procedures with acids have following problems:

(a) Safety and environmental concerns.
(b) Solutions of acids cause some metal loss of equipment (even in spite of use of corrosion inhibitors). Therefore, acidic solutions intended for chemical cleaning must be evaluated for corrosiveness toward all materials which may contact during the cleaning, appropriate corrosion inhibitors and their concentrations must be chosen.

Passivation

After the acid cleaning and rinse steps, pure metal surface is in active state. If left and exposed to the atmosphere or immediately introduce in processing, the metal would oxidize with uncontrolled rust formation. In order to prevent this, passivation is applied. Passivation is an artificial process creation of passive protective (against corrosion) layer on metal (usually carbon steel) surface. This layer consists of tightly adhered metal oxide (Fe_3O_4) or salts ($Fe_3(PO_4)_2$, $FePO_4$) non-dissolved in water and hydrocarbons. The term 'passivation' is also applied to cleaning from contaminations (iron dust, grease, dirt) and creating passive oxide Cr_2O_3 on stainless steel equipment [60, 61]. For this, 20–25 wt% HNO_3 solution is used after degreasing and rinsing. Passivating solution for carbon steel is alkaline (NH_4OH. pH = 9.5) and contains oxidizing agent (H_2O_2). If equipment is not used immediately after cleaning, inner space should be filled with inert gas (usually dry nitrogen) and keep it till start-up. Passivation procedures and solutions for boilers and steam lines differ from that for cooling water system. Detailed procedures are described in Appendices L and I.

Appendix L
Boil-Out Procedure (Chemical Cleaning and Passivation of Inner Surfaces of Boiler and Steam Pipelines)

This procedure is based on the proprietary standard of several oil refineries. The system (an example) needed for cleaning consists of new steam lines, flue gas cooler and two evaporators made of carbon steel. In order to decide about the application of all steps of the procedure, examination of inner surface of this system should be done. The procedure is described below.

1. Examination of inner surface of steam lines, flue gas cooler, and evaporators. Visual examination and by means of a boroscope.
2. If the surface is relatively pure (oil and grease contamination), without mill scale and rust, only alkaline cleaning is recommended (see 3.2).
3. If the mill scale and rust are present on inner surface, the steps of chemical cleaning and passivation are following:

   3.1. Wash with tap water containing suspended solids not more than 5 ppm and total dissolved solids not more than 500 ppm.

   3.2. Alkaline cleaning from oil contaminants:

      3.2.1. Soda ash (1.8–2.2 wt%).

      3.2.2. Tri-sodium phosphate (0.15–0.5 wt%).

      3.2.3. Liquid detergent (0.15 wt%).

      Boiler feed water with hardness not more than 1 ppm $CaCO_3$ is used for the preparation of the solution. Alkaline cleaning is carried out by circulation at 50–60 °C for 6 h.

   3.3. Wash with the tap water (two rinses) and then with boiler feed water for 2 h. Washing process can be finished when the quality of drainage water would be: conductivity is less than 50 μS/cm, 0.5 ppm PO_4^{3-}, 1 ppm SiO_2, 5 ppm suspended solids and 2 ppm oil. Drainage water can be discharged into separators at the oil refinery (ecological facility). Operator staff at this facility should be informed about the disposal.

A. Groysman, *Corrosion Problems and Solutions in Oil Refining and Petrochemical Industry*, Topics in Safety, Risk, Reliability and Quality 32,
DOI 10.1007/978-3-319-45256-2

4. Acid cleaning (pickling) with aqueous solution of:

   4.1. Hydrochloric acid 5 wt%.
   4.2. Corrosion inhibitor Rodine 213 (0.2 wt%). It is a proprietary corrosion inhibitor based on the mixture of amines, nonionic surfactants, isopropanol, propargyl alcohol, acetone, and HCl.
   4.3. Ammonium Hydrogen Fluoride (0.5 wt% $NH_4F{\bullet}HF$).

   Pickling is carried out by circulation of acid solution at 50 °C for 6 h.
5. Pretreatment (prepassivation) with:

   5.1. Citric acid (1.2–1.5 wt%).
   5.2 Ammonia solution (till pH = 9.5).

   Period—1.5 h till the constant concentration of iron in the solution.
6. Passivation with:

   6.1. Citric acid (1.5 wt%) at 40 °C.
   6.2. Ammonia solution (till pH = 9.5).
   6.3. Hydrogen peroxide (30 vol%) till positive redox potential (+30 mV). Period—2 h at 40 °C.

7. Drainage with nitrogen pressure and storage under atmosphere of dry nitrogen (overpressure 0.3–0.5 bar).
8. Examination and inspection of inner surface—quality of chemical cleaning and passivation (visual and with boroscope).

Conclusion and Insight into the Future

Arriving at one goal is the starting point to another.
John Dewey (1859–1952), an American philosopher

We have arrived to the end of our long journey among corrosion problems and their solutions at refinery and petrochemical units. I have tried to explain corrosion problems like a huge risk for the existence of oil refineries. It is impossible to cover all corrosion problems in one book. We had to find prioritizations in this large sea of problems.

Fires and destructions at oil refineries occurring because of corrosion phenomena distort the image of oil refining and petrochemical industry. Probably there is no refinery where corrosion did not lead to small or large catastrophes. Corrosion is an imminent danger and present explicitly or implicitly. Corrosion rates can be diminished, but never reduced to zero. This situation creates illusion that if corrosion is controlled, there is no need in corrosion monitoring. We should do it at any corrosion control method. Our task is to predict behavior of metallic equipment and as a result to stop using it in time, to examine, and to repair or replace before the failure. The main role in solutions of corrosion problems belongs to human.

We have reached a 'singularity' point in amount of published information and new knowledge. Corrosion specialists cannot track all publications. Only computer treatment will be able to search, find and analyze all the news. However, the final evaluation and decision should be made by specialists.

Some futurologists forecast that the period of crude oil will be finished in 15–30 years. Only nuclear, solar and wind energy will be used in 2050s. We should not forget that besides fuels, crude oil is the source of numerous materials which occupy and play significant role in life. So far, we will deal with oil refineries and petrochemical plants in near 50 years. Therefore, I see this topic is significant and perspective in foreseeable future. I hope everyone who has read this book found it useful, important, and new.

References

1. Gary JH and Handwerk GE (2001) Petroleum Refining, 4th edn, Marcel Dekker, Inc., New York
2. Wu Q (2010) Processing high TAN crude: part1, PTQ Q4. http://www.digitalrefining.com/article_1000524.pdf
3. Lu T (2012) Challenges in opportunity crude processing. NALCO, An Ecolab Company
4. Hatch LF and Matar S (1977) From Hydrocarbons to Petrochemicals. Part 2. Hydrocarbon Process 56(6):189–194
5. Groysman A (2014) Corrosion in systems for transportation and storage of petroleum products and biofuels. Springer, Dordrecht
6. Nelson WL (1958) Petroleum Refinery Engineering, 4th edn. McGraw-Hill Book Co., New York
7. Manning FS, Thompson RE (1995) Oilfield processing of petroleum. Volume two: crude oil. Penn Well Publ Company, Tulsa, Oklahoma, USA
8. Matar S, Hatch LF (1994) Chemistry of petrochemical processes. Gulf Publishing Company, Houston, Texas, USA
9. Dowd WT, Kuuskraa VA (1988) A technical and economic assessment of domestic heavy oil. Final Report. DOE Report, DOE/BC/10840-1
10. Rossini F (1960) Hydrocarbons in Petroleum. J Chem Educ 37(11):554–561
11. Toxicological Profile for Hydrogen Sulfide and Carbonyl Sulfide (2014) U.S. Department of Health and Human Services, Public Health Service, Agency for Toxic Substances and Disease Registry, USA
12. The Science of Smell Part: Odor perception and physiological response (May 2004) Iowa State University Extension
13. USEPA Health and Environmental Effects Profile for Hydrogen Sulfide (1980)
14. Zenz C, Dickerson OB, Horvath EP (1994) Occupational medicine. St. Louis, MO, USA
15. Lindenmann J et al (2010) Severe hydrogen sulfide poisoning treated with 4-dimethylaminophenol and hyperbaric oxygen. Diving and Hyperbaric Medicine: J South Pacific Underwater Medicine Soc 40(4):213–217
16. Clarke PH (1953) Hydrogen Sulfide Production by Bacteria. J Gen Microbiol 8:397–407
17. Nekrasov BV (1973) Fundamentals of General Chemistry, vol 1. Khimiya, Moscow
18. Shreider AV, Shparber IS, Archakov YI (1976) Vliyanie vodoroda na khimicheskoye I neftyanoye oborudovaniye (Influence of hydrogen on oil and chemical equipment). Mashinostroyeniye, Moscow
19. Volinski NP (1971) Thiosulfur acid. Polythionates. Wackenroder Reaction. Nauka, Moscow
20. Groysman A (2010) Corrosion for everybody. Springer-Verlag, Dordrecht, pp 97–98

21. NACE Standard SP0170-2012 (2012) Protection of austenitic stainless steels and other austenitic alloys from polythionic acid stress corrosion cracking during a shutdown of refinery equipment. NACE International, Houston, Texas, USA
22. Port RD, Herro HM (1991) The Nalco Guide to Boiler Failure Analysis. McGraw-Hill, Inc., New York
23. Heidersbach R (2011) Metallurgy and corrosion control in oil and gas production. John Wiley & Sons Inc., Hoboken, New Jersey, USA
24. ASTM A234 WPB (2015) Standard specification for piping fittings of wrought carbon steel and alloy steel for moderate and high temperature service. ASTM International, West Conshohocken, PA, USA
25. ASTM A209-03 (2012) Standard specification for seamless carbon-molybdenum alloy-steel boiler and superheater tubes. ASTM International, West Conshohocken, PA, USA
26. ASTM A217/A217M-14 (2014) Standard Specification for Steel Casting, Martensitic Stainless and Alloy, for Pressure-Containing Parts, Suitable for High-Pressure Service. ASTM International, West Conshohocken, PA, USA
27. Griffith WP, Morris PJT (2003) Charles Hatchett FRS (1765–1847), Chemist and Discoverer of Niobium. Notes and Records of the Royal Society of London 57(3):299
28. Rayner-Canham G, Zheng Z (2008) Naming elements after scientists: an account of a controversy. Foundations of Chemistry 10(1):13–18
29. Magee JH (1996) Two Galling Resistant Stainless Steels used for bridge Hinge Pins. https://www.cartech.com/techarticles.aspx?id=1666
30. ASTM B111 (2011) Standard specification for copper and copper-alloy seamless condenser tubes and ferrule stock. ASTM International, West Conshohocken, PA, USA
31. ASTM B148-14 (2014) Standard specification for aluminum-bronze sand casting. ASTM International, West Conshohocken, PA, USA
32. ASTM B338-14 (2014) Standard specification for seamless and welded titanium and titanium alloy tubes for condensers and heat exchangers. ASTM International, West Conshohocken, PA, USA
33. ASTM B348-13 (2013) Standard specification for titanium and titanium alloy bars and billets. ASTM International, West Conshohocken, PA, USA
34. Johnson WH (1873) On some remarkable changes produced in iron and steel by the action of hydrogen in acids. Iron 1:291, 452–453.
35. Jones RH (Ed) (1992) Stress-corrosion cracking. ASM International, USA
36. NACE International Publication 8X194 (2006) Materials and fabrication practices for new pressure vessels used in wet H_2S refinery service. Houston, TX, USA
37. NACE Standard MR0103-2007 (2007) Materials resistant to sulfide stress cracking in corrosive petroleum refining environments. NACE International, Houston, TX, USA
38. ANSI/NACE Standard TM 0284-2003 (2003) Standard test method evaluation of pipeline and pressure vessel steels for resistance to hydrogen-induced cracking. NACE International, Houston, TX, USA
39. API RP 941 (2008) Steels for hydrogen service at elevated temperatures and pressures in petroleum refineries and petrochemical plants. 7th edn. American Petroleum Institute, Washington, DC., USA
40. ANSI/NACE MR0175/ISO 15156 (2009) Petroleum and natural gas industries—materials for use in H_2S-containing environments in oil and gas production. NACE International, Houston, TX, USA
41. Messer B et al (2008) Role of nascent hydrogen in refinery corrosion. Paper no. 08549 presented at the NACE Corrosion Conference, NACE International, Houston, TX, USA
42. Compendium of Chemical Terminology (1997) IUPAC recommendations. In: MaNaught AD, Wilkinson A (eds) 2nd edn. Blackwell Scientific Publications, Oxford, Great Britain
43. Bunnett JF, Jones RAY (1988) Names for hydrogen atoms, ions, and groups, and for reactions involving them. Pure Appl Chem IUPAC, Great Britain 60(7):1115–1116

44. Wach S, Miodownik A, Mackowiak J (1966) The diffusion of hydrogen through pure iron membranes. Corr Sci 6(6):271–278
45. Lasocka M, Raczynski W (1967) Bull de l'Acad Polonaise des Sci, Serie de Sci Chim 15 (1):31–36
46. Shively JH, Hehemann RF, Troiano AR (1966) Hydrogen permeability in a stable austenitic steel. Corrosion 22(9):253–256
47. ASTM E1351–01(2006) Standard practice for production and evaluation of field metallographic replicas. ASTM International, West Conshohocken, PA, USA
48. API 570 (2003) Piping inspection code: inspection, repair, alteration, and rerating of in-service piping systems. American Petroleum Institute, Washington, DC., USA
49. Nondestructive Evaluation and Quality Control (1989) Vol 17, Metals Handbook, Ninth Edition, ASM International, USA, pp 52–56
50. Groysman A, Kaufman A, Feldman B, Balali R (2011) Hydrogen damage and prevention in the oil refinery. Paper no. 11295 presented at the NACE Conference CORROSION 2011, NACE International, Houston, TX, USA
51. Korb LJ, Olson DL (eds) (1987) Metals Handbook, vol 13: Corrosion. ASM International, USA, pp 1277–1280
52. White RA, Ehmke EF (1991) Materials selection for refineries and associated facilities. NACE, USA
53. Failure Analysis and Prevention (1986) Vol 11, Metals Handbook. 9th edn. ASM International, USA, pp 469, 645
54. Parthasarathy TA, Shewmon PG (1987) Hydrogen attack behavior of the heat affected zone of a 2.25Cr–1Mo steel weldment. Metall Mater Trans A, Springer 18(7):1309–1312
55. Müller-Steinhagen H (2000) Heat exchanger fouling: mitigation and cleaning technologies. PUBLICO Publications, Germany
56. Watkinson AP, Li YH (2009) Fouling characteristics of a heavy vacuum gas oil in the presence of dissolved oxygen. In: Müller-Steinhagen H, Malayeri MR, Watkinson AP (eds) Proceedings of International Conference on Heat Exchanger Fouling and Cleaning VIII-2009. June 14–19, 2009, Schladming, Austria, pp 27–32
57. Asomaning S, Watkinson AP (1999) Deposit formation by asphaltene-rich heavy oil mixtures on heat transfer surfaces. In: Bott TR (ed) Understanding heat exchanger fouling and its mitigation. Begell House Inc., NY, pp 283–290
58. Coletti F, Hewitt JF (eds) (2015) Crude oil fouling: deposit characterization, measurements, and modeling. Gulf Professional Publishing, Elsevier, USA
59. Paikert P (1983) Verschmutzung von Kondensatoren und Kühltürmen. GVC Weihnachtstagung, pp 371–390
60. ASTM A380 (2013) Standard practice for cleaning, descaling, and passivation of stainless steel parts, equipment, and systems. ASTM International, West Conshohocken, PA, USA
61. ASTM A967 (2013) Standard Specification for Chemical Passivation Treatments for Stainless Steel Parts. ASTM International, West Conshohocken, PA, USA

Recommended Literature

62. ASTM G35-98 (2010) Standard Practice for Determining the Susceptibility of Stainless Steels and Related Nickel-Chromium-Iron Alloys to Stress-Corrosion Cracking in Polythionic Acids. ASTM International, West Conshohocken, PA, USA
63. Garveric L (ed) (1994) Corrosion in the petrochemical industry. ASM International, USA
64. Nagashima E, Matsumoto K, Shibata K (1998) Effects of sensitization and service fluid chemistry on polythionic stress corrosion cracking of 18-8 stainless steel. Paper no. 592 presented at the NACE CORROSION conference, Houston, TX, USA
65. NACE 6A192/SSPC-TR3 (2000) Dehumidification and temperature control during surface preparation, application, and curing for coatings/linings of steel tanks, vessels, and other enclosed spaces. Joint Technical Committee Report
66. Tamaki K, Nakano S, Kimura M (1987) Application of CrNi stainless steel weld metals to polythionic acid environments. Mater Perform 26(8):9–13
67. Dove D, Messer B, Phillips T (2001) An austenitic stainless steel, resistant to high temperature creep and naphthenic acid attack. Paper no. 01523 presented at the NACE CORROSION conference, Houston, TX, USA
68. Brophy AJ (1974) Stress Corrosion Cracking of Austenitic Stainless Steels in Refinery Environments. Mater Perform 13(5):9–15
69. Nishida H, Nakamura K, Takahashi H (1984) Intergranular Stress Corrosion Cracking of Sensitized 321 SS Tube Exposed to Polythionic Acid. Mater Perform 23(4):38–41

Glossary

Aldehyde An organic compound containing a functional group CHO with the general formula R–CHO

Alkanes (paraffins, saturated hydrocarbons) Chemical compounds consisting only of carbon and hydrogen atoms and are bonded exclusively by single bonds. The general formula: C_nH_{2n+2}

Alkenes (olefins, unsaturated hydrocarbons) Chemical compounds consisting only of carbon and hydrogen atoms and containing one or more pairs of carbon atoms linked by a double bond. The general formula: C_nH_{2n}

Alkyl A functional group R– (radical – C_nH_{2n+1}), e.g., CH_3–, C_2H_5–

Amides Organic compounds with the functional group $RY(O)_xNR'$ where R and R′ refer to H or radical; Y = carbon or sulfur or phosphorous atoms

Amines Organic compounds; derivatives of ammonia, where in one or more hydrogen atoms have been replaced by an alkyl or aryl (C_6H_5–) group

Amine carboxylates Carboxylate salts of amines (amine salts of carboxylic acids)

Amphoteric metals Metals that corrode in acidic and alkali aqueous solutions

Antifoulant Chemical substance injected into organic media (crude oil, residue) to prevent, reduce, or eliminate fouling

API gravity Used in the petroleum industry as a density scale. Named after the American Petroleum Institute (API)

Aromatic ring The configuration of six carbon atoms in aromatic compounds; is known as a benzene ring

Aromatic solvents (aromatics) Aromatic compounds based on benzene ring

Asphalt (bitumen) A sticky, black and highly viscous liquid or semi-solid material (mixture of high molecular weight hydrocarbons)

Asphaltenes Heterocyclic aromatic compounds containing N, S and O atoms.

A. Groysman, *Corrosion Problems and Solutions in Oil Refining and Petrochemical Industry*, Topics in Safety, Risk, Reliability and Quality 32,
DOI 10.1007/978-3-319-45256-2

Atmospheric gas oil The mixture of liquid hydrocarbons obtained under atmospheric distillation below 350 °C

Atmospheric residue The undistillable fraction that remains after atmospheric distillation of a crude oil

Bacteria (microorganisms) Large domain of microorganisms; a few microns in length, bacteria have a wide range of shapes, ranging from spheres to rods and spirals

Benzene An aromatic hydrocarbon with the molecular formula C_6H_6; a natural constituent of crude oils

Biocide A substance for killing microorganisms

Biodegradation Destruction of materials by microorganisms

Biofouling (slime, sludge) Biological fouling, the accumulation of microorganisms, plants, algae or small animals on wetted surfaces

Bitumen Naturally occurring highly viscous hydrocarbons of viscosity greater than 10,000 cSt at reservoir conditions and API gravity less than 10°

Blog A web log using for broadcast content created by a single author across an entire organization or the internet

Boiling range The range of temperature over which a fuel, or other liquid mixture of compounds, distills

Brass An alloy consisting of copper and zinc (15–50 wt% Zn)

Bronze An alloy consisting primarily of copper and tin ($\sim$10 wt% Sn) as the main additive

Carbide An intermetallic compound containing carbon (e.g., Fe_3C, TiC). There are many possible combinations of carbon and other atoms (such as iron, titanium, niobium, vanadium) that combine to form carbides in steel

Carbon steel An alloy containing iron (Fe) and carbon (C) at concentrations from 0.008 to 2 wt%, and small amounts of other elements

Carboxylic acids Organic acids containing at least one carboxyl group –COOH

Carcinogenic Producing or tending to produce cancer

Cathodic protection A technique used to control the corrosion of a metal surface by making it the cathode (which does not corrode) of an electrochemical cell

Cementite (see carbide) An intermetallic compound iron carbide Fe_3C

Cladding A process used to join dissimilar metals together to form a single metal piece

Coalescence A process of uniting small droplets of one liquid preparatory to its being separated from another liquid (separation of emulsion)

Coalescer A device performing coalescence

Coating disbondment The destruction of adhesion between a coating and the surface coated

Coke Organic carbonaceous solid deposits (fouling)

Colloid A substance microscopically dispersed evenly throughout another substance

Composite materials (composites) Materials made from two or more components with significantly different physical and chemical properties, that when combined, produce a material with characteristics different from the individual components

Conventional crude oil The type of crude that can be recovered from the reservoir by conventional means (e.g., by pumping)

Conventional heavy oil Hydrocarbons in the range of 10–21 °API that can be recovered along with the conventional oil

Corrosion inhibitors Chemicals that, when present in low concentrations (1–15,000 ppm) in a corrosive environment, retard the corrosion of metals

Cracking A series of reactions that takes place by breaking bonds (e.g., C–C, C–S, C–N, or C–H) of heavier molecules into smaller molecules

Crude oil A naturally occurring liquid mixture of different hydrocarbons that exist in the Earth's crust

Cyclic amines Organic compounds with N atoms inside the cycle

Cycloalkanes (cycloparaffins, naphthenes) Types of saturated hydrocarbons that have one or more rings of carbon atoms in the chemical structure

Cycloparaffins (see cycloalkanes)

Dehumidification The removal of the moisture from air

Demulsifiers (detergents, surfactants, emulsifiers, emulgents, wetting agents) Polar compounds that cause a marked reduction in the interfacial tension of liquids

Dew point The temperature, at which the moisture content in the air will saturate the air and condense on nearby surfaces

Diesel fuel (diesel oil, gas oil, heating oil, or petrodiesel)—a liquid mixture of hydrocarbons C_{12} to C_{24} distilled in the range 180–370 °C

Dispersant A surfactant additive designed to hold particulate matter dispersed in a liquid

Distillation (rectification) A process of separating a liquid homogenous mixture into fractions based on differences in boiling points of its components

Electrolytes The substances whose water solutions or molten states conduct electric current on account of free ions

Emulsion A two-phase system of a mixture of two or more immiscible liquids

Ester Organic compound containing the group COO combining with two radicals

Ether Organic compound where two radicals are bonded through oxygen atom

Ethyl mercaptan An organic compound C_2H_5SH (ethanthiol) added to the propane–butane gas in order to detect the leakage of the latter according to its specific unpleasant odour

Extraheavy oil (unconventional heavy oil) Heavy oil that cannot move by itself and cannot be recovered by conventional means. Extraheavy oil falls below 10 ° API, with viscosity less than 10,000 cSt

Fatty acids Saturated monocarboxylic acids

Fiberglass A composite material, a glass reinforced plastic

Fuel oil A liquid mixture of hydrocarbons ($>C_{20}$) with boiling point >340 °C

Fungi Microorganisms including yeasts and molds (more familiar as mushrooms)

Gas oil (see Diesel fuel)

Gasoline (Gas, Petrol) A liquid mixture of hydrocarbons (C_4 to C_{12}, with the most prevalent C_8) boiling between 20 and 210 °C

Grease A semisolid lubricant

Gum Polymerized organic materials of high viscosity formed during fuel storage

Heat-affected zone (HAZ) The area of base material (either a metal or thermoplastic), which is not melted and has its microstructure and properties altered by welding

Heavy oil In general, heavy oil refers to a crude petroleum hydrocarbon mixture of below 21 °API

Hydrogen-induced cracking (HIC) Stepwise internal cracks that connect adjacent hydrogen blisters on different planes in the metal or to the metal surface. No externally applied stress is needed for the formation of HIC. HIC is commonly found in steels with high levels of impurities, especially sulfur

Hydrocarbons Compounds composed only of hydrogen (H) and carbon (C) atoms

Hydrocracking The process of cracking the C–C bond accompanied by the addition of hydrogen, where bigger molecules in the heavy oil are transferred to the lower-molecular-weight liquid oil

Hydrodesulfurization The process of removing hydrogen sulfide (H_2S) and other sulfur-organic compounds from petroleum products at the oil refineries

Hydrogenation The process of adding hydrogen to remove contaminants (e.g., S, N, and metals) and to saturate the unsaturated olefins or aromatics

Hydroperoxides Organic compounds R–O–O–H

Hydrophilic Water accepting. Hydros (from the Greek) means water; philia means love

Hydrophobic Water repelling. Hydros (from the Greek) means water; phobos means fear

Hydrotreating Treatment with hydrogen

Immiscible Liquids which are mutually insoluble

Ketones Organic compounds where two radicals are bonded with the group C=O

Kerosene (jet fuel, aviation kerosene) A liquid mixture of hydrocarbons C_9 to C_{16} boiling at 150–290 °C

Lean amine Aqueous solution of amine containing very small concentrations of H_2S (regenerated amine, recycled for reuse in the absorber)

Make-up The water that must be added to the circulating water system in order to compensate for water losses such as evaporation, drift loss, blow-out, and blow-down

Membrane A thin, continuous sheet of nonconductive synthetic material used to contain and/or separate two different media

Mercaptans A sulfur-containing organic compound where radical is combined with the group –SH

Microbial metabolism The set of life-sustaining chemical transformations within the cells of living organisms

Miscible Liquids which are mutually soluble

Mold (mould) A fungus that grows in the form of multicellular filaments

Naphthenates Salts of naphthenic acids

Naphthenes (see Cycloalkanes)

Naphtha The lightest and most volatile distillate fraction of the liquid hydrocarbons in crude oil

Neutralization Number (see Total Acid Number)

Non-electrolytes Substances that do not dissociate into ions and do not conduct electric current being dissolved in some solvent

Non-polar hydrocarbons Molecules which have symmetry

Nutrients Chemical substances that organisms need to live and grow

Octane number The percentage (by volume) of iso-octane in a combustible mixture

Oil shale (kerogen shale) An organic-rich fine-grained sedimentary rock containing kerogene from which liquid hydrocarbons can be produced

Oil sands Bitumen-soaked sand, located mostly in Alberta (Canada)

Olefins (see alkenes)

Paraffins (see alkanes)

Peroxide A compound containing an oxygen–oxygen single bond named peroxide (or peroxo) group (–O–O–)

Phenols Organic compounds containing aryl combining with one or more group OH

Pig A device that moves through the inside of a pipeline for the purpose of cleaning, dimensioning, and inspecting

Pyrophoricity Spontaneous combustion of a material upon exposure to air

Plankton Microscopic organisms that float in liquids

Polar hydrocarbons Molecules which have no symmetry

Polymer A material consisting of repeating group of atoms

Porphyrins Nitrogen containing compounds derived from chlorophyll and occur as organometallic complexes of vanadium and nickel in crude oils

Produced water Water in the oil industry that is produced as a byproduct along with the oil and gas

Rectification (see Distillation)

Relative humidity The percentage of water vapor present in air, relative to the maximum amount of water that the air (saturated by water) can hold at the same temperature

Residue/residuum/resid Nondistillable portion remaining after atmospheric or vacuum distillation of crude oil or heavy oil

Rich amine Aqueous solution of amine containing large concentrations of H_2S (after absorber)

Schedule (Sch) Non-dimensional number specified pipe size. It was invented in 1927 to specify the nominal wall thickness of pipe

Secondary containment A device, system or construction used to control the accidental escape of a stored product so it may be properly recovered or removed from the environment

Slime (see Biofouling)

Slops Liquid wastes (emulsion) containing mixtures of various fuels, organic solvents and water

Soda ash (washing soda, sodium carbonate) Na_2CO_3

Sodium naphthenate Surfactant

Stainless steel An alloy of iron with chromium content above 12 wt%

Stress-oriented hydrogen-induced cracking (SOHIC) A stacked array of small blisters joined by HIC, aligned in the through-thickness direction of the steel as a result of high localized tensile stresses

Sulfidation The reaction of a metal or alloy with sulfur-containing compounds

Sulfide stress cracking (SSC) Cracking of a metal or an alloy under the combined action of tensile stress and corrosion in the presence of water and H_2S

Surfactants (surface active agents, see Demulsifiers)

Suspension A heterogeneous mixture containing solid particles in liquid

Synthetic crude oil (also called syncrude) a product resulting from the upgrading of heavy oil or bitumen that does not contain residue

Tar Highly viscous material resulting from destructive distillation of bitumen or coal

Tar sands Term incorrectly used to refer to oil sands, which, by definition, are not a tar

Teflon Brand name of polytetrafluoroethylene (PTFE)

Tempered water Mixture of hot (BFW) and cold water via a tempering valve. A tempering valve mixes enough cold water with the hot water to keep the outgoing water temperature fixed at a moderate temperature $\sim$50 °C

Toluene Organic aromatic solvent

Total Acid Number (TAN; Neutralization Number) A measure of the numbers of milligrams of potassium hydroxide (KOH) needed to neutralize 1 g of crude oil or its distillate fraction

Upgrading Conversion of heavy oil into lighter oil

Waxes Chemical compounds that are plastic at ambient temperatures

Wiki A short for 'wiki wiki web', from the Hawaiian expression 'wiki wiki' meaning 'fast'. It is a web-site where anybody can create and edit a web page

Yeasts Microorganisms in the kingdom Fungi

Different words are used, in different countries, to describe the same piece of equipment (Table 1).

Table 1 Chemical Engineering Terms[a]

United States	United Kingdom
Accumulator	Reflux Drum
Agitator	Mixer or stirrer
Check valve	Nonreturn valve
Clogged (of filter)	Blinded
Consensus standard	Code of practice
Conservation vent	Pressure/vacuum vent
Dike, berm	Bund
Discharge valve	Delivery valve
Downspout	Downcomer
Expansion joint	Bellows
Explosion proof	Flameproof
Faucet	Tap
Fiberglass reinforced plastic (FRP)	Glass reinforced plastic (GRP)
Flame arrestor	Flame trap
Flashlight	Torch
Fractionation	Distillation
Gasoline	Petrol
Gauging (of tanks)	Dipping
Generator	Dynamo or alternator
Ground	Earth
Horizontal cylindrical tank	Bullet
Install	Fit
Insulation	Lagging
Inventory	Stock
Manway	Manhole
Mill water	Cooling water
Nozzle	Branch
OSHA (Occupational Safety and Health Administration)	Health and Safety Executive
Plugged	Choked
Shutdown	Permanent shutdown
Sieve tray	Perforated tray
Sparger or sparge pump	Spray nozzle
Spigot	Tap
Spool piece	Bobbin piece
Stack	Chimney
Tank car	Rail tanker or rail tank wagon
Tank truck	Road tanker or road tank wagon
Torch	Cutting or welding torch
Tower	Column
Tray	Plate
Turnaround	Shutdown
Utility hole	Manhole

[a]Trevor Kletz (1996) Dispelling chemical industry myths. CRC Press

Index

A. Groysman, *Corrosion Problems and Solutions in Oil Refining and Petrochemical Industry*, Topics in Safety, Risk, Reliability and Quality 32,
DOI 10.1007/978-3-319-45256-2

The manufacturer's authorised representative in the EU is Springer Nature Customer Service Centre GmbH, Europaplatz 3, 69115 Heidelberg, Germany. If you have any concerns regarding our products, please contact ProductSafety@springernature.com

Printed and bound by CPI Group (UK) Ltd, Croydon, CR0 4YY
25/09/2026
02222841-0009